005

To A.G.,

My father's good friend and my esteemed colleague and friend.

With best wishes for the New Year,

David

*Yale Agrarian Studies Series*

JAMES C. SCOTT, SERIES EDITOR

"The Agrarian Studies Series at Yale University Press seeks to publish outstanding and original interdisciplinary work on agriculture and rural society—for any period, in any location. Works of daring that question existing paradigms and fill abstract categories with the lived-experience of rural people are especially encouraged." —James C. Scott, *Series Editor*

Christiana Payne, *Toil and Plenty: Images of the Agricultural Landscape in England, 1780–1890* (1993)

Brian Donahue, *Reclaiming the Commons: Community Farms and Forests in a New England Town* (1999)

James C. Scott, *Seeing Like a State: How Certain Schemes to Improve the Human Condition Have Failed* (1999)

Tamara L. Whited, *Forests and Peasant Politics in Modern France* (2000)

Peter Boomgaard, *Frontiers of Fear: Tigers and People in the Malay World, 1600–1950* (2001)

James C. Scott and Nina Bhatt, eds., *Agrarian Studies: Synthetic Work at the Cutting Edge* (2001)

Janet Vorwald Dohner, *The Encyclopedia of Historic and Endangered Livestock and Poultry Breeds* (2002)

Deborah Fitzgerald, *Every Farm a Factory: The Industrial Ideal in American Agriculture* (2003)

Stephen B. Brush, *Farmers' Bounty: Locating Crop Diversity in the Contemporary World* (2004)

Brian Donahue, *The Great Meadow: Farmers and the Land in Colonial Concord* (2004)

J. Gary Taylor and Patricia J. Scharlin, *Smart Alliance: How a Global Corporation and Environmental Activists Transformed a Tarnished Brand* (2004)

Raymond L. Bryant, *Nongovernmental Organizations in Environmental Struggles: Politics and the Making of Moral Capital in the Philippines* (2005)

Edward Friedman, Paul G. Pickowicz, and Mark Selden, *Revolution, Resistance, and Reform in Village China* (2005)

Michael Goldman, *Imperial Nature: The World Bank and Struggles for Social Justice in the Age of Globalization* (2005)

Arvid Nelson, *Cold War Ecology: Forests, Farms, and People in the East German Landscape, 1945–1989* (2005)

Steve Striffler, *Chicken: The Dangerous Transformation of America's Favorite Food* (2005)

Lynne Viola, V. P. Danilov, N. A. Ivnitskii, and Denis Kozlov (editors), *The War Against the Peasantry, 1927–1930* (2005)

ARVID NELSON

# *Cold War Ecology*

## FORESTS, FARMS, AND PEOPLE IN THE EAST GERMAN LANDSCAPE, 1945–1989

*Yale University Press*
*New Haven*
*& London*

Set in Sabon type by Keystone Typesetting, Inc.
Printed in the United States of America.

Library of Congress Cataloging-in-Publication Data
Nelson, Arvid, 1951–
Cold war ecology : forests, farms, and people in the East German landscape, 1945–1989 / Arvid Nelson.
p. cm. — (Yale agrarian studies series)
Includes bibliographical references and index.
ISBN-13: 978-0-300-10660-2 (alk. paper)
ISBN-10: 0-300-10660-2 (alk. paper)
1. Forests and forestry—Political aspects—Germany (East)—History. 2. Forest policy—Germany (East)—History. 3. Agriculture and politics—Germany (East)—History. 4. Human ecology—Germany (East)—History. I. Title.
II. Yale agrarian studies
SD195.N35 2005
333.75′0943′1—dc22
2005013231

A catalogue record for this book is available from the British Library.

The paper in this book meets the guidelines for permanence and durability of the Committee on Production Guidelines for Book Longevity of the Council on Library Resources.

10 9 8 7 6 5 4 3 2 1

To my mother,
Aurora Carvajal Nelson

It was with a feeling of ecstasy that the Contessa recaptured the memories of her early childhood and compared them with her sensations at that moment. "The Lake of Como," she said to herself, "is not surrounded, like the Lake of Geneva, with wide tracts of land enclosed and cultivated according to the most approved methods, calling up ideas of money and speculation. Here, on every side, I see hills of unequal height, covered all over with clumps of trees that chance has planted, and which the hand of man has never yet spoilt, and forced to *yield a return.* Among these hills so admirably shaped, pressing on toward the lake down slopes of a strangely curious formation, I can preserve all the illusions of Tasso's and Ariosto's descriptions. Everything is noble and tender, everything speaks of love; nothing recalls the ugly spectacles of civilization. The villages half-way up the slopes are hidden in tall trees, and above the tree-tops rises the charming architecture of their graceful spires. If some tiny field, fifty paces across, comes here and there to interrupt the clumps of chestnuts and wild cherries, my satisfied eye sees growing on it plants more vigorous and happier than elsewhere."
—*Stendhal,* The Charterhouse of Parma

*Contents*

List of Figures ix

Conversions x

Preface xi

Acknowledgments xv

A Note on Terminology xix

1. Prologue 1
2. Landscape and Culture 10
3. Initial Conditions and Reparations 29
4. "A Law Would Be Good": Land Reform 53
5. The Landscape's "Socialist Transformation" and Flight from the Countryside (1949–1961) 77
6. The Landscape Transformed (1960–1961) 99
7. Cybermarxism and Innovation (1961–1971) 118

8. The Grüneberg Era and the Triumph of Industrial Production Methods (1971–1989) 141

9. Reunification 171

Notes 189

Glossary 259

Bibliography 265

Index 299

*Plates follow pages 52 and 140*

*Figures*

Figure 1. Natural forest cover type map of East Germany 18
Figure 2. Current (1986) forest cover type map of East Germany 19
Figure 3. Block diagram of the Central German Uplands viewed from the south 24–25
Figure 4. Allied occupation zones, 1945 35
Figure 5. Fuelwood and timber harvests, 1935/6–1950 70
Figure 6. Political and administrative map of East Germany, 1970s–1980s 130
Figure 7. Aerial fertilization of the East German forest, 1955–1980 156
Figure 8. Pulp and sawlog harvest plotted against total yield per hectare 161
Figure 9. Age class distribution, 1927–1989 162
Figure 10. Stocking growth, 1956–1989 163
Figure 11. Emergence of the German nation, 1937–1990 164–65
Figure 12. State harvest plotted against afforestation and TSI investment, 1946–1985 166
Figure 13. Red deer kill and stocking, 1960–1989 168

# *Conversions*

### *Volume and Yield*

1,000 board feet (mbf) = 2.36 cubic meters ($m^3$)
1 cubic meter ($m^3$) = 35.31 cubic feet ($ft^3$) = 423.78 mbf
1 cord = 3.62 $m^3$ = 128 $ft^3$
1 $m^3$/hectare = 14.5 $ft^3$/acre ~ 89 mbf/acre

### *Distance and Area*

1 foot (ft.) = .305 m
1 hectare (ha) = 10,000 $m^2$ = 2.471 acres = 107,639.10 $ft.^2$
1 kilometer (ki) = 3,280.84 ft. = 0.6214 miles
1 meter (m) = 3.281 ft.
1 mile (mi.) = 5,280 ft. = 1,760 yd. = 1,609.34 m
1 yard (yd.) = 0.914 m
1 acre = 43,560 $ft^2$ = 0.4047 ha = 4,046.86 $m^2$
1 square kilometer ($ki^2$) = 0.3861 square mile ($mi^2$) = 247.105 acres
1 square mile ($mi^2$) = 259.00 ha = 640 acres

# *Preface*

Mixed blessings confronted social scientists working in the 1980s and 1990s. They witnessed two rare events: the collapse of a great empire and the reordering of the global economy. Their failure to foresee the global revolutions of the 1990s also confronted them. For despite all the rigor and resources of cold war analysis, social scientists failed so completely to understand the quality and sustainability of the Soviet empire as to drive "a nail in the coffin of Western political science," as Mark Mazower noted. Western social scientists paid more attention to official data and goals than they did to physical reality. How else could they have squared their direct experience of massive pollution, much less their witness of more than two hundred deaths at the Berlin Wall, with their vision of East Germany as a disciplined, efficient industrial power of indefinite tenure?

The collapse of the Soviet empire surprised the world as have few other events. One has to go back two hundred years, to the late 1780s and the French Revolution, to find comparable contradictions between perceptions of power and reality of decline. The Soviet Union seemed to have reached a climax stage of stability, strengthened by the reforms and policies of its youthful leader, Mikhail Gorbachev. The riches and material accomplishments of Soviet power, with its great natural wealth, geographic expanse, and intimidating strategic and conventional forces, made dramatic change hard to imagine. Even though few would have exchanged Soviet for Western problems and opportunities, the scope and dynamic of Soviet geopolitical and military power itself inspired awe.

The mystery today is how Westerners could not have foreseen the collapse of the Soviet empire given the abundant telltales of decline. Anyone traveling in East Germany in the 1970s and 1980s could have read the instability of the Marxist-Leninist Ponzi scheme in the Party's cruel and arbitrary human rights policies and in the waste in the country's economy. Pollution was a constant fellow traveler, an outward and visible sign of decline. There was a near total lack of even primary sewage facilities, no controls on smokestack emissions, backward transportation and communications networks, and primitive living conditions outside the ruling elite. Nitrates and heavy metals fouled much of even the subsurface water, and the noise pollution was almost intolerable. In East Berlin, the showcase "capital of the German Democratic Republic," coal furnace waste and industrial gases rendered the air palpable with a sour, sulphurous pall of brimstone. Men and women suffered from abnormal levels of chronic asthma, bronchitis, and rare respiratory and blood diseases. A racket of coughing and sneezing stirred up from the street through the thick, fouled atmosphere, competing with the background noise of city streets, the roar of truck exhaust, and the grinding of battered machinery. Sufferers discharged ounces of dust, particulate, ash, and sulfur dioxide into their handkerchiefs. East Berlin's decay is even more remarkable given the Party leadership's disproportionate infrastructure investments there, starving other cities and the consumer economy to polish and fine-tune the capital.

The people most directly harmed by Marxism-Leninism, those who had to live in the poor world of the "workers' paradise," read the state's instability in forced collectivization, and then in forest decline and pollution. They first voted with their feet by fleeing between 1949 and 1961 in numbers that bled the state white until the Party leadership imprisoned them behind the Wall. People then emigrated internally away from pollution centers in the industrial uplands and later joined to protest environmental destruction under the protection of the church. Western observers ignored these signals, seeing toughness and discipline in the leadership's willingness to bear such high levels of pollution damage for the sake of industrial production, perversely reading discipline and sacrifice into the dysfunction and inefficiency of pollution.

Most Western observers came to East Berlin and Leipzig, saw the heavy pall of smoke, soot, and ash, and the material gloom, and still parroted back the data fed them, data drawn from late-Ptolemaic accounting systems whose epicycles, eccentricities, and retrogradings bestowed false harmony upon the gathering disorder. Jane Kramer, writing in the *New Yorker* eight years after reunification, recalled the poverty that official data obscured: "The West had imagined East Berlin as the cutting edge of the Soviet Empire—the efficient, disciplined, technological, German edge—but East Berlin turned out to be a deteriorating place, a shabby police-state capital not so different from most of the other Eastern-bloc capitals." A walk through a crumbling and dirty East German city, or the smells of animal waste and chemicals wafting off poisoned collective farm fields, could have exposed the official data as fraudulent.

West German forest scientists and historians knew before many of their colleagues in economics and political science that the East German forest economy was collapsing, the consequence of forty years of the reduction of forest structure and the Party's peculiar economic ideology: Industrial Production Methods. Bowing to pressure from the West German government, the Party leadership agreed in the mid-1980s to open their forests to West German foresters. West and East German foresters then walked together in the forests east of the Elbe River for the first time since the early 1950s. Soon the forests of eastern and western Germany knitted back together seamlessly, leaving no mark of their joining apparent to an outside observer.

Western observers and analysts missed invaluable indicators of East Germany's (and of the Soviet empire's) instability in the natural landscape. Forest structure manifests political and economic history spanning many previous generations. How closely forest structure emerges from ecological conditions, and how foresters, managers, and owners plan for the forest of the future, offer physical expressions of historical influences and events. The qualities of the forest, its diversity, structures, resilience, and flows, revealed the essential political and economic structures of the East German polity and were the most accessible and accurate leading indicators of its future. The citizens of the former East Germany, even without formal training, knew this and read in forest death the Party's inner corruption and incompetence, which most Western analysts and observers missed. Indeed, one could have read the instability and brittleness of the East German regime from satellite photographs, which illuminated a landscape of extreme artificiality, suffocating under low levels of diversity and a linear structure. What one saw in the forested ecosystem one also saw in the East German society and economy.

Landscapes are excellent leading indicators of change. The clearing of Central Europe's primeval Hercynian Forest, so vast that "Germans whom Caesar questioned had traveled for two months through it without reaching the end," was the great social and economic task of the early Middle Ages, presaging the rise of feudalism and early village and urban communities. Enclosure of forest and agricultural land prepared the way for the Industrial Revolution. Lenin's and Stalin's brutal collectivization of private land foreshadowed forced industrialization and the Great Terror. The new Soviet zone government in 1945 took control first of the countryside, a foreshock of the Party leadership's assertion of its power. Finally, pollution damage to farms, forests, and fresh water was a constant and growing telltale of the inefficiencies and corruption of the East German political leadership. Landscapes are a bellwether of change, and when the Party leadership simplified and stressed the rural landscape, it signaled what it was doing in society and what was to come.

Forest policy offers three simple benchmarks to evaluate the Party's stewardship: to reverse forest decline endemic to East Germany's dominant pine and spruce plantations, to protect forests from industrial and agricultural pollution, and to cut back pollution and the oppressive game wildlife stocks. East Germany's impoverished forest ecology meant that long-term stability and basic health concerns, rather than fiber output and cash flow, had to be management's principal goals. When taken together with geographic and ecological constraints, these tasks call for restoration of a forest of diverse species and ages and complex structure, or a close-to-nature forest. The Party leadership ignored these goals, preferring instead to maximize gross production and impose structures consistent with Marxist-Leninist theory. Just as East Germany's political collapse was never inevitable, neither was forest collapse. But as long as forest management was directed by political and ideological concerns rather than by concerns for productivity and long-term stability, the East German forest was bound to end in decline and economic ruin, just as the East German state was destined to dissolve into oblivion.

Finally, no policy, whether grounded in Marxist-Leninist hypermaterialism or in ultra-green environmentalism, could have succeeded without integrating economic and ecological policies on an equal footing. Forest health in East Germany was critically dependent on a diverse rural economy and an open political system, a priori conditions of overwhelming importance. Yet ecological health and stability are also absolute goods independent of politics and economics. And the concerns of ecosystem analysis—for high levels of diversity, for low levels of control and uncertainty, for healthy information and resource flows, for individual autonomy and populations that interact freely with their environments—are equally worthwhile in economic and political life and cannot be compartmentalized. Integrating these values when thinking about political, cultural, and economic systems will improve analysis, make policymakers better stewards, and give a practical and ethical basis to policymaking. So the process and tools of ecosystem analysis not only aid in policymaking and problem solving but also serve vital ethical and moral concerns.

# *Acknowledgments*

No writer can have been more blessed in his teachers, readers, friends and family than I. Two extraordinary forestry scholars from opposite ends of the cold war spectrum, Professor Dick Plochmann of the University of Munich and Dr. Hans-Friedrich Joachim of the Eberswalde Forest Research Institute northeast of Berlin, gave unsparingly of their support and advice. Dick Plochmann introduced me to Central European forest history and policy through a German Academic Exchange Service (DAAD)–sponsored field trip to West German forests in 1986, and then guided and encouraged my study of East German forest management. Dr. Hans-Friedrich Joachim took enormous pains to support my research in the field, my interviews, and my work in the archives. His guidance and patronage won me access to the top levels of the East German forest bureaucracy and opened the doors to many once secret archives. His personal example of not only enduring, but thriving under "real, existing socialism" stands as a victory of an individual in a collectivist regime. Dr. Robert Hinz, chief forester of the Protestant Church forest in Brandenburg and chief of all Brandenburg forests after 1989, also gave me invaluable help. Their generosity and kindness made the debt American foresters owe to German foresters personal and immediate.

There may be no better place to study applied ecology and policy or an intellectual atmosphere more encouraging to interdisciplinary and international work than Yale University's School of Forestry and Environmental Studies. I have had many mentors: Professor Stephen Kellert encouraged me to use ecology to frame history with humans as an integral part of the landscape. More than a generation of Yale foresters have learned how to read the natural landscape from Tom Siccama, whose analytic structures I still apply when thinking about complex systems. Professor David M. Smith's teaching in silviculture gave his students a grounding in applied ecology and forest history which transcends the context of the United States. Professor Garry Brewer directed my dissertation, "Acid Rain and Romanticism," and his teaching in the policy sciences grounds my analysis of East German policymaking. Professor Henry Ashby Turner, Jr., was a critical influence on my work, and I came to love historiography through his seminars on modern German history. Professor Bill Cronon's research seminar in environmental history, his writing on environmental history, and his thinking on narrative and on how humans fit into the natural landscape have also influenced me greatly. I am deeply grateful as well to former dean John C. Gordon for his support, to Dr. Paul Draghi for his sympathetic reading, and to my colleagues at the School of Forestry and Environmental Studies for their support. Dr. Jim Scott's scholarship and support have been very important, and I'm particularly delighted that he accepted this book for his Agrarian Studies Series.

Dr. Claus Dalchow of Brandenburg's Center for Agricultural Landscape and Land Use Research (ZALF) and Monika Thiele at the Brandenburg Land Survey helped me greatly in tracking land use changes. I am also grateful to Dr. Dieter Vorsteher and Christine Zeidler at the German Historical Museum in Berlin for their generous permission to reproduce parts of their wonderful collection. Bill Nelson worked with me to create new maps of Central Europe and Germany of elegance and voice. The final manuscript benefited greatly from the insights and criticism of Professor Emeritus Paul V. Ellefson of the University of Minnesota, Dr. Thomas E. Lovejoy, Nancy Geary, and Professor John Lewis Gaddis. Patricia van der Leun read this manuscript at every stage and gave me invaluable support and encouragement.

I am most thankful to the directors of Yale University Press for taking a chance on a first-time author, particularly one emerging from a dissertation, for their support of this new perspective on cold war history, and for supporting environmental history. Environmental history helps us see how "everything is connected to everything," in Lenin's words, and to perceive the complex relation among humans, the natural landscape, culture, and ideas. I thank particularly Jonathan Brent, a scholar of vast range and a great editor, for his encouragement and Eliza Childs for her invaluable advice and skill in editing this manuscript.

This work would have been far different without the constant stimulation and refreshment I reap from my work with the Smith Richardson Foundation (SRF). It was a privilege to work with H. Smith Richardson, former chairman of the foundation, in the 1980s and 1990s as the cold war era closed. His example endures as a model of the highest personal and business ethics. Peter Lunsford Richardson designed and oversaw the foundation's transition into the post–cold war era, deepening the foundation's work in international security studies and foreign policy and is a constant friend. Donna Walsh, Peter's executive assistant, has been a steady support and confidant, and SRF would be a lesser institution without her. Dr. Marin Strmecki, SRF's director of programs, patiently and generously worked through many thorny problems of political philosophy with me, helping me to resolve my ideas more than he knows. To my colleagues on SRF's trustees and boards of governors, I can only say that as East German farm and forest collectives were "the greatest schools of socialism," so SRF's grants meetings have been extraordinary schools of policy analysis and scholarship.

I've been particularly blessed to have steady friends and family. So much of the time any book demands comes from the margin, taken from family and friends, yet I never felt anything but their unconditional support and approval. Maggie Daly, a garden designer of unsurpassed grace and judgment, sparked my interest in forest ecology twenty years ago. During our walk together one early spring afternoon, the magic of natural structures and complexity emerged from the disorder of the rough, chaotic pasture below my garden, and I was off to forestry school within the year. I am deeply indebted to Maggie and to her husband, Michael Daly, for their wisdom and friendship. Emily Kruger's interest and intelligence lifted me to the final push to complete this book, and I drew deeply from Dick Richardson's and Bob Falkenhagen's encouragement and friendship.

My sister Julia Nelson has given me unstinting support. I've depended on the continuity and depth of the family life my sons and I share with my sisters, Julia, Susan Nagle, Louisa Dysenchuk, and Mary Anne Grammer. Our father, Lawrence Nelson, was the most insightful and patient teacher I've ever had. No businessman ever had greater imagination or a better understanding of market risk or saw more clearly the relations between human populations, geography, and agriculture than he did.

My three sons, Malcolm, Eric, and Arvid, have passed through their youth to become young men in the time it has taken me to complete this project. They uncomplainingly shared my travails and joys and forgave me my moods. Their grandmother Marjorie Meacham Meek shares and amplifies my delight and pride in these fine young men. She has been an astute reader of my work and an unfailingly constructive councilor. When I started my dissertation many years ago Marjorie gave me a small cast-bronze bull, legs planted and head lowered. It has been on my desk ever since, and its form and memory strengthened me whenever the goal seemed far away. I treasure her judgment, aesthetics, and love as I cherish my sons' goodness, intelligence, and affection.

# *A Note on Terminology*

Political language enriches narratives of cold war history. Soviet bloc and Chinese propaganda in particular bristled with unintentional humor: ritualistic catchphrases such as "running dogs of capitalism" and "fascist warmongers and imperialists" contrasted invidiously with praise for "the peace-loving socialist camp" and "freedom fighters." Naturally, U.S. Air Force pilots wore "Yankee air pirate" patches on their flight suits to show their disdain for North Vietnamese propaganda and abuse of captured fliers. The publisher Malcolm Forbes had "capitalist tool" painted in high letters down the length of the fuselage of his jet. The jet often sat at Sheremetyevo Airport during his visits to Moscow as stone-faced paramilitary guards and customs inspectors rifled tourists' bags, oblivious to the challenge to state censorship outside on the tarmac. Soviet propaganda marked the general corruption of Soviet political life that George Orwell singled out in "Politics and the English Language."

As a researcher, however, I almost despaired at the numbing rhetoric of triumphalist and opaque speeches, journals, books, and internal reports written in Parteichinesisch (Party Chinese). Official language and economic data often contradicted reality. Henry Kissinger, cleverly paraphrasing Napoleon's comment about the Holy Roman Empire, observed that the East German state, the German Democratic Republic, was "neither German, democratic, nor a republic." And how could East German propagandists anathematize the timid West German leadership as "imperialist" when the East German People's Army had marched into Prague in 1968 at the head of the Warsaw Pact tank columns? The weight and repetition of the propaganda alone made it as impossible to ignore as it was difficult to decipher.

Once I traveled within the soon-to-dissolve East Germany and spoke to Party officials at their stations, I saw that official language encoded policy debates fought at the level of the Politburo and the Party's Central Committee, such as whether politics or economics were the primary tool of policy. Not politics in the sense that Aristotle or economics as Keynes or Schumpeter understood them; "politics" and "economics" were code words for ideological positions within the narrow margins that Marxism-Leninism afforded. When the Party leadership declared "real, existing socialism" or the realization of the "union of workers and peasants," they signaled the victory of orthodox over technocratic Marxist-Leninists. The distinctions are without a difference, however: both schools believed in the absolute truth of Marxism-Leninism, the sole role of the Party, and in their unerring ability to read the future and direct society toward communism. The more this secular utopia faded—as the leadership's stunning incompetence seemed to destroy more than it produced—the more insistent and apodictic the Party leadership's language became. With "real, existing socialism" the leadership abandoned hope of winning popular approval or even of achieving socialism, much less communism.

When I write of "socialist forestry" I repeat the coded language of the East German state. I could as easily have termed it "Marxist-Leninist," "Nazi," "First Five-Year Plan," or "industrial" forestry, but never the enlightened forest management for which West German Social Democrats fought. Socialist forestry, in any event, may be an oxymoron, for politics and forestry fit together as appropriately as politics with ethics and religion. Similarly, when I write of the "Party leadership," I'm speaking of the Soviet-installed, narrow Marxist-Leninist hierarchy despite the sham of the "Socialist Unity Party": again, neither "socialist" nor a party of "unity." Naturally, their lack of legitimacy haunted them. "Socialism" in the East German context means: a command economy, shortages, central control, and suppression of individual rights, not liberal democratic socialism.

Geographic language is equally rich in meaning. Europe's boundaries, and with it our sense of geographic equilibrium, underwent radical change in 1945 as Soviet political control extended as far as Stalin's armies and Germany's center of balance slipped decisively to the West. In 1937 Berlin lay roughly 650 kilometers from East Prussia's eastern frontier along the Nieman (Memel) River. After Stalin's 1945 border adjustments, the center of Berlin was only 80 kilometers from Germany's eastern frontier—now along the Oder River with

Poland, not Russia, on the far bank. Cold war "East Europe," the lands the Soviet Union controlled through satraps, was more a political than a geographic idea. Europe's geographic center lies in the modern Czech Republic, in the Bohemian massif. Cold war rivalries — and political realities — blinded us to the fact that Stalin took most of Central Europe in 1945.

What we think of today as East Germany historically is central Germany, *Mitteldeutschland,* the middle Elbe Basin south of Magdeburg between the Czech border and the Harz range to the north. "Eastern Germany" was a geographic term weighted toward Germany east of the Oder River, particularly to the great province of East Prussia and Kant's city of Königsberg, now the Russian oblast of Kaliningrad (see figure 11). "East Germany" refers only to the rump of Prussia wedged tightly between the Oder and Elbe rivers, the Soviet zone of occupation briefly dignified as a nation-state (see figure 6). When I refer to "eastern Germany" in the first chapters, I mean essentially the bulk of the Prussia that the Allies abolished at Potsdam: the New Mark, eastern Pomerania, Silesia, and the great geographic outlier and fortress-home of the Teutonic Knights — East Prussia.

# I

## *Prologue*

*It is base to receive instructions from others' comments without examination of the objects themselves, especially as the book of Nature lies so open and is so easy of consultation.*
*— William Harvey*

In March, Berlin can be cold, open as it is to northeastern winds that blow in unchecked from the Urals across the lowlands of the North European Plain. Overcoats and shoes comfortable in Hamburg or Frankfurt did not keep out the freezing wind sluicing across the deserted train platform despite the morning's brilliant sunshine and dryness. But at least travelers shivered above the street, clear of its pall of furnace gases and greasy exhaust fumes. And the street noises dulled as they drifted up against the constant wind, which almost drowned out the coughing of car and truck engines, the pounding of burnt-out bearings, and the sputtering complaints of jackhammers and generators deployed in desultory repairs below. And yet, despite the cold and filth—the sheer poverty of prospect and habit—I was exhilarated just to be in the East's Lichtenberg Station, on my way to eastern forests. Westerners could not have made the slow trek from the West to this poor district a few months earlier, and certainly could not have traveled unescorted into the forests around Eberswalde some twenty-five miles northeast of Berlin, my destination.

The Lichtenberg station agent hesitated to sell me the cheap passage to Eberswalde, checked unconsciously by the institutional memory of control. East German bureaucrats had always stood out among the East bloc for their prim coldness, their intolerance of dissent, and the sly, sharp business practices of their finance and trade officials, such as the shadowy "louche Stasi Colonel," Alexander Schalck-Golodkowski, so different from the Hungarians' openness and humor, the Czechs' gentle world-weariness, or the Poles' independence and anger at Russian, rarely "Soviet," power.[1] The station agent hesitated, visibly trying to remember if it was still forbidden to sell such a train ticket to a foreigner, then, with a shrug, grudgingly passed my ticket under the window grille.

"You must go to Eberswalde. You must talk to Dr. Joachim," urged Professor Richard Plochmann, a senior West German forest historian and policy scholar widely respected in West and East Germany. Many assumed that the Party leadership had created a radically new forest, even if it was on the brink of collapse, changing East German foresters themselves. Dick Plochmann thought differently, as did most older West German foresters who remembered their shared schooling before the war and exchanges in the early 1950s before the Party leadership forbade contact. Most of all, they remembered that modern forest ecology was born at Eberswalde in Prussia, the intellectual center of forest science before the Second World War. They also recognized that forests are long-lived ecosystems with high levels of inertia. So change was unlikely to be as absolute as the Party leadership's rhetoric and ideology forebode.

I made my way to Eberswalde on a cold, early March morning in 1990, memories of the cold war and the Wall still powerful. An ancient coal-fired locomotive pulled the short line of carriages to Eberswalde, past old brick roundhouses, water towers, and coaling stations which graced the rail yards. Narrow, cobbled roads lined closely with poplar and lime trees ran through fields and small towns unmarred by the garish advertising posters, brightly lit filling stations, and convenience stores which blight West German villages. The modern world seemed far away, not immediately recognizable even in the monotonous fields of the collective farms or in the serried ranks of the People's Forest.

Arriving in the industrial town of Eberswalde itself meant an abrupt return to urban grime. I supposed I would find the Research Institute on the town's edge, where the Eberswalde forest began. Local maps were useless; as in most East bloc countries, they often did not show important government buildings. The station and dingy buffet were deserted, so after waiting a few minutes for a taxi, I trekked through Eberswalde's streets looking for signs to the Forest

Research Institute, meeting only glum Soviet soldiers. I found the institute compound fairly soon, at the end of a once-grand boulevard lined with shabby villas and a deserted inn, the remains of the resort Berliners thronged to before the war for weekends in the forest.

Razor wire, a high fence, and a cleared strip circled the institute grounds, the only entrance running past a guard post and formidable gate where a uniformed armed guard demanded my papers. I gave him Dr. Hans-Friedrich Joachim's letter of invitation, and he withdrew to announce me. In a short time I saw Dr. Joachim's spare frame; he walked quickly across the institute's dusty forecourt, dressed in civilian jacket and tie rather than the Forest Service's military-style uniform. The guard unlocked a narrow side wire gate, and I passed through to meet Dr. Joachim, who, perhaps out of habit, waited on the inside to greet me and take me to his small office behind the institute's main building.

Dr. Joachim, who had planned to retire, was instead taking over as the institute's codirector and moving into an office in the main building. A generous, kindly man of high intelligence, he agreed to help with my research. Many had already judged that the Party leadership and Marxism-Leninism were responsible for East German forest decline; certainly their aggressive management and industrial and farm pollution had pushed the forested ecosystem to the brink of economic and ecological collapse. Although their responsibility was clear, their pretensions of control and mastery of radical change would not prove out—they never had the competence. A better explanation of the cause would follow from study of East German forestry in its historical and cultural contexts, as appropriate for its long biological and economic lives. Two hundred years of human agency led to the near collapse of the East German forest, with the Party leadership unwittingly playing a minor, if destructive, role in the struggle. More important, the landscapes of East German forest and farm, and the vast frontier and forbidden zones, witnessed the true qualities of "real, existing socialism" and testify to the costs to the East German people in the constant dearth of the "thousand little things" and in East Germans' forty years of penance for National Socialist crimes, penance served as the Soviet Union's frontline state and its principal Warsaw Pact ally. Months later as I walked in the forest with Robert Hinz, the new director of Brandenburg's forests, we came upon members of a local men's chorus who nervously asked him, "Herr Forstmeister, is it true that the communists destroyed our woods?" Hinz replied, "The Church's Brandenburg forest is over 400 years old—What does forty years of communism mean next to that?" Comforted, the men respectfully touched their hats and continued their Sunday walk.

Over the next year I saw the barriers and controls left over from the Party's rule disappear—the brittle guard at the institute's main gate evolved into a cheerful porter dressed in a misbuttoned sergeant's surplus tunic dating, clearly, from the time of Elvis Presley's army service. People could move in and out of the institute compound freely, the closed gate and razor wire gone. Foresters no longer locked their inventories and management plans in the safes standard in every office. Copier locks and logs also disappeared. Gradually Party members lost their defensiveness, no longer rationalizing their hard management by appealing either to the principles of Marxism or to their true goals of a Swedish-style socialist state. One idealistic Party member explained with unintentional irony his service to the rigid and destructive system: "I had to be optimistic, I was a communist." Sheepishly, senior foresters recovered their footings in science, sometimes pulling from a bottom drawer or the back of a bookcase prewar natural histories and romantic tracts on forestry. The stiff East German forest bureaucracy and the forest itself blended into the liberal West German structure with remarkable smoothness.

Two years after I first traveled to Eberswalde, I walked through the institute's pine forest to Dr. Joachim's house for a last Sunday afternoon with his family before I returned home. East Germans cherished Sundays with family and friends, often sharing a Haydn concert on an old radio set after breakfast followed by a long walk on forest roads and late afternoon tea, a retreat into their most intimate circle. I was not accepted at once, and in time only because they took pity on an outsider marooned without the family and friends necessary to survive in the cold, monochromatic East German society. My life gradually fell into a rhythm of long workweeks punctuated by privileged, warming Sundays with generous colleagues and their families. Such reflective, intimate Sundays endured in the early days after reunification, the call of Western consumer culture still distant.

The forest roads and streets were empty, but on this Sunday I did not miss the roar of struggling engines and grinding gears from Soviet army trucks and jeeps clambering in, out, and across the oil-dressed ruts they ground deep into the old cobbled forest roads. In my daily walks through the forests that girdled the institute I was more likely to surprise a motorcycle courier or a short, heavily clothed soldier from Central Asia or Mongolia than I was to meet an East German. These soldiers always seemed more surprised than I at meeting, turning back abruptly to their barracks deeper in the woods. By 1992 these sightings were less frequent as the 400,000 Soviet, now Commonwealth of Independent States, soldiers beat their awkward retreat east.

The Eberswalde forest I walked through, with its tall, old trees and clear understory, mirrored the classic German forest of myth—park-like openness, dense ranks of uniform stems clear of lower branches, high canopies of loosely bundled needles puffing out soft green against the brilliant blue sky. Outwardly, the trees looked healthy, no worse than those in West Germany. Apart from forests on higher elevations in the Erzgebirge on the Czechoslovak border or in the worst industrial areas in the south, it was difficult to see evidence of forest death, *Waldsterben*, or of mismanagement.

Dr. Joachim and his wife, both in their mid-sixties, lived alone on the second floor of the villa of Alfred Möller, dean of the institute in the early 1920s and the most influential forest scientist of the twentieth century.[2] A fine villa in the Biedermeier style, the house sat comfortably back from the street. Tall conifers of the institute's arboretum rose above the claret-colored tile roof and the white mullioned windows set into earth-colored stone, framing the garden and house. The finery of most prewar buildings in Eberswalde's once affluent suburbs had faded through neglect and the lack of materials, but the serenity of the old villa's setting, its footings on the arboretum margins, and the Joachims' care lent it luster.

Now, late in spring and my work and research in the former East Germany nearing its end, I walked quickly out of the security of the orderly pine forest and approached Möller's villa in bright sunshine. I had visited the Joachims often in the past two years, but always in the suddenly darkening twilight of the cold and wet north European winter. I took great comfort from their warmth, living as I was in a seedy and expensive rented room above a shuttered-up pub. As I came nearer in the midday sun, for the first time from the main street and not from the arboretum, I saw a waist-high meadow filling the front garden, a shapeless sea of flowing textures and earth colors. The meadow's lack of formal structure after weeks in East Germany's pine plantations and well-ordered archives startled me. I felt rootless, bereft of the clear signs and comforting borders which ordered the forest behind me.

But as I walked a bit farther and stood square against the formal, worn façade, a four-foot-wide, closely mown path of rich green-blue turf revealed itself as a safe passage through the center of the meadow's disorder. Dr. Joachim saw me at the same time, rose up from the small garden at the side of the house where he was working, and came forward to lead me toward his front door, moving athletically despite his age. The pleasure of garden work in the sudden warmth after weeks of rain and sleet showed in his smile and pale blue eyes. Over the past two years we had seen together the uncertainties of reunification melt away as Western institutions swept before them the old system. The new liberal market economy doomed East Germany's industry, freeing the landscape overnight from its Augean loadings of gas, dust, and particulate. The new free market in forest products also meant the end of harvesting in the exhausted stands, and foresters could resume the century-long task of restoring a close-to-nature forest ecology.

We stopped halfway down the path to look deeper into the meadow, the high sun behind our backs illuminating the suddenly airy and spacious interior reaches as we leaned forward. Taller grasses and sedges, wood millet and fescue filled the overstory. Purplish poverty grass spoke to the dryness of the northern lowlands while native weeds, such as the beautiful, invasive sand grass *Calimagrostis* and knapweed, added rich texture and color as well as the resonance of a wilder habitat. Wildflowers of all heights thrived in the sunlit understory—columbine in violet-blue, white, and red; white-flowered wood anemone; and yarrow burst chaotically among the gray-green shafts of grass and sedge. Oxeye daisy, yellow archangel, and kidney vetch added diverse shades of yellow. Wind and changing sunlight created a constantly emerging structure of height, texture, and color as diverse wildflowers and grasses native to woodland and heath mixed and clashed. Our small effort had called forth a new, deeper picture.

My first impressions, after all, were not far from the mark. There were no real borders within the woodland meadow; plants mixed by chance into random patterns of color, height, and species, each blending seamlessly into the other. Where the suburban garden pleases the eye with its ordering of nature and clearly defined borders—the English enclosure—this meadow was more truly a community of plants given space to grow together rather than a collection of plants shoehorned into discrete niches to fulfill the designer's aesthetic vision.

We withdrew to the house, both elated at the pleasure we took in his garden within the relentless order of the state forests surrounding us. We went on with our afternoon walk and early supper, no longer talking of meadows and the natural forest but of Dr. Joachim's experiences as a young student of ecology at the University of Berlin before the war. After army service on the Russian front and homecoming to the terrible winter of 1946 and near starvation, he resumed his studies at Eberswalde and began a promising career as a forest ecologist and professor. In the late 1950s the Party selected him and other promising young scholars to join East Germany's delegation to the UN's International Union of Forest Research Organizations (IUFRO) conference in Rome. The Party leadership viewed the Rome conference as an opportunity to promote their political and economic achievements, and it therefore issued orders not to discuss problems in East German forestry or to criticize the government and to report to the secret police all contacts with West Germans. Dr. Joachim gave a West German colleague a lift to the train station after the conference ended and did not report it. But one member of the East German delegation did not fail in his duty and turned him in. Dr. Joachim was ordered to account for himself upon his return from Rome. In the aftermath of the 1956 Hungarian Revolution and in the midst of one of the Party's episodic spy hunts and purges, Dr. Joachim could not satisfy his interrogators.

The Party withdrew his Western travel privileges, took away his teaching duties, and limited his research and publication work, his professional life essentially over until he agreed to cooperate. "But we are not monsters," they told him, "you may keep your housing and continue to work, but you will never advance in our society until you join us." Dr. Joachim decided that, in fact, he did not want to be a part of the Party's increasingly intolerant system. The heavy conformity of socialist forestry repelled him. He hated the regulation that professors wear uniforms and badges of rank modeled on those of the National People's Army, or the expectation that he would ostracize fallen colleagues as he had been ostracized. He would not share in any of the other countless petty cruelties of the East German bureaucracy. And so the institute's dean banished him to the flimsy trailer where I first met him, put up as temporary space many years before but still in use in 1989. For the rest of his professional career he endured steady, low-grade harassment and professional isolation without hope of rehabilitation. Yet he lived a full life with his family in Alfred Möller's house, working at his research on hardwoods and tending his garden outside the main pulse of East German society.

As we sat with his wife later in the evening by the coal fire in the second-floor parlor overlooking the front garden, I confessed that I did not know if I would have had the courage to resist the Party, which was indeed the sole legitimate government and recognized as such by the Western powers. Few, East or West, thought the Party's rule would not last forever. Dr. Joachim, momentarily confused, replied in a puzzled tone, "But I had no choice." His wife added with pride, "And I could not have loved a man who was a Party member."

When the Wall fell in 1989, just short of Dr. Joachim's retirement, his career resolved itself at the eleventh hour as West Germans reached out to eastern Germans untainted by hypermaterialist socialist forestry. The church foresters for the Protestant Church in Brandenburg and the Catholic Church in Saxony, the only land spared compulsory collectivization, guided their states' return to ecologically based forest management and rebuilding of the Forest Service. Dr. Joachim took over as codirector at the institute and directed IUFRO's 1992 centennial celebrations at Eberswalde, the birthplace of IUFRO in 1892. Thus, his career had come full circle since he came home to disgrace from the Rome IUFRO conference thirty years earlier. Far from a quiet retirement in 1990, Dr. Joachim celebrated German forest ecology's return to its home at Eberswalde and the resumption of the German forest's century-long process of restoration.

Later that week, I returned to Berlin and then, home. Dr. Joachim had just taken delivery of a bright red Lada automobile, a relict of the Soviet empire's evergreen days, ordered and paid for seven years earlier. He offered to drive me to the airport, to share one last field trip together—his own homecoming to the Berlin Botanical Garden in West Berlin where he had studied botany before the war. He did not talk about his life then or his service on the Russian front, but memories of his teachers, of lesson and example in the garden precincts, clearly filled his thoughts. As with most homecomings after a long absence, he reasonably may have feared disillusionment, that the memories of his youth and of the persistence of scholarship which had sustained him for the past forty years may have been false.

So, in a mood of nervous anticipation we drove to West Berlin, easier, we thought, than taking the lurching old steam train. Cartographers had optimistically resutured East and West Berlin on their post-Wall maps, using bright colors to refashion broken subway, road, and rail lines. But we found that the prewar avenues which once crossed Berlin were still nearly impassible, and we soon were mired in a tangle of dead-end streets and struggling Trabis and Wartburgs in a crumbling district of dingy apartment blocks and rundown factories. It took three hours to cover the thirty miles from Eberswalde to the Botanical Garden in West Berlin's prosperous Dahlem district; we arrived with only forty-five minutes for Dr. Joachim to make his homecoming and tour the world's largest botanical garden.

Dr. Joachim, hesitating only a moment while he got his bearings, seemed to recall the layout of the garden's forty-three hectares perfectly, the grand buildings housing the great library and conservatories, classrooms, galleries, and cabinet rooms, all descendants of the Prussian monarchy's herbarium and botanical collections founded in 1815. Without speaking Dr. Joachim took my arm to lead me deeper into the interior, toward the fourteen-hectare arboretum. We sped past complexes of greenhouses, planting sheds, cold frames, and plant beds, all energy and order behind glistening glass and wood frames enameled glossy white, purpose and order written in every structure. As we hurried over the crest of a hill suddenly we saw spreading below us a vast wildflower meadow that filled the shallow valleys and ran up the lower slopes of far sandy ridges where native hardwoods, magnolia in flower, birch and beech, maple and oak mixed thinly. Hawthorn and common rose grew underneath and just below, too loosely to form hedges in the dry soil. Great drifts of color and texture were constantly articulated anew as high-flying white clouds raced across the sky, and sun and wind swept up through the swales and over the gently rolling, sandy ridges which laced the arboretum. Not a word was needed; the wildflower meadow's infinite gradations of form and structure, its subtle shadings, and diverse plant communities instantly recalled Dr. Joachim's woodland meadow in the front garden of Möller's villa in Eberswalde.

For a while we stood silently. Then Dr. Joachim, still holding my arm, spoke softly, "It's as I remembered it," and I felt, suddenly, that I had returned home myself. And the cold war division of Germany became an illusion as images of woodland meadows and landscapes in East and West Germany fused. And rather than we from the east coming as insolvents begging for redemption, we came bearing ecological science and older values evolved in eastern Germany, values that had triumphed over Marxism-Leninism and now challenged the materialist culture of consumption.

# 2

# *Landscape and Culture*

*If you stand right fronting and face to face to a fact, you will see the sun glimmer on both its surfaces, as if it were a cimeter, and feel its sweet edge dividing you through the heart and marrow, and so you will happily conclude your mortal career. Be it life or death, we crave only reality. If we are really dying, let us hear the rattle in our throats and feel cold in the extremities; if we are alive, let us go about our business.*
*—Henry David Thoreau,* Walden

So I came to East Germany in the early days after the Wall fell, drawn to the Neiße River marshes and Tacitus' "bristling forests" of remote Thuringia in the south and the broad wetlands of the Oder River valley, the *Oderbruch*, and the old Prussian "sand and pine" landscape east of the Elbe River.[1] I was drawn by reports of crisis in the eastern forest—the end result of Marxism-Leninism's forty-year collision with the natural landscape. This struggle offered a natural experiment, how a long-lived ecosystem with a 1945 provenance weighted toward the *Kaiserzeit* before 1918 changed under rule of the Plan. The Party leadership devoted particular attention to forest products and valued timber, the most valuable and abundant raw material in eastern Europe, as a vital domestic asset in their otherwise impoverished natural resource economy.[2] The qualities of forest stewardship under the Party's rule testify to its competence and power. Documenting its avowed war on nature is a critical question, in Willy Brandt's words, "of political hygiene."[3] The quality, direction, and scale of change in the German forest, itself a palimpsest of two centuries' political, cultural, and ecological history, also reveal underlying social and political change. The qualities the Party strove for in the forest, I suspected, mirrored its leaders' core values and illuminate the history of East Germany, the most secretive of the East bloc satellite countries.

For centuries Germans have found symbols in nature to express a common political and cultural identity, in particular reaching out to the Rhine River, the Brocken, and the forest. The Rhine sutures the German landscape, rising in the Alps, cutting through the uplands, and then flowing across the lowlands to the North Sea, an organic thread uniting Germany. The great composer Richard Wagner used the Rhine's myths, of the Nibelungen and gods, as a symbol of purity and the German soul. The Rhine maidens celebrate its power at the close of *Das Rheingold,* singing, "Goodness and truth dwell but in thy waters," and the Rhine purifies the ring, representing the incorruptible essence of the German people.[4] Johann Wolfgang von Goethe wove another major feature of the German landscape, the Brocken, tightly into the high culture of literature, poetry, and music. "The Brocken," wrote Heinrich Heine, "is a German" full of tolerant rationalism conflicting with the wild-student madness of the "real German romantic."[5] Goethe set the Walpurgisnacht climax of *Faust,* which Heine celebrated as "the great mystical German national tragedy," on the Brocken summit, where Mephistopheles tempted Faust with the "marvelous view of Mammon's blazing light" below while Druid spirits and witches rioted about them.[6] Goethe returned often to the Brocken as a symbol of the clash of modernity with untamed nature.

The forest resonates through German culture with a force unmatched even by these powerful symbols. Germans still find shared cultural and political identity through their "particularly intimate and mystical relationship to the forest."[7] Tacitus marveled that the ancient Germans "consecrate whole woods and groves, and by the names of the gods they call these recesses; divinities these, which in contemplation and mental reverence they behold."[8] "Tree worship," mused Frazer in *The Golden Bough,* "is hardly extinct today. Nothing," he concluded, "could be more natural." Jacob Grimm, for whom forests played a major role in the fairy tales he collected with his brother, wrote of central Germany's "immense primeval forest stretching eastward from the Rhine for a distance at once vast and unknown."[9] Forest covers less than 30 percent of the modern German landscape, but it claims greater space in the national consciousness.

Modernity's conflict with custom defined and shaped the European landscape as much as economics and the forces of ecology and geology. Symbols and myths drawn from nature, the Rhine, the Brocken, and the forest reveal more than what the landscape looks like and is; their resonance in culture also marks the landscape's transformations and tells why it changed. From a symbol such as the Brocken we can see the physical as well as the cultural landscape and understand not only what happened to the East German forest, but why — and how to imagine its future.

Goethe first climbed the Brocken in December 1777, drawn by its unusual natural history and mythic power: a massive, old plutonic core of granite robed in gneiss and crystalline schist sculpted and scrubbed clean by the Vistula glacier. The twenty-eight-year-old Goethe came to the Brocken looking for a sign that his personal destiny was for greatness. He told only his closest friends of his climb, asking particularly that they keep it secret.[10]

The climb was unusual for its time, as mountain climbing was not considered a sport until later in the nineteenth century. The forester J. C. Degen, who had not climbed to the summit in the past thirteen years, led Goethe up the unmarked summit path over one meter of snow with a hard frost crust, the harsh conditions and heavy fog almost turning them back several times. They took three hours to climb three hundred meters and reached the top just as the sun broke through on the summit and clouds filled the valleys far below them. Goethe wrote in his journal, "Bright, magnificent moment, the whole world in clouds and mist, and on top everything bright. What is man that thou are mindful of him?"[11] Goethe, almost exhausted by the struggle, filled with reverence for nature's power and meditated on the dangerous ascent as if in prayer. But he also took careful notes of the summit's geology and plant life and made pen and ink drawings for later study, taking intellectual as well as emotional sustenance. This experience with untamed nature came at a critical stage in Goethe's life, and afterwards he worked with greater confidence and insight. The Brocken and the Walpurgisnacht myth powered *Faust*'s climax, central symbols in his work of the conflict between custom and modernity and the cultural anxiety which arose from it.[12] This sensibility crossed the Atlantic to influence the early American conservation movement.

Ralph Waldo Emerson developed Goethe's theme as "the primal antagonism, the appearance in trifles of the two poles of nature."[13] Americans of the mid-nineteenth century looked to Germany first for models of scholarship in the natural sciences, philosophy, and religion, and for professional forestry.[14] John Muir followed Emerson in his admiration of Goethe and kept works of the great German naturalist von Humboldt by his bedside. In December 1874 Muir climbed a one-hundred-foot spruce on a ridge in the Sierra perched over a deep valley in "one of the most bracing wind-storms conceivable," the setting echoing Goethe's 1777 hazardous ascent of the Brocken. Delighting in his carefully calculated risk and the surrender of personal identity at the height of the storm, Muir exulted in the power and sublimity of raw nature, writing, "In its wildest sweeps my tree-top described an arc from 20–30 degrees, but I felt sure of its elastic temper. I was therefore safe, and free to take the wind into my pulses."[15] Muir blended a mechanic's common sense and appetite for utility with a passion for natural history and wildness, responding primally to his sense of nature's complication of reason and intuition which Goethe first expressed — more elegantly but no less passionately.

Goethe again picked up the Brocken's symbolism of the conflict of custom and modernity and innovation in the last decade of his life in partnership with the young Felix Mendelssohn. Their friendship was born when Mendelssohn's teacher brought the twelve-year-old boy, widely acknowledged as a composer whose genius might surpass even Mozart's, to Weimar to meet the aging Goethe in 1821.[16] Goethe instantly befriended the youth, soon taking him on as a protégé. Six years later Goethe asked Mendelssohn to compose a cantata based on his 1799 poem *The First Walpurgisnacht,* whose battle climax on the Brocken between Christian zealots and Druid priests Goethe declared in a letter to Mendelssohn to be "symbolic in the highest sense": "For in the history of the world, it must continually recur that an ancient, tried, established, and tranquilizing order of things will be forced aside, displaced, thwarted, and, if not annihilated, at least pent up within the narrowest possible limits by rising innovations. The intermediate period when the opposition is still possible and practicable, is forcibly represented in this poem, and the flames of a joyful and indestructible enthusiasm once more blaze high in brilliant light."[17] Emerson's "primal antagonism" would play out over the Brocken summit on May Day Eve, Goethe's libretto soaring on the force and majesty of Mendelssohn's music.

Before Mendelssohn began work on the cantata he determined to climb to the Brocken summit as Goethe had done. But a violent spring storm rose up suddenly, and Mendelssohn turned back short of the summit, not helped by his stumbling, drunken guide who had never before been on the Brocken. Mendelssohn later described the mix-up to his family in letters suffused with humor and charming self-deprecation.[18]

Heinrich Heine also tried to emulate Goethe's analytic eye when he followed him to the Brocken summit, scorning the view as no more than "an accurately designed and perfectly colored map" — artificial and lacking the spirit and genius of German myth and poetry.[19] But Heine missed the point implicit in Goethe's understanding: the German landscape is artificial, the product of conflicting human needs over centuries.

Mendelssohn's life and work linked the eighteenth-century world of Goethe's natural history, philosophy, and poetry with the modern world of the natural sciences, Hegelian philosophy, and Marxism. Mendelssohn studied for one year—in 1828—at the University of Berlin, where he heard von Humboldt lecture on natural history and Carl Ritter teach geology. Eduard Gans, also Karl Marx's teacher, taught Mendelssohn modern history and the great G. F. W. Hegel lectured to the nineteen-year-old Mendelssohn on the aesthetics of music.[20] Hegel may have known of Mendelssohn's presence in the lecture hall; he was a frequent visitor, along with von Humboldt, at the Mendelssohn family's Sunday teas and concerts. Mendelssohn had also started his systematic study of Bach's work, which restored Bach's reputation.

But Hegel was not fazed by the famous young composer's presence in his lecture hall; undaunted, he lectured of a complex but alien art with which Mendelssohn was not familiar. Mendelssohn afterwards parodied Hegel's ponderous style to his family's and friends' delight. Nevertheless, Hegel commanded respect; his philosophy was the dominant intellectual force in German universities and would capture the intellectual passion and fierce focus of the young Karl Marx, a more diligent student of the great man than Mendelssohn.[21]

Mendelssohn left the University of Berlin after one year despite his parents' pleas and set out on a walking tour in Switzerland to prepare to write the score for Goethe's *The First Walpurgisnacht.* As he climbed in the Alps he thought deeply about the conflicting philosophies of Goethe and Hegel. He wrote his sister Fanny of the sublimity of the Alpine landscape and his awe at the sudden storms, avalanches, and floods which transformed the landscape overnight: "I know Hegel says, 'that every single human thought is more sublime than the whole of Nature;' but here that strikes me as hardly modest. It is a fine idea, but a confounded paradox nonetheless. For once I will be on the side of the whole of Nature, which, indeed, is likely to be the safe side."[22] The timbre is romantic, yet Mendelssohn scorned romantic pieties, and although Berlioz praised his score for *The First Walpurgisnacht* as "a masterpiece of Romanticism," Mendelssohn congenitally disliked Berlioz' music and Romanticism.[23] Mendelssohn, like Goethe, would not surrender either reason or intuition for an abstraction, either Hegel's or Berlioz'.

Mendelssohn came down from the Alps and pressed on to Rome where he wrote again to Fanny as he neared the end of his work on the cantata, "Listen and wonder!"[24] Mendelssohn returned to Weimar a year later in 1832 to play *The First Walpurgisnacht* score for the dying Goethe. Full of emotion, Goethe delayed his departure, commissioning a portrait of Mendelssohn which he kept in his study until his death.

One of Mendelssohn's most accomplished works, *The First Walpurgisnacht* is a soaring secular cantata incandescent with Goethe's passion for wildness.[25] As in *Faust,* the climax rang out on the Brocken in the driving *Kommt mit*

*Zacken* final movement where Druid guards disguised as devils battled the Christian missionaries, guarding nature and custom against modernity and innovation. Perhaps, as Mendelssohn confessed later, he gave the Druids' "pagan strain" the better tune.[26] The Druids' resistance against the present, irresistible force of modernity fascinated Goethe and Mendelssohn, not opposition to modernity or progress itself.

A profound interest in change permeated Goethe's thinking about the relation of humans with nature. Nature offered a model for thinking about the dynamics of change as a random process governed by disturbance beyond human power, anticipation, or control. Hegel, and his disciple Marx, saw change as following a single, nonrepetitive process governed by discoverable laws, historicism. Whereas Marx saw rational power as unlimited, Goethe was concerned with marking the limits to human power and control. Mendelssohn was too good a student of Goethe's to accept either Idealism or historicism, just as he was too accomplished and intuitive a musician to accept Hegelian aesthetics.

Marx, the avatar of revolution and absolute truth, arrived at the University of Berlin shortly after Mendelssohn as an ardent Young Hegelian. Even though he eventually rejected Idealism, Marx remained a "convinced, consistent and admiring follower of the great philosopher" for the rest of his life.[27] Marx's economic theories and philosophy are innovation writ large, if nothing else, and the polar opposite of custom and nature. Marx was the consummate Christian missionary bearing an ideology of progress locked in inevitable triumphant conflict with custom and nature.

Fifty years passed between Goethe's 1777 and Mendelssohn's 1821 pilgrimages to the Brocken, yet they both looked out upon the same landscape: the fragmented political scene of small kingdoms, petty princedoms, duchies, and Hanseatic cities which characterized the weak post-Napoleonic German nations of the early 1800s. Great swaths of a low forest of oak, maple, and beech coppices, thickets, and groves blanketed the landscape, a mirror to the late-feudal rural economy. This medieval forest of coppices survives in isolated remnants in the great sea of farm fields and high forests of pine and spruce. Coppicing relies upon hardwoods' ability to sprout from cut stumps or "stools." Growth from these stumps is more reliable and faster in the first years than seedlings, trees grown from seed. Coppiced trees enjoy an immediate advantage over seedlings of an existing mature root structure, and they often have little competition for growing space. Peasants could take from such growths an almost infinite volume of poles and saplings and at far more frequent intervals than would be possible from a high forest regenerated from seed. The health of coppiced woods declined rapidly with time, but owners and peasants preferred the products of the coppiced woodland. Coppicing suited the preindustrial economy, yielding dependable volumes of small-dimension round wood for charcoal burning, fence and barrier production, and as an osiery. Yet the low feudal forest of coppices was natural only in the broadest sense of the word and it stagnated under the weight of peasant abuses and the overstocking of game. Despite the emotional appeal of the idea of a virgin forest or *Urwald,* the mixed hardwood coppices and thickets of Goethe's youth were as much an artifact of human management as the pine and spruce monocultures which followed. But romantics had a hard time seeing this.

The principal technical task of foresters was to provide fuel wood, mostly charcoal, an indispensable raw material for the preindustrial iron and steel industries and the most important commercial product taken from the forest before the Industrial Revolution.[28] Peasants enjoyed limited rights, such as mushroom and honey collection, gleaning of small wood and fallen branches for firewood, and trapping of hare and vermin for the pot. But the more valuable peasant rights damaged long-term forest health and dictated forest structure; litter-raking, or taking of fallen leaves and straw for livestock bedding, and pasturing of swine and cattle on beech and oak mast, robbing forest soils of nutrients and damaging tree roots and stems. Elimination of these peasant abuses and exercise of complete control over the forest moved landowners to refabricate a forest where livestock foraging was forbidden and litter-raking theft. And the quickening pulse of economic activity in the early nineteenth century suggested that a different, modern forest structure might soon be preferable to the medieval low forest.

So Mendelssohn's and Goethe's view from the Brocken changed entirely in the early years of the nineteenth century because of an ecological coup d'état —the liquidation of Germany's old, preindustrial forest and its refabrication as a modern industrial forest within a generation. Yet they anticipated in *The First Walpurgisnacht* the approaching Industrial Revolution conflict between custom and innovation. This battle is still fought in the Central European forest. German landowners, drawing signals from increasing industrial demand and the new forest, financial, and statistical sciences, clear-cut the native hardwood coppices covering almost 80 percent of the total forest area and planted in its stead pure pine and spruce plantations at a staggering initial cost and without hope of economic returns for generations.[29] This ecological revolution foreshadowed both the Industrial Revolution and the creation of the German empire in 1871: thickets, groves, and coppices gave way to the modern industrial forest just as the myriad of small polities gave way to the unified German state of the Wilhelmine Empire, the ultimate goal of Hegelian historical process and progress.

Goethe's encouragement of natural history and his empiricism and skepticism prepared the way for this ecological revolution in the forest. A forest scientist of comparable genius in his field, Heinrich Cotta, completed in Goethe's lifetime the first systematic study of forest economics and silviculture, *Anweisung zum Waldbau,* influenced by Goethe's blending of classical and romantic forms and Goethe's natural science.[30] The abused ecology of the German forest in the early 1800s cried out for rationalization of forest management and closing the forest to peasants and their livestock. Cotta proposed clearing the feudal coppiced forest and planting a pioneer forest of pine and spruce to replace the hardwoods as an intermediate stage toward the final goal: restoration of a high forest of mixed, native hardwoods managed through natural regeneration—a forest closer to nature. Cotta's plan demanded an ecological intervention on an unprecedented scale, accomplishing in a few decades what would take centuries naturally.

Clearing the old coppices at a stroke generated cash flow to finance the heavy planting expenses, the greatest direct financial burden in plantation forestry. Conifers transplant more easily and tolerate degraded soils better than hardwoods; this solved the immense technical problems of replanting the vast clear-cuts, even though conifers were not native. Felled trees fed the young coal industry with pit props, beams, and construction lumber and yielded railroad ties and building materials for the railroads. Liquidating the low forest also ended the peasants' intimate and ancient relationship with the coppiced forest, increasing owners' control over their property and improving forest health and vigor at a stroke. Despite the long-term economic and ecological problems of plantation forestry, and Cotta was clearly aware of them, he called for pine and spruce plantations as an intermediate stage to bring the stagnant and degraded low forest quickly into full production and to shut out the peasants. After one rotation foresters would gradually restore an uneven-aged, mixed hardwood forest suited to local site conditions through natural regeneration, Cotta's restoration plan.[31]

The forest commons closed before the commons as landowners throughout Europe abolished peasant rights in the rural landscape in the nineteenth century. Enclosure and clearances came first in Britain, just as Britain's Industrial Revolution came before that of the Continent.[32] In the first decades of the nineteenth century German forest landowners enclosed their forests and took back from the peasants their feudal rights to forage for fallen wood and gather honey, to collect mushrooms and take small game, and to pasture their livestock on oak and beech mast. The Central European forest evolved from a medieval into a modern industrial forest of property rights, clear boundaries, and intensive management. Baron Georges-Eugène Haussmann's (1809–91)

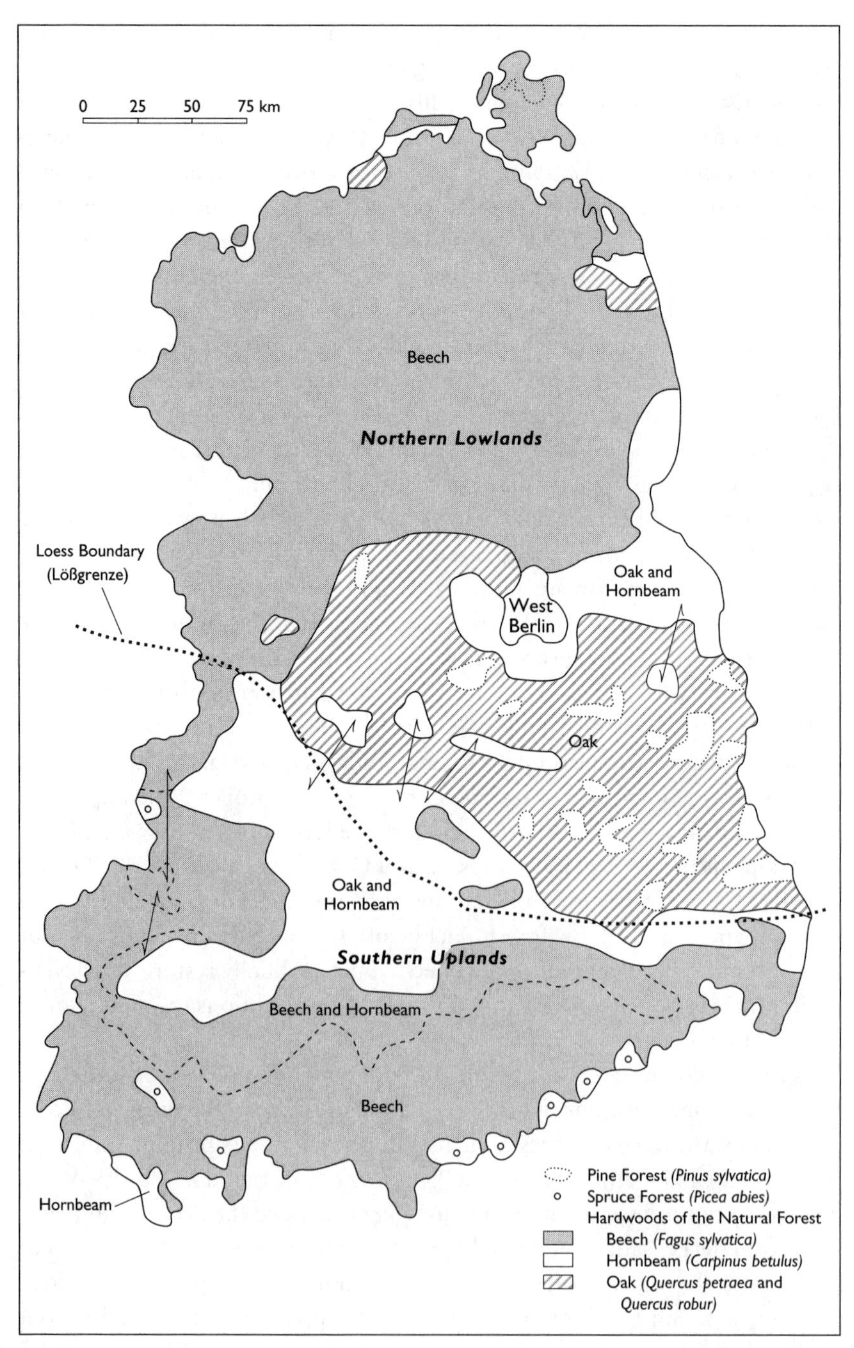


Figure 1. Natural forest cover type map of East Germany. Data gathered from Hofmann, "Vergleich der potentiell-natürlichen und der aktuellen Baumartenanteile auf der Waldfläche der D.D.R.," 119.

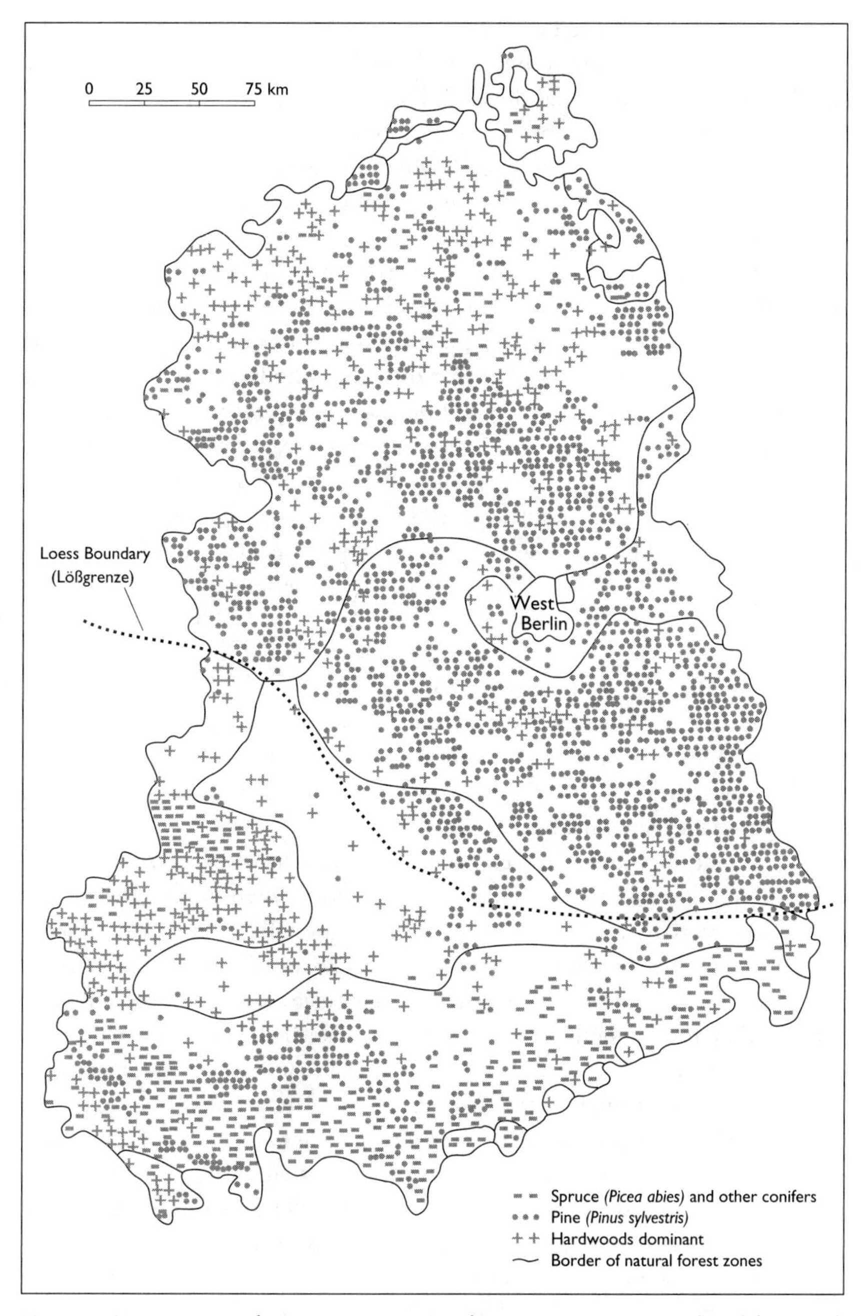


Figure 2. Current (1986) forest cover type map of East Germany. Data gathered from Hofmann, "Vergleich der potentiell-natürlichen und der aktuellen Baumartenanteile auf der Waldfläche der D.D.R.," 119.

mid-century razing of medieval Paris to make room for the broad boulevards, squares, and Second Empire buildings brought similar changes to the great city. Haussmann modernized the city's business and transport and gave it an aesthetic soon seen as distinctly Parisian and timeless. In a similar manner the new German forest was quickly seen as both more modern and more natural and "German," all at the same time, an impressive feat of mass cognitive dissonance.

Peasants bitterly fought the closing of the forest commons, enlisting the young Karl Marx as an unlikely ally. A bill introduced in the Rhenish Diet forbidding peasants from gathering fallen wood galvanized the twenty-four-year-old Marx, a newly minted doctor of philosophy from the University of Jena and editor-in-chief of the liberal *Rheinische Zeitung* as of October 1842. Government came to the aid of landowners and liberal reformers and forest scientists, an action which Marx condemned as part of a larger campaign to abolish communal rights, granting rights to trees while denying them to people. Marx declared in a "passage of exhilarating eloquence," as Edmund Wilson wrote in *To the Finland Station,* that even though feudal justice was essentially the law of the jungle, kill or be killed, at least "among the bees it was the workers that killed the drones and not the drones that killed the workers."[33] His fury at the government's suppression of peasant rights in the forest signaled a new emphasis in his writing on politics and economics. For the first time he came up "against problems for which no solution had been provided for him by Hegel."[34] This catalyzed the evolution of his thought and writing from Hegelian Idealism to materialism.

Despite custom, protest, and cost, the forest was enclosed and remade with stunning speed and assurance. From where did forest owners, particularly the Junker squirearchy, the *Rittergutsbesitzer* traditionally seen as reactionary, draw such confidence and foresight? They anticipated the Industrial Revolution by two generations in a remarkable, progressive act of economic courage. Perhaps they took a signal from Britain's Industrial Revolution, well under way in the early 1800s when German industrialists began importing coal mining technology from England, freeing the forest from charcoal production. Coal mining and steam transport created enormous demand for higher-value products from the forest: large dimension, straight-stemmed wood for lumber, pit props, poles, and railroad ties, products far more lucrative than those drawn from the cottage industries of the charcoal burners and firewood gleaners. Additional demand from chemical industries for cellulose and from the rapidly growing paper industry for long-fibred spruce pulp increased the cash flow and profitability of the new industrial forest by the mid-1800s.[35] The release from charcoal production together with unanticipated strong demand from new industries confirmed the wisdom of enclosure and refabrication of the old coppiced forest as a modern, industrial forest.

Cotta's restoration plan, gradually to reintroduce a healthy, high forest of mixed hardwoods native to local conditions, stalled in the intermediate stage of artificial conifer plantations. Cotta could not have foreseen that industrial demand would radically change the economics of forest management, making owners reluctant to lose the control and predictable cash flow benefits of the new industrial forest.[36] The costs of restoring a mixed hardwood forest, which were far off in the early 1800s, were unacceptable by the end of the nineteenth century. Industrial forest management was as well suited for the real, existing forest of simple structure in the late nineteenth century as it was antithetical to the complex natural forest.

And Cotta could not have expected that the German people would so readily embrace the new artificial forest, paradoxically coming to see in its park-like, uniform structure the sublimity of nature and the expression of German identity. Even though the long-term sustainability of the industrial forest was poor, and most managers understood this, in the near term people settled into the new, clean forest of dense spruce and pine which soon acquired the aroma of ancient forest in the popular imagination. Pure pine and spruce forests became the guest that wouldn't leave, as deeply entrenched in popular culture as in the modern economy.

A scholarly and cultural counterattack on industrial forestry mounted after the First World War as romantic conservativism, rather than the liberalism of the 1848 generation, increasingly appealed to many foresters who were recoiling from the manipulation of nature and utilitarian cast of industrial forest management and disillusioned by the slaughter and waste of the First World War.[37] The call for the reestablishment of a naturally regenerated forest based on native species, a German forest, mirrored the broader appeal of *völkisch* themes and romantic conservatism, such as Rudolf Steiner's biodynamism and anthroposophism, the *Naturschutz* (nature protection) movement, and anti-modernism. The forestry profession split in a debate, as much philosophical as scientific, over whether forested ecosystems were quasi-organic entities, creatures of custom, or whether the forest was an array of material assets, a creature of innovation.[38] Thus, even though Goethe and Mendelssohn might not have recognized the modern industrial forest, they would have understood the debates about human control over nature and the dynamics of change.

Growth rates and vigor began falling, slowly at first, then alarmingly after the First World War, as predicted for forests planted outside their natural range and stressed by surging industrial and agricultural pollution. Saxony's chief forester lamented in 1923, the year of catastrophic hyperinflation, "Saxon state forestry is bankrupt." The 1922–23 harvest collapsed 42 percent from 1913 levels, seriously damaging the rural economy. He cited clear-cutting, "artificial" plantations' ecological fragility, and "Saxon soil sickness," or soil exhaustion caused by pure spruce stands, as the principal agents of forest decline. He forecast that in thirty years, by 1954, "the stands of spruce to be tapped by thinning and cutting will be much reduced in size."[39] Indeed, conditions in Saxony's "People's Forest" of the 1950s were even worse than the old chief forester could have imagined.

These scholarly and cultural concerns culminated in Alfred Möller's seminal *Dauerwald,* "permanent forest," movement.[40] *Dauerwald* had a broad cultural resonance, particularly appealing to such "green" National Socialists as the Reich chief forester, Hermann Göring.[41] Möller based *Dauerwald* on the insight that forests were an "integrated organic entity," appealing "fundamentally to many foresters' romantic conception of the forest as a living, even spiritual, organism" threatened with annihilation by clear-cutting. Möller asserted, without experimental evidence, that abolishing clear-cutting would free untapped reservoirs of productivity and vigor. Mainstream foresters scented a laissez-faire aroma behind Möller's *Dauerwald* philosophy. Its very simplicity and focus on a negative act, abolishing clear-cutting, implied neglect of basic management. Möller's reluctance to call directly for a close-to-nature forest is puzzling, particularly since a mixed hardwood forest almost inevitably would follow an end to clear-cutting. *Dauerwald* supporters, led by the charismatic autodidact Hermann Krutzsch, carried *Dauerwald* from National Socialist forestry into the new East German state in 1946, where the Party leadership would prove the wisdom of *Dauerwald*'s skeptics.

From the beginnings of scientific forestry in the early 1800s, then, industrial forest management conflicted with ecological, close-to-nature forestry, a surrogate for the larger cultural conflict of custom with innovation, of romantic conservatism versus liberalism. Recknagle's "great dividing point" of clear-cutting versus natural regeneration defined East German forestry, not the Party's orthodox version of a clash between "socialist" and "bourgeois" forestry. As Rudolf Rüffler, orthodox Party member, Honecker's deputy minister for agriculture, and director of Eberswalde's Forest Research Institute, confessed in 1991, "the main, constant fight centered on *Dauerwald* versus industrial forestry."[42] Goethe's "primal antagonism," then, not capitalism versus communism or class conflict, defined East German forest management debates despite the Party's suppression and denial of ecology and pretensions to revolutionary power and effect.

Forest decline was not unique to East Germany, nor was the clash of custom and innovation. The great ecological revolution of the early nineteenth century consumed all of Europe, destroying and then remaking farm and forest landscapes from the Mediterranean to the Urals. It persists long after the structures and ideologies of the Soviet and National Socialist political revolutions have faded. This ecological revolution stalled before its final stage, however, trapping the European landscape in an ecological limbo of artificial forests — a phenomenon of stagnation not uncommon to many revolutions in middle age. But Europe's industrial, artificial forest is dying. Forest decline, attributed to pollution since the mid-nineteenth century, stems from the forest's artificial, simple structure and reduced system qualities more than from pollution. Management options are binary: either introduce a close-to-nature, complex forest structure and end forest decline, or continue with industrial forestry and watch forest death accelerate. This was true throughout Europe, not just on the sandy plains of Brandenburg or in the Saxon uplands.

East Germany's weak political geography and ecological and material deficits compounded forest decline — a close-to-nature forest therefore was particularly critical to the health of the East German forest. East Germany was a fragile construct, as its European and regional contexts reveal. Compared to West Germany, the East German landscape was dry, nutrient and resource poor, and small. The stark difference is readily seen from the Brocken summit, interior Germany's highest mountain. Goethe, Heine, and Mendelssohn, and legions of wanderers after them, climbed the Brocken for its unmatched views across the German landscape and for its window on Germany's past and culture. An English traveler of the early 1800s marveled at its cynosural power:

> Although the height of the Harz Giant, the Brocken, is only three thousand four hundred feet above the level of the sea, yet being isolated from its diminutive brethren, and no other mountains of equal altitude intervening to obstruct the view, we have consequently a more extensive prospect than from one of a much greater height in the Tyrol, or Switzerland; but those persons who tell us that the Baltic, the German Ocean, and the Vosges mountains, are visible, we may be assured are most imaginative tourists. However, Magdeburg, with the Elbe, Erfurt, Gotha, and Wilhelmshöhe, at Cassel, may be distinctly seen, together with the towns of Brunswick and Wolfenbüttel. The surrounding country is also exceedingly interesting as it affords numerous excursions, and at the same time a fund of amusement to the imaginative tourist; for he is now in the region of enchantment, and every hill, glen, and wood, has been the theater of some supernatural legend.[43]

The Brocken's outlook was so advantageous that the Soviet army fortified it as a forbidden military zone bristling with antennae, watchtowers, and weapons from which they spied on East and West Germany. The Russians surrendered the Brocken summit only in 1994, five years after the fall of the Berlin

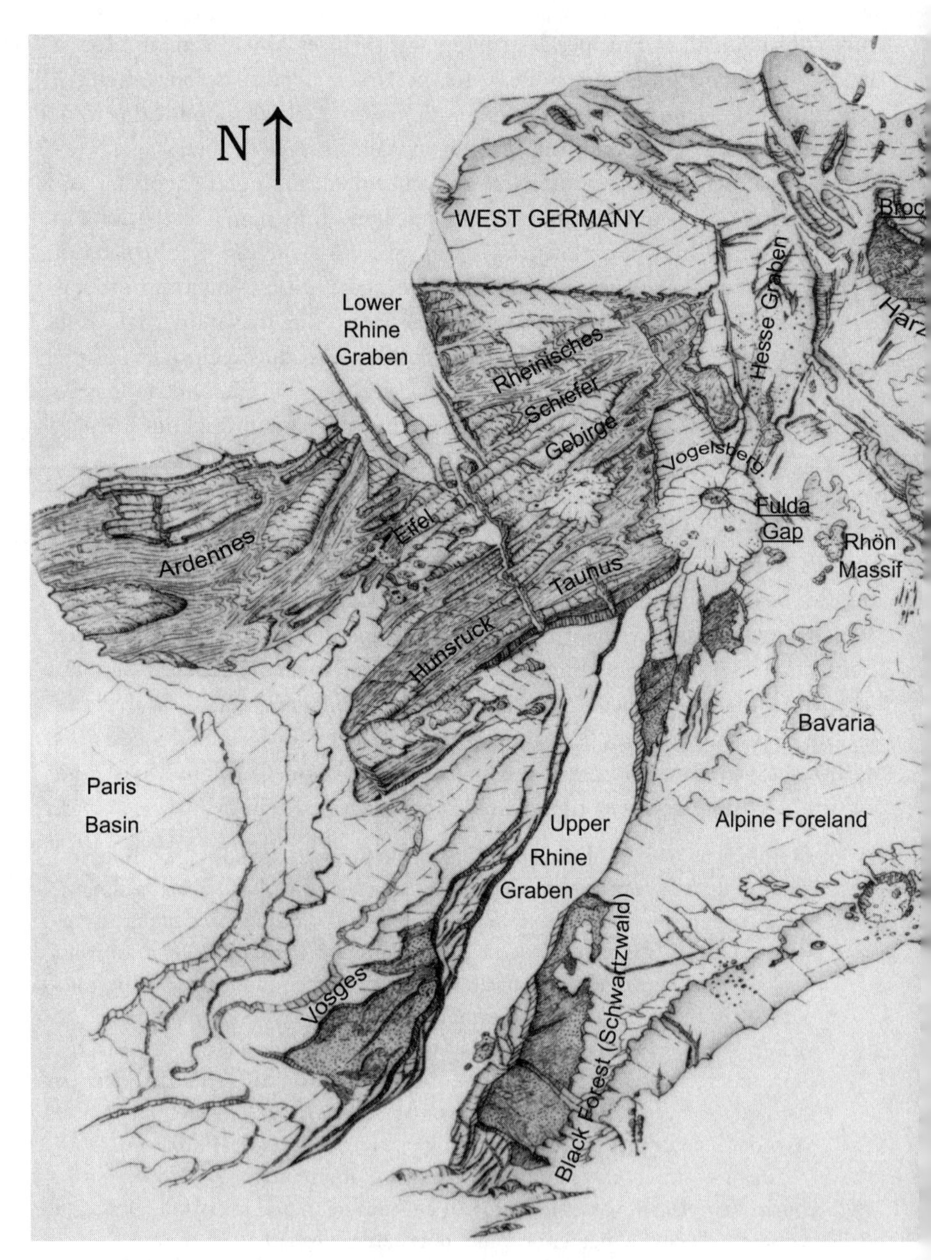


Figure 3. Block diagram of the Central German Uplands viewed from the south. Data gathered from Cloos, "Ein Blockbild von Deutschland," and Rutten, *Geology of western Europe,* fig. 38, "Block diagram of Germany," 68.

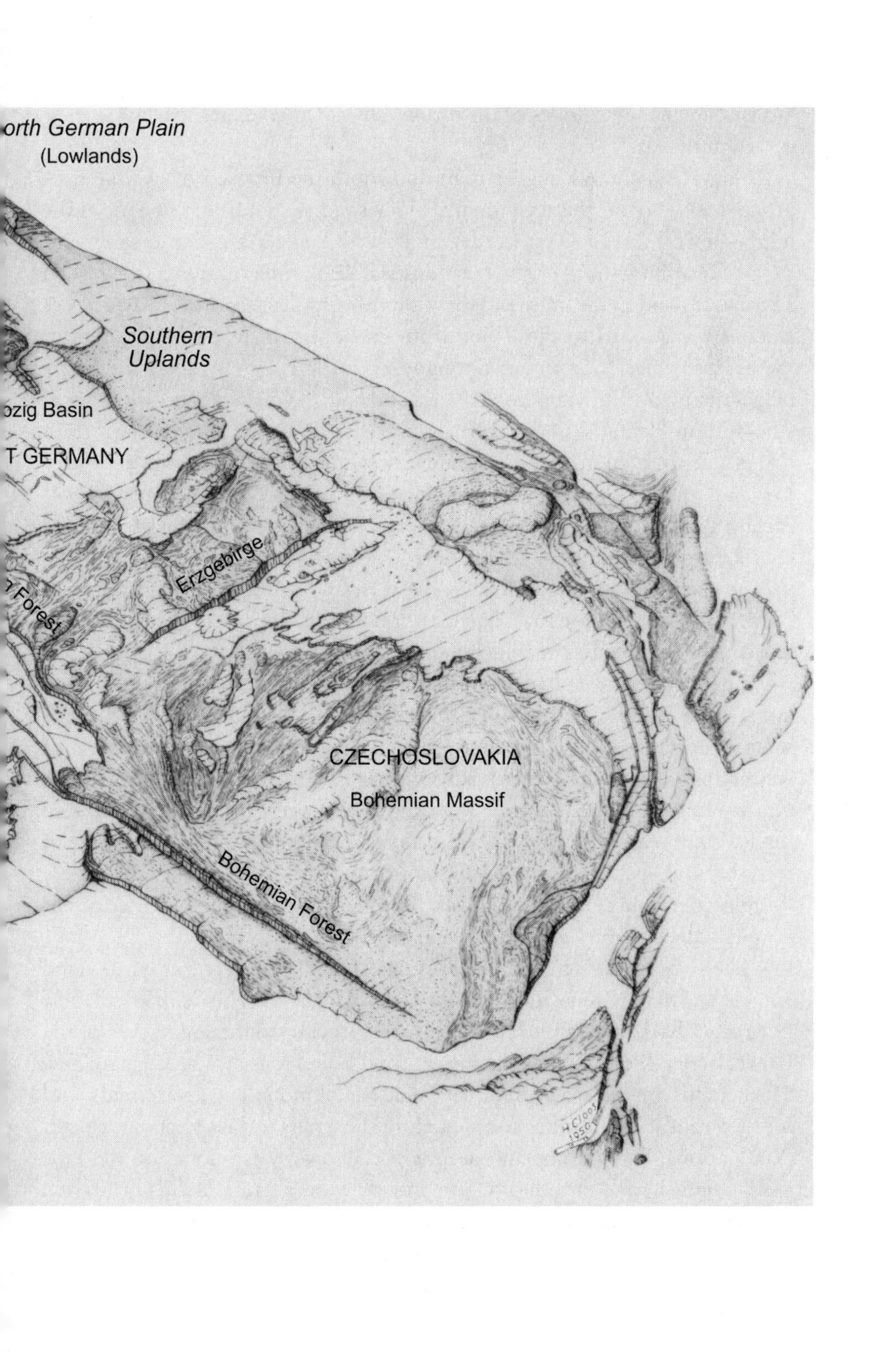

orth German Plain
(Lowlands)
Southern
Uplands
ozig Basin
T GERMANY
Erzgebirge
Forest
CZECHOSLOVAKIA
Bohemian Massif
Bohemian Forest

Wall, when the last soldiers of the former Soviet Union reluctantly made their eastward retreat.[44]

Rising 1,141 meters, higher than Snowdon, the Brocken marks the geographic center of northern Germany.[45] Views radiate out from this aerie at the midpoint on the East-West border in a 225-kilometer arc comfortably embracing Hamburg to the north, the industrial Ruhr Valley to the west, Dresden southward, and Berlin to the east. Within this radius, the Main River flows sinuously westward to the Rhine from its Bohemian forest headwaters in the eastern uplands, marking Germany's narrow waist. Southern Germany bellies out below the Main into Bavaria and Baden-Württemberg into the rich Rhein-Main Plain, the Danube Basin, and the Alpine foreland.[46] The terrain slopes steeply away to the north and east of the Brocken toward the sandy lowlands of the North German Plain where most of East Germany lay.[47] Setting aside curvature of the earth, a person standing on the Brocken could see clear from the Aquitaine Basin, the Mediterranean, and the Pyrenees all the way to the Pennines in England in the west and to the Urals and Europe's border with Asia in the east.[48] There are few natural barriers to movement and few geographic features on which to base stable frontiers.[49]

A small country even by European standards, East Germany was not much larger than Bavaria and smaller than Tennessee. Its poor land and hardscrabble farms yielded the lowest returns in Central Europe—half those of western German farms.[50] The Vistula glacier 20,000 years earlier scraped the North German Plain, leaving behind poor, dry soils, glacial outwash, and sands. Trees recovered only 12,000 years ago, limiting the lowland forest to forty natural species, many fewer than on similar North American sites.[51] Postglacial terrain defined two-thirds of the East German landscape: a dry, flat island floating on the vast lowland sea.

Western Germany's richer, lusher lands spread westward across lowland and plateau to the Rhine, mostly in the Central Uplands. The Central Uplands, Germany's backbone, stretch out below the Brocken to the south, part of the Great Barrier Arc which rises across Europe from France to the Carpathians. The uplands are better watered, with block-fault mountains, scarplands, and forested sandstone plateaus forming natural barriers to east-west movement. Wind storms, heavy rains, and deep winter snows sweep down off the Eurasian continent and drop heavy rain and snow on western Germany, leaving Germany east of the Elbe to suffer chronic drought.[52] Frustrated eighteenth-century Huguenot silk-weaver emigrants, seeing their carefully planted and tended mulberry plantations fail on the dry lowland plains near Berlin, christened them "Moabit," the Bible's barren and dry trans-Jordan desert.[53]

As the great Vistula ice sheet retreated northward vast floodplains and outwash basins formed along its margins, laden with the shrinking glacier's gritty runoff engorged with rich clays, silts, and loams and supercharged with organic matter from shells. Wind picked up these rich soils as the meltwater basins and glacial floodplains drained, depositing them in a broad band of deep, fertile, and fine-grained loess soils along the glacier's southern margins, forming the *Lößgrenze* (loess boundary).[54] This great arc of loess soils forms a border region (*Börden*) of rich farmland stretching across Germany from Essen at the Rhine and Ruhr into East Germany's best farmland, the Leipzig Basin northeast of the Thuringian Forest. The loess boundary also marked a major change in ecology and the boundary between pine and spruce silviculture. Scots pine (*Pinus sylvestris)* tolerates drought better than spruce and therefore is the dominant species in the lowlands. Norway spruce (*Picea abies*), "the drunkard of the forest," is widely planted in the uplands south of the loess boundary where soil quality and climate are more favorable, creating a forest ecology similar to Bavaria's.[55]

Below Germany's narrow waist along the Main the Central Uplands' arms open to shelter the rich scarplands, plateaus, and terraces of the densely populated and prosperous upper Rhine valley and the Rhein-Main Plain, the Alpine foreland, and the Danube Basin.[56] Other than the Rhine, there is only one natural pass through the Central Uplands' folded blocks and volcanic highlands: where the uplands converge at the town of Fulda near Frankfurt am Main. A narrow thirty-five-kilometer-wide pass, the Fulda corridor squeezes between two great isolated volcanic horsts, the Vogelsberg and the Rhön Massif. Prosperous merchants, factors, and tradesmen, drawn by the thriving north-south trade moving through Fulda since the Middle Ages, made Fulda a wealthy entrepôt.[57]

Warsaw Pact planners noted the natural pass at Fulda into the industrial and financial heartlands of western Germany just where the East German border was closest to the Rhine. Frankfurt beckoned one hundred kilometers, just over sixty miles, away.[58] This pass at Fulda became known within NATO as the "Fulda Gap" by the 1980s as a Soviet sallyport from the formidable glacis of the steep Thuringian Forest into the rich Rhein-Main Plain.[59] The Soviets placed a powerful force within striking distance; the Eighth Guards Army's armored and mechanized infantry divisions lurked only fifty kilometers east of Fulda under thick forest canopy. But a Soviet invader moving south through the Fulda Gap would face U.S. Army forward units armed with tactical nuclear weapons to halt any Soviet army breakout into the Rhein-Main Plain and to prevent the capture of Frankfurt.[60] No, if the Soviets attacked their drive would come through the Lüneberger Heide in the north, the Lüneburg Heath on the North German Plain where only the Elbe blocked east-west movement and natural defensive positions were weak.

Thus we've seen that East Germany lay predominantly in the lowlands, dry, poor farmland with weak natural borders. Since the Teutonic Knights first settled Prussia, the "sandbox of the Holy Roman Empire," and the lowlands of the valley of the Vistula in the late twelfth century, German rulers struggled to maintain a stable, German population.[61] Eastern German towns were founded much later than those in the west, and the agricultural economy suffered from episodic pulses of depression and mass desertion.[62] Eastern Germany had always been dependent on western Germany and chronically on the brink either of being overrun from the east or of collapsing into bankruptcy under the weight of population, resource, and ecological deficits. East Germany's population loss to the West in the 1950s, *Republikflucht,* reflected these long-term difficulties as well as the economic and cultural wreckages left in the wake of Walter Ulbricht's erratic and harsh economic and social policies.

West Germany, in contrast, not only was larger in area and blessed with formidable natural borders but also was endowed with a richer, more productive ecology. Then, on top of these historic constraints, Stalin seized Silesia and Prussia east of the Oder-Neiße line in 1945, taking almost thirteen million hectares, driving out twelve million Germans, and killing two million more in a ruthless act of ethnic cleansing. This left the future East German state with only the rump of historic Prussia, a landscape impoverished geographically and ecologically.[63] The "great givens," as Timothy Garton Ash termed them, of geography and ecology became even more critical with the end of direct Soviet rule in 1949 as the Party leadership pursued autarky and control.[64]

The political revolution of 1989 granted the German forest only a reprieve; its health has always depended on restoring a close-to-natural species distribution of mixed hardwoods, an ecological revolution. The fall of the Wall removed economic and political barriers; now popular perceptions and culture would also have to change. As long as Germans see the artificial forest as natural and distinctly German, until they abandon the aesthetics of order and conformity and embrace a messier aesthetic, the forest will remain in decline. Ideas of what is beautiful and natural are as important as the knowledge of what is natural and what is efficient. Aesthetics, as Hazlitt observed, precede ethics, or, in this case, forest management.

# 3

## *Initial Conditions and Reparations*

*Chaos was the law of nature; Order was the dream of man.*
—*Henry Brooks Adams,* The Education of Henry Adams

In the closing months of the Second World War, the broad northern plains between the Oder and Elbe rivers settled the flood of more than eight million refugees from Germany's eastern provinces. The refugees found there, in the core of the future East German state, a fragile calm.[1] They had raced westward blindly at first, put to flight like wildlife driven by beaters from thicket and earth, slowing only as they flowed into Brandenburg's sandy lowlands. The European landscape had never seen such a shift of population—the liquidation and replacement of an entire people in a few weeks' time through the "largest mass expulsions in European history."[2] The London *Times*' special correspondent vented the anger many felt at Soviet ethnic cleansing: "It is surely not enough to say that the Germans brought these miseries upon themselves; brutalities and cynicism against which the war was fought are still rife in Europe and we are beginning to witness human suffering that almost equals anything inflicted by the Nazis."[3] Sympathy slowly grew for the helpless refugees, only recently seen as Nazis, as news of the Soviet atrocities leaked out from the East.

The refugees took flight from ancestral homelands in East Prussia, driven across the marshes of the Masurian Lake District, crossing the Vistula River valley which Teutonic Knights and German colonists had first entered in the late 1100s. They were joined by the stream of refugees from Prussia's New Mark and Pomerania and by Germans from the uplands of Silesia, breaching the Warthe River and finally crossing the broad Oder River, the last natural barrier to the heartland of central Germany.

The Brandenburg landscape must have been soothing in its familiar pattern of sandy plain and glacial lake, with thick pine forest layering over the end moraines, drumlins, and glacial ridges left in the margins of the Vistula glacier's retreat. Early in the spring of 1945, with Zhukov's armies coiled on the Oder's east bank, German civilians could feel protected in this lowland landscape between the Soviet and Allied armies. Refugees from east of the Oder and Neiße rivers took shelter at night in farmers' barns and houses, or camped in woodlot and forest, taking food and firewood to warm themselves and comfort from Brandenburg's farmers and villagers. The generous landscape absorbed the flood from the east, granting these people temporary respite from the storm to come.

Rudolf Binsack, a thirteen-year-old boy living on his family's 208-hectare farm some 60 kilometers northeast of Berlin, stood outside the farm gates on a clear early-March morning in 1945 and watched as women, old men, and children streamed past on their way west.[4] He heard a refugee grimly tease a gawking farmworker: "Enjoy the war, peace will be much worse!" The Brandenburgers listened, puzzled—they had little direct experience of the war. Still, even in the morning's calm, the refugee's caution rang true. More than five million German soldiers were dead or severely wounded, millions more languished in POW camps, and hundreds of thousands of women and children were dead in the Allies' day and night bombing. Almost every German had lost one or more family member in the war.

When Rudolf looked back over the walls of his mother's garden to the fields and woodlots beyond, spreading in patterns of sandy brown fields and soft-green Scots pine stands on the low, rolling hills, he saw a healthy, if somewhat frayed, landscape promising recovery. The Binsack farm was tattered, the family lacking labor, fertilizer, and petrol. Still, like most German farmers, the Binsacks kept the German people fed throughout the war at the same level as the British population. Germany's farms and forests were in good condition when the war ended two months later in May 1945.[5]

Western officials and reporters who ranged through the four occupation zones found that Germany "had *not* been nearly as badly damaged by air bombardment as had been reported" and that war damage was concentrated in western Germany (emphasis in original).[6] Cities, and housing, took far greater damage than industry or the countryside. British and American airmen bombed the transportation hubs clustered in and near cities with particular ferocity in the last eighteen months of the war—neglecting the munitions and armaments factories which the Nazis had dispersed in the countryside.[7] Only 20 percent of Allied bombs hit within the generous "target areas" allowed, 1,000 feet around the aiming point of attack, about one-third of a kilometer. "The safest place," joked Allied airmen, "is on the target." And the bombers dumped the heaviest tonnages on Hamburg, Bremen, and the Ruhr in western Germany. Bombing destroyed less than 12 percent of Soviet zone housing, but more than 25 percent of western Germany's.[8] Eastern German industry also took less heavy damage from aerial bombardment. Bombing cut western German industrial productivity by 21 percent compared to a 15 percent cut in eastern Germany.[9] After the war the Soviet zone was more than self-sufficient in food whereas western Germany, as it had historically, depended on imports from the east.[10] Germany's physical and capital assets, particularly in the Soviet zone, survived the war far better than anyone expected.

The war ended only two months after the clear March day when Rudolf watched the slow stream of dusty refugees move past his family's farm. The Soviets immediately set about remaking the countryside under the guise of land reform. The apologetic mayor of the local village warned Rudolf's father in October 1945, "the Russians demand your farm's break-up," seizing all farms over one hundred hectares to create new, small farms of seven to eight hectares. One of the Binsacks' older farmworkers, knowing what was happening at neighboring farms, gently advised young Rudolf to hide his old bicycle from the land reform "functionaries." Rudolf wheeled his bike behind a shed and later that afternoon gave it to a friend. Several days later land reform functionaries, backed by Red Army soldiers, split up the Binsacks' farm among thirty or so "new peasants." Resigned to the loss of their family farm, the Binsacks watched the ceremony, cheered by the hope that the Soviets would let them keep a small parcel so they could stay on as neighbors of the new peasants.

The prospects of farming their small plots alone, alienated from the larger farm structure—and their lack of experience—perplexed the new peasants. An old farmworker who got a small parcel from the Binsacks under land reform begged Rudolf's father to manage the estate as before, but as a cooperative. Everyone would need the barns, sheds, pastures, and machinery of the whole estate to succeed, assets which were not divisible. And the new peasants needed Mr. Binsack's experience and skill. So it seemed, for the moment, as if the farm economy and rural society might emerge in a new, stable form and the Binsacks could remake their lives and farm in a cooperative farm community.

But five days later Soviet zone police came and, without warning, arrested and shipped the Binsack family to a former Russian POW camp near Dresden, splitting their personal property among themselves.

The Binsacks shared one room at the Dresden prison with twenty other families—without toilets and running water—and were fed a starvation diet. After a week, guards locked them inside cattle wagons for five days and nights, freezing and starving, with only straw and a pail for comfort. The Party sentenced father, mother, young boy, and sister without trial to the former Nazi concentration camp on Rügen in the Baltic. The family made a narrow escape to the West just before the Soviets could deport Rudolf to a forced labor camp in the Soviet Union, leaving their family farm, community, and life on the North German Plain behind forever.[11] They were lucky; many tens of thousands of German farmers, civilians, and political prisoners, particularly Social Democrats whom the communists hated, died of starvation and disease in the Soviet camps.[12]

Had Rudolf been able to return to his family's home and his mother's garden two years later, when land reform had run its course and Soviet reparations were peaking, he would have seen the thrifty family farm destroyed, the machines and livestock gone for reparations, the fields underplanted or fallow, and once well-stocked forest stands clear-cut and left unplanted. But the agricultural wealth of the Soviet zone made the initial prospects for its population brighter than in the chronically food-deficient and war-ravaged states of western Germany.

Thus, in the first months of Soviet occupation most eastern Germans lived better than Germans in the West. Times were far harder on civilians in all four zones than during the war years, but British and French zone civilians suffered hunger and deprivation that exceeded even the notorious years after the First World War.[13] Herbert Hoover, head of President Truman's Economic Mission to Germany and Austria, reported that western German conditions had sunk to "the worst modern civilization has seen, the (people living at) the lowest level known in a 100 years of human history."[14]

Throughout 1945 it was not unusual for Western correspondents to show disdain for what they perceived as German apathy, in reality the onset of famine from near-starvation diets and heavy labor, and to dismiss German reports of Soviet atrocities.[15] But the suffering of civilians, in particular the starvation of children and their suffering from such diseases of famine and chaos as edema, cholera, and tuberculosis, changed British and American opinion; Westerners began to see the former German enemy as human and increasingly the Soviets as hostile and brutal.[16]

The advantages civilians in the Soviet zone enjoyed in May 1945 soon evaporated due "above all" to Soviet occupation policies: to reparations and human rights crimes, to land reform, and to the imposition of Soviet-style political and economic systems.[17] East German personal income fell to half West Germany's by 1950, although in 1939 they had been equal. West Germany reached 1936 levels of production in the early 1950s; East Germany did not regain even this low level for ten more years.[18] Reparations, which lasted from 1945 to 1954, crippled the East German economy in its infancy. German fixed assets, cash and production, food, intellectual property, and labor funded Soviet reconstruction and leveled Germany in preparation for a "People's" economy. Land reform, Stalin's first postwar political initiative in his "parliamentary road to socialism," crushed economic and ecological diversity in the rural landscape to wage class warfare against the conservative estate owners, farmers, and peasants.

The outward forms and purposes of land reform and reparations were different—land reform democratic in form, reparations classic *Machtpolitik*—but they were prosecuted in tandem and cemented Soviet power. Both of these policies were first made manifest in the natural landscape, in eastern Germany's farms and forests, and both have marked indelibly the countryside for generations. Stalin's "border adjustments" prepared the ground for land reform and Soviet hegemony by pushing the Soviet security frontier to the margins of Central Europe and by flooding and then almost overwhelming all of Germany with refugees. Land reform and reparations followed to complete the leveling Stalin began with ethnic cleansing, conquest, and reparations.

At the end of the war Germany lost to Stalin all its lands east of the Oder-Neiße line, one-third of the prewar Reich: 129,500 square kilometers, provinces almost one-third larger than the future East German state. German economic experts valued the Oder-Neiße territories at $26 billion, almost $210 billion in current value.[19] Different estimates of the value of lost Oder-Neiße territories are distinctions without a difference, however; no value can be placed on such a loss, or on the deaths of millions in ethnic cleansing, or on the liquidation of the ancestral homes and memories of an entire people. (See figure 11, "Emergence of the German nation, 1937–1990," for the areal loss.)

Stalin also held fast to most of the Polish land east of the Narew, Vistula, and San rivers, roughly the area of Spain, that he had grabbed as Hitler's partner in 1939 under a secret protocol of the Molotov-Ribbentrop treaty. Stalin's 1939 terror in Poland foreshadowed his ethnic cleansing of German civilians in 1945 as his secret police "cleansed" the twelve million Byelorussians, Ukrainians, and Poles living in the eastern reaches of Poland.[20]

Stalin thus changed the political geography of Europe. He incorporated eastern Poland into the Soviet Union in 1939 and if Germany reunited after the war's end, as most expected, it would be a shadow of its former self, its center of gravity shifted west to the Rhine. No one imagined in mid-1945 that Germany would divide into zones of occupation, or that the East-West fault line would run along the Elbe River. The border adjustments left the future East German state an artificial and weakened geographic construct, the rump of Prussia, only slightly larger than Bavaria, but impoverished ecologically and economically.

Stalin's "border adjustments" also cost Germany more than 3 million hectares of forestland, a quarter of Germany's 1937 forest along with most of its pulp supply and half of its forest products industry. Pine and spruce plantations within the future East Germany's narrow boundaries rose to 82 percent of total forest cover, increasing the urgency of restoring a close-to-nature forest. Thus Stalin's political calculus hobbled eastern Germany and its forest economy right at the start. Land reform and reparations then crippled the rural economy, creating unfavorable initial political, economic, and ecological conditions from which East German forestry never recovered.

The Party leadership blamed the severe forest decline revealed in the early 1950s on their inheriting a 1945 forest devastated by "fascist looting" and war damage.[21] Early postwar working inventories supported this claim by underestimating stocking levels and skewing the age-class distribution toward the younger classes, presenting a forest depleted of its most valuable, mature timber—a forest that needed conservation.[22] Klaus Schikora, director of East Germany's forest inventory, echoed Ryle's astute observation that there was "a big temptation to underestimate the quality classes and stocking percentages to strengthen the arguments in favor of a rigid conservation policy. It is fairly certain that the figures are heavily underestimated."[23] Instead, the working inventories unintentionally lent cover to Soviet reparations harvests by supporting the myth of fascist looting. By undercounting, German foresters gave headroom to the Soviets to take mature timber "off the books," as so much volume had been squirreled away in hidden reserves. The stratagem was naïve in any event: maximum reparations deliveries alone determined harvest levels, not the forest's biological capacity.

But what was the forest structure in 1945? The last complete inventory was more than twenty years old in 1945, and individual forest records were either poorly kept or lost in the war.[24] The whole forest inventories needed to set forest conditions are highly complex and require an infrastructure of experienced foresters, technicians, clerks, and support staff working from reliable, recent databases, all impossible luxuries in 1945. Instead, we can reconstruct

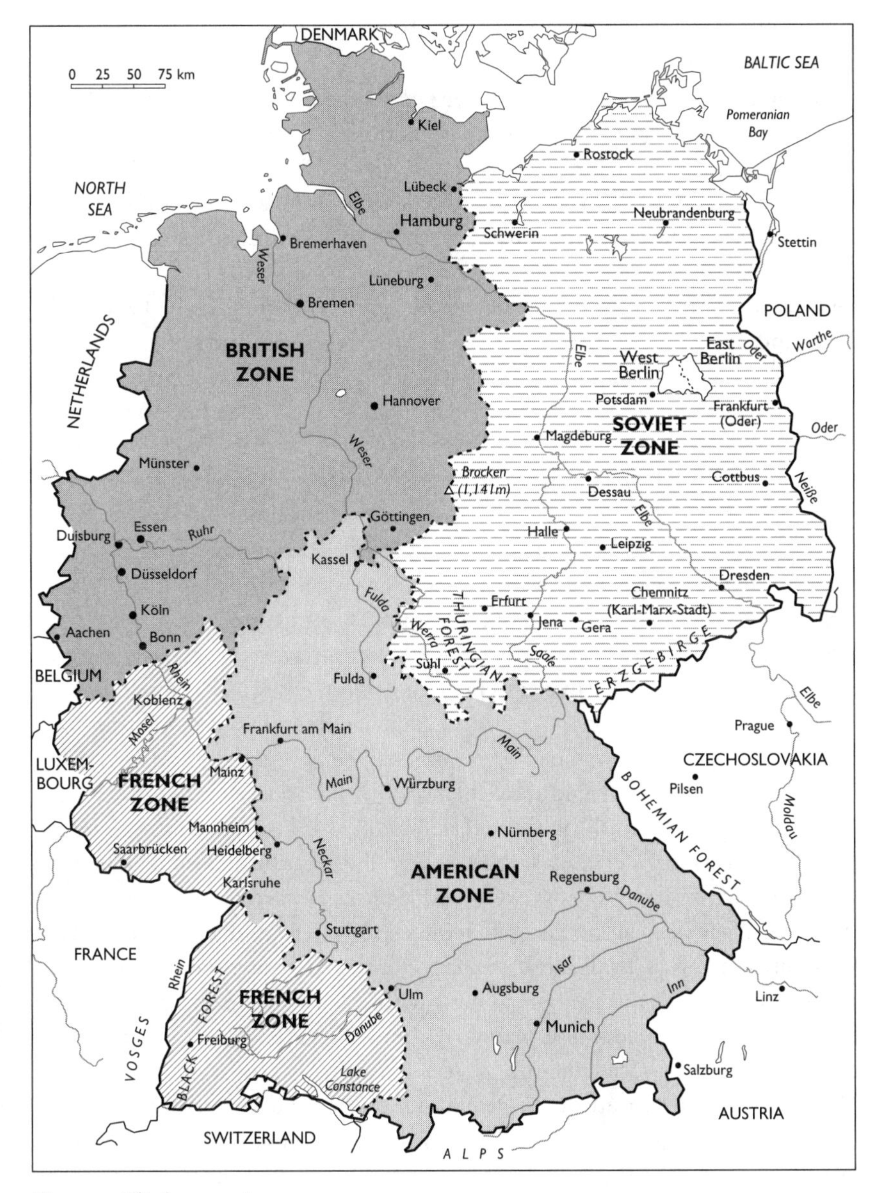


Figure 4. Allied occupation zones, 1945.

the 1945 forest structure from study of forest history and comparison to postwar western German conditions.

Germany's forest and farm land were in good condition in 1945, as British and American foresters, Prussian foresters fleeing westward, and internal reports witnessed.[25] Postwar observers agree that the Second World War harvest only exceeded annual growth, a measure of sustained yield, "by not more than thirty percent," leading one surprised Western forester to remark that the German forest was "not badly overcut."[26] Stocking in 1945, the volume of wood in the forest, was down only 10 percent from 1930 levels, leading the Food and Agriculture Organization (FAO) to comment: "German forests were well husbanded throughout the war years and now are in excellent condition. It is safe to say that not more than the increment of the next two to five years has been cut, and in northern Europe the growing stocks remain practically unimpaired."[27] A U.S. forester who toured Soviet zone forests reported that war damage was limited to 2 to 3 percent and that "the forests are in an excellent state."[28] Stephen Spurr, a young forest ecologist and future dean of the Yale School of Forestry, observed that "the visitor cannot but be impressed with the overall good condition of the German forest. Coming from Scotland, where perhaps seventy percent of the merchantable forest was clear-cut during the war, the small acreage of clear-cut areas in Germany seems insignificant."[29]

Against all expectations, Nazi forest and wildlife management was thoroughly "green." Environmentalism suffused Nazi ideology and the Nazis managed the forests and wildlife in their conquered lands, even in the occupied Soviet Union, ecologically, with remarkable care for long-term values and the natural landscape.[30] As late as 1944, Alfred Heger, a future senior East German forester, still vigorously promoted close-to-nature forest management throughout the Nazi empire despite looming defeat.[31] Conflict between close-to-nature and industrial forest management, a debate reminiscent of the *Dauerwaldstreit* of the 1920s, endured within the Forest Service and within the offices of Nazi economic planners, yet National Socialist forestry never strayed far from close-to-nature forest management.[32] Even Reichsführer Heinrich Himmler, head of the S.S. and Gestapo and warden of the death camps (and an agronomist by training), preached environmentalism in his late-war decree "On the Treatment of Land in the Eastern Territories": "The peasant of our racial stock has always worked steadily to increase the natural powers of the soil, plants, and wildlife, and to conserve the balance of the whole of nature. If, therefore, the new *Lebensräume* [living spaces in eastern Europe and the Soviet Union] are to become a homeland for our settlers, we must without deviation manage the landscape as a close-to-nature organism. It is the critical base for fortifying the German *Völk*."[33]

Germany's forests suffered in the war mostly from neglect. German foresters not taken up by the army, mostly older men, were too few and forest labor was too scarce to undertake thinning, road repair, and inventory work. Most work except for harvesting and replanting was deferred. Further damage came from the postwar chaos as millions of starving refugees and displaced people, as historically happened in the aftermath of war and revolution, fell upon the forest for shelter and sanctuary, for food and fuel.

Yet within several years of the end of the war the Soviet zone forest neared collapse.[34] Kurt Hueck, a senior Soviet zone forest scientist, described the forest of the late 1940s as blighted by "huge clear-cuts of unimaginable size, with depressed stocking and thin, poorly tended stands."[35] One-third of the state forest was highgraded, its battered stands stripped of all valuable timber and "without a forestry future." The rest of the forest area was divided between immature stands and vast, unplanted clear-cuts.[36]

Storm breakage, forest fires, and infestations of insects and pathogens, damage common to industrial forests, surged in the immediate postwar years following an unusual series of natural calamities. A severe northeast storm on 13 June 1946, the "worst in 100 years," destroyed overnight 2 million cubic meters of valuable sawtimber, a third of the annual harvest. A bark-beetle infestation of terrible ferocity quickly followed in 1946–47: "the greatest insect plagues which have ever hit Central Europe, made worse by the monoculture," as a contemporary forest scientist recorded.[37] By the time the infestation ran its course more than 20 million cubic meters, equal to several years' growth, was lost.[38] In addition, salvage requirements forced the Party to shift foresters and woodsworkers to the south, further crippling afforestation and thinning for several years. But the unstable artificial forest structure, the Soviet's purge of the Forest Service, land reform and reparations destruction of forest structure, and the lack of workers—not the specific intensity of the June 1946 storm—were responsible for the severity of damage.[39] Even without such natural calamities salvage volumes still ran between 1.5 and 2 million cubic meters per year through 1950 and regularly ran at 25 percent of annual harvests through 1989.[40]

Unlimited Soviet reparations harvests and the artificial forest structure were the direct cause of the ecological and economic collapse of the eastern German forest between 1945 and 1950. Soviet forestry experts worked backwards from reparations quotas to set harvest levels, not from growth rates or stocking. Maximum reparations deliveries, not physical counts, determined the stocking and volume levels in forest inventories.[41] Thus, while whole forest stands vanished into clear-cuts before foresters' eyes, stocking volumes on the Soviet Military Administration's books doubled between 1946 and 1947, just at the time when Soviet reparations harvests were at their highest.[42] No forester dared to contradict Soviet experts or reveal the political secret of Soviet destruction of the forested landscape.[43]

Reparations withdrawals, both industrial and from the farm and forest economy, were the dominant Soviet policy goal. The Soviet officers charged with meeting the reparations deliveries were in no doubt as to their masters' priorities. When reparations deliveries faltered in the first six months of 1948, the Soviet Military Administration issued an edict holding key Soviet personnel personally responsible. Reparations, the edict declared, had the "top priority" of all Soviet policy in Germany.[44] An unidentified senior Soviet zone official, possibly Tulpanov, declared in an internal memorandum that reparations withdrawals were the Soviets' primary policy in Germany, followed by the creation of Soviet Joint Stock Companies, the notorious *Sowjetische Aktiengesellschaften,* or "SAGs," which ran eastern German factories to produce exclusively for the Soviet economy. "Without resolution of these two issues [reparations and the SAG]," the author concluded, "there can be no democratization of eastern Germany. But how can we explain to the German worker the significance of the SAGs?"[45]

The Allies had agreed at Potsdam to consider Soviet reparations demands of $10 billion, $100 billion in 2002 terms, by taking German fixed assets from each zone. But the Potsdam limits to reparations, even though never formally ratified, were irrelevant to Soviet withdrawals, which were "as severe as could be devised and tolerated," more than gross investment in the economy between 1945 and 1954.[46] The British and Americans, conscious of the disastrous consequences of the Versailles Treaty's cash reparations of 132 billion gold marks ($34 billion in 1947 value), wanted to limit reparations to fixed assets and to make reparations "short and sharp." The three chief Allies (Russia, Britain, and the United States) agreed at Yalta to bring Germany down to a "middle European standard for some time." The Americans and British argued that the Allies should not risk again the political chaos which followed from the punitive cash reparations of Versailles.[47]

Cash reparations and reparations taken from current production not only would slow global economic recovery by suffocating the German economy, but would create suffering and economic hardship that could only favor the communists. The Soviets, as they did with the Potsdam agreements about coordinating land reform and forest management, ignored British and American cautions. The Soviets did not see a difference between cash and fixed asset reparations, except that cash reparations were far more valuable and efficient than shipping German plants and factories to the Soviet Union: much better to take the goods, not the plant and equipment.[48] Forest reparations combined the worst features of fixed asset and cash reparations, stripping production out of the local economy without payment and slashing capital asset values.[49]

But the specific form of reparations meant less than who paid. As with sanctions, innocent women, children, and men paid the bill: individual human beings, not a government or class. Although the British and Americans argued against cash reparations, they accepted the reductionist orientation at the heart of Soviet reparations policies. They accepted the idea that an abstraction, "Germans," would pay the reparations bill, whether out of fixed assets or in cash. Similarly, no political leader questioned taking forced labor from civilians or former POWs as a legitimate form of reparations, a modern form of slavery.

As Churchill said, there is no way to draw up an indictment of an entire people. A. G. Dickens, an English scholar of the Reformation and a military journalist in Germany in 1945, countered a Soviet trade attaché's (a KGB agent under cover in Britain) defense of Soviet reparations: "There exists no genuine moral argument for imposing so utterly barbarizing a load upon precisely those German individuals—women, children, manual workers—least deeply responsible for the crimes of German militarism and probably most capable of moral reorientation. From our democratic viewpoint let us moreover be quite clear on one point: that the enemies of free institutions will exploit most unfairly the subhuman standards entailed by this proposed catastrophic drain upon western Germany. Any student of current German politics could exemplify, *ad nauseam,* how effectively Communist demagogues take advantage of hardship in the western zones in order to discredit democracy as opposed to Communism."[50] Individual Germans paid the Soviets' reparations bill, and so too, indirectly, did British and American taxpayers.

Reparations dominated relations between the British and Americans and the Soviets in the early postwar years.[51] Soviet reparations shifted massive financial burdens onto the British and American governments as resources planned to be available for all of Germany were spirited away to the Soviet Union instead. The Soviet refusal to ship food to the western zones cost the United States $200 million in 1947 and the British a staggering $320 million, more than $2.5 billion in 2002 terms, to meet the barest needs of their zones.[52] And despite this expense, British zone civilians were still near starvation. Ernest Bevin, British foreign secretary, reported to the House of Commons on 16 May 1947:

> The Prime Minister and myself never for a moment imagined that they [Soviet reparations] would involve the imposition of added taxation on the British people. The Soviet Government had taken from their own zone reparations

> from current production. They had consistently refused to give any information about those unauthorized removals which, with removals of capital equipment and the effect of Soviet cartels of German enterprises, amounted in value to well over $7 billion. Yet at the same time the taxpayers in Britain and America had to find money to keep Germany from starvation. Since the end of the war the United Kingdom contribution to the re-establishment of German economy had been over £200 million [$806 million, almost $6 billion in 2002 terms].[53]

The editors of the London *Times,* looking forward to the Council of Foreign Ministers' meeting in Moscow in April 1947, echoed Bevin's impatience, writing: "[The Soviets scheme] to use Germany as a channel through which Allied assistance would flow to her benefit."[54] Carl J. Friedrich and Henry Kissinger reported: "At the end of the Second World War, it was the manifest desire of the Soviet Union to employ reparations from Germany as a device to siphon resources from the United States."[55] Soviet reparations' destructive force and hostility led to an early break with the Western Allies sooner than the Soviet Union wanted.[56]

The *Times'* Berlin correspondent marveled at the central role reparations played in Soviet policy and the opportunism that damaged the Soviet's long-term strategic interests, writing: "The only discernable policy of the Soviet occupation—namely, that the Russian zone should be exploited for Russia's benefit, regardless of German interests and of what on a longer view might have seemed to be Russian interests also. [Russian exploitation has] been carried farther than had been thought probable or profitable."[57] Bruno Gleitze also marked the aroma of opportunism in Soviet policy: "Any policy seemed correct (to the Soviets) as long as it revived the German economy and made possible the withdrawal of reparations."[58] Stalin designed reparations to compensate the Soviet Union for its grievous war losses, to pay for reconstruction in the Soviet Union, to reduce Germany's "war potential," but mostly to advance Soviet power at the expense of British and American power. Yet in the final analysis the material benefits of reparations were swamped by the counterreaction his harsh policies sparked in his former Allies and the loss of popular approval throughout Germany.

Despite their harm to Soviet strategic interests, reparations did destabilize the European economy and fomented economic chaos favorable to Soviet-style communism. The Soviets had a lot on their side in 1945. In the early postwar period they were widely seen as managing their zone more wisely than the Allies. Many in the West saw central planning and controls as superior to the liberal market economic policies which had failed in the 1930s. Economists coming out of the war environment of planning, of wage and price control and rationing, such as John Kenneth Galbraith, favored central planning for occupied Germany and were pessimistic about the global economic outlook. The great accomplishments of planning in mobilizing for victory also disposed people to favor planned economies, particularly since many feared a future of stagnant global food production and uncontrolled population growth.[59]

Even the editors of Henry Luce's *Fortune* magazine, a bastion of American capitalism, called for continued economic controls after the war and for concentrating control of the global food supply in the newly minted United Nations, opinions which ten years later *Fortune*'s editors would have labeled communist.[60] French leaders as late as 1957 planned to centralize European Community decision making within an administrative elite, a strategy not far from French designs for the modern European Union: "The French hope that the new community will pursue a 'dirigiste,' or at least a Keynesian policy, regulating and guiding investment on a European scale."[61] Thus, the cold war's contest of ideas began with a distinct advantage for the Soviets.[62] Yet the Soviets threw away this advantage, along with the moral credibility they had gained from the sacrifices of their population during the war, to follow an opportunistic strategy. Reparations, particularly cash reparations, were intimately connected with central planning and a command economy: the connection is practical and intuitive. When Westerners reacted to Soviet reparations with anger for their inefficiency and harm to global economic well-being, central planning and control as ideas also suffered.

Raw materials played a central role in Soviet reparations demands, and they "took everything from the forest which could be turned into hard currency or other value."[63] Forest products were equal to coal and food in importance to postwar European economies.[64] Although industrial reparations claimed more than 25 percent of total eastern German production, timber reparations took almost 100 percent of the eastern German postwar harvest: most of the 44 million cubic meters officially harvested between 1945 and 1953.[65] And these harvests came from the forest's muscle and sinew, its best quality, oldest timber and capital, destroying stand structure and forest health. Perhaps the only form of reparations more destructive was holding former POWs as slave labor in the Soviet Union for years after the war ended. Two to three million German POWs were unaccounted for in May 1945, most in Soviet camps. Soviet Foreign Minister Molotov admitted in mid-1947 to keeping more than 890,000 former German POWs in forced labor camps, giving at best a low estimate of the total.[66]

All the Allies, particularly the French and the Soviets, mined the German forest for timber. Timber harvests were set "without regard to growth" in the first two postwar years, ostensibly to eliminate "war potential" but also to take cheap timber for reconstruction and to let their own forests recover. International forestry experts determined that Germany must "make a major contribution" to alleviating shortages of pine and spruce pulp and wood forecasted to last until 1960. Germany could do this: its standing timber was "in good order and well managed."[67]

Timber markets, and markets for most other basic commodities, exploded in the postwar economic environment of released demand and shortages. The global economy until early 1948 was desperately short of everything—food, coal, steel, and timber. Population problems and perceptions of uncontrolled population growth created a Malthusian atmosphere, with demand seeming to grow faster than food production. After food, basic raw materials were in greatest demand. The Germans desperately needed pit props to rebuild coal mines and railroad ties, "sleepers," to mend the broken rail lines. A senior German forester reported that even in 1949 "delivery of pit props to the pits and mines continues to be forestry's critical economic task."[68] The economy needed timbers, planks, and beams to rebuild bridges, harbors, workshops, farmsteads, and homes. A senior German forester prayed in 1947 that "one day pure economic, and particularly pure forestry considerations, will determine harvest levels, not political or 'war potential' considerations."[69] Pressure on the forest was far heavier in the postwar years than it ever had been under the Nazis.

The Soviets cut, mostly from state forests, more than thirteen years of growth as reparations between 1945 and 1949.[70] As an example of the intensity of reparations harvests, the Social Democratic news service reported that in 1949, 7,500–8,000 railcars loaded with timber "roll each month to the east or to hard currency lands."[71] Manifests of trains running through Frankfurt am Oder to the Soviet Union offer a snapshot of Soviet plundering (although they do not include wheat, coal, and timber taken out by ship and truck, or valuables taken in private baggage, such as Marshal Zhukov's notorious personal loot).[72] These manifests reveal a startling array of plunder: potash, vehicles, and precious farm horses; sugar, grain, and pigs taken from the starving Soviet zone population to the food surplus Soviet Union. One train's manifest listed fifty-two wagons of "literature"—were they texts from German libraries or archives? Another train of thirty-four cattle cars held nothing but political prisoners bound for the gulag.[73]

The Soviets even broke down German houses, taking 3 million cubic meters of first-quality timber between 1947 and 1952.[74] More than one thousand railcars carried dismantled housing timbers and furniture off to the Soviet Union in June 1954 alone. The Soviets sold much of their timber reparations to the West by way of Murmansk and Leningrad, or to the U.K. through Königsberg/Kaliningrad after reparations ended, paying for them in low 1944, fixed east mark prices while taking payment at world market prices in U.S. dollars.[75] With forest reparations harvest running at such levels, forest stocking in 1948, or capital invested in standing timber, plunged one-third from 1933 levels to 78 cubic meters per hectare, almost two-thirds below normal levels.[76] "The damage is so severe," noted a special 1949 forest report, "that it won't be overcome in a man's lifetime."[77] In 1965 an East German forester looking back on the waste and damage described Soviet harvests as "reaching shocking dimensions."[78]

The Soviets kept the catastrophic results of their reparations harvests a Party and state secret until formal reparations ended in 1954.[79] Soviets reparations policy flowed from Stalin's opportunism, from uncertainty as to the government of a reunited Germany—better to take everything possible now, before a peace treaty was signed and the opportunity lost—and to cover up the vast human rights crimes and the economic and ecological catastrophes which followed in the wake of Soviet occupation. Bruno Gleitze, a leading observer of Soviet economic policy, termed the period of direct Soviet reparations from 1945 to 1954 the "period of improvisation and covering up."[80] The Soviets banned Western observers and refused to conduct joint forest surveys or share data as agreed at Potsdam.[81] Secrecy was vital to hide the Soviet failure to observe the Potsdam agreements on reparations and to avoid unfavorable criticism at a time when the Soviets still courted popular opinion. The perfunctory forest inventories the Soviets did sporadically supply to the Allies were useless, so "primitive and incompetently manned as to be unacceptable." Social Democratic forestry experts at the time concluded: "One glance at Soviet forest inventories and practice shows at once that everything is intentional camouflage—a wordy paper swindle."[82] The Soviets ordered eastern German foresters to deny the intensity and scale of their reparations harvests and hide plummeting growth and health data from their western colleagues while learning as much as possible about western forest conditions.

The Soviets failed to replant the vast clear-cuts left in the wake of reparations, the most basic charge of plantation forestry. All four zones had large areas of unplanted clear-cuts after the war, between 5 and 6 percent, not remarkable under wartime conditions.[83] Yet by 1948, the first year afforestation data were available, the Soviet zone's unplanted forest area soared to 13 percent, whereas in the U.S. zone it had dropped to less than 4 percent and by 1950 closed below 2 percent, within striking distance of the normal 1 percent unplanted area. The area of unplanted clear-cuts in the Soviet zone almost doubled between 1945 and 1949, a staggering economic loss to the eastern

German economy that persisted into the late 1960s as the Forest Service struggled to clear the huge overhang of unplanted forest.[84] A forestry expert for the Social Democratic Party linked the destructive force of reparations with the chaos left in the wake of land reform: "Efforts to replant these clear-cuts have been so ineffective that it will take one hundred years to clear away the damage. All the Soviet zone power's [the Soviet's] policies, particularly the land reform, end up with the disappearance of the forest into the machinery and schemes of the monopolistic command economy."[85]

The rapid recovery of the U.S. zone forest owed much to the influence of U.S. foresters and the prompt inclusion of German foresters, politicians, and civilians in forest policy planning. Following the advice of foresters, American diplomats argued unsuccessfully against the punitive harvests the French and Soviet governments demanded at Potsdam.[86] The Americans refused to implement the Potsdam harvest in Bavaria, itself almost the size of the future East Germany, and convinced British occupation authorities in 1948 to combine their forest operations with the United States to rebuild the western German forest.[87] The U.S. military government drew on the traditions of U.S. forestry in Europe after the First World War when Lt. Colonel Henry Graves, the first dean of Yale's School of Forestry, Gifford Pinchot's successor as chief of the U.S. Forest Service, and provost of Yale University, led the 10th Engineers (Forestry) Battalion in Europe. Lt. Colonel William Greeley, also dean of the Yale School of Forestry, succeeded him. Graves' and Greeley's successors in the Second World War came to their work already in tune with their German colleagues by virtue of the origins of U.S. forestry in Prussian forest science and silviculture and army experience after the First World War. The traditions of the U.S. Forest Service also conditioned them to think in terms of multiple use, and to consult the local populace and government leaders in forest planning. Unlike in the Soviet zone, western German politicians and citizens complained with great effect about food shortages and reparations, and particularly about Allied over-harvesting in the German forest.[88]

Sympathy for the German people was not uniform. General Lucius Clay, commander in chief of the U.S. forces in Europe and military governor of the U.S. zone, empathized with the Germans, perhaps because of his Georgia roots.[89] Clay scathingly derided State Department officers seconded to the U.S. military government as "carpetbaggers" for their indifference to civilian hunger and suffering. Charles Kindleberger, who became a prominent economist, was one of these "carpetbaggers," and yet he stood in awe of Clay's energy and ideas. He wrote to his wife, Sara, in August 1946: "Yesterday I saw General Clay, who is a very strong-minded and strong-willed man, not a little scary to talk to for us State Department types whom he regards as very unreasonable. Did I tell you that the day before, when I had only an opportunity to shake his hand, he recommended orally to Ambassador Murphy, in my hearing and partly addressed to me, that it would be appropriate to put me on the 1,225 calories per day diet which everybody in Washington thought it was enough for the Germans to eat? It was a joke, of course, but a grim one with lots of irony in it."[90] Officers in the Soviet army of occupation may also have felt empathy for German civilians, but there was little they could do to express it and no tolerance for civilian protest. Above all, reparations deliveries were inflexible and failure to deliver carried direct, dire consequences for Soviet reparations officials.

The Soviets quickly imposed central control over their zone's forests to ensure maximum delivery of timber reparations and to anchor the Party's rule in the traditionally conservative countryside. The Soviets rechristened state-owned forests as *Volkswald* (People's Forest) and placed them under the management of Unified Forest Districts on 1 April 1946.[91] The Soviets based the Forestry Division in Potsdam with Soviet "experts" directing operations; Germans contributed workers and transport.[92] National Socialist forest administration forms fit well with Soviet economic policies, so they continued.[93] The Soviets issued "Forest Management Regulations for the Soviet Zone" on 13 October 1945, decreeing that "the entire forest economy in the Soviet zone will be managed and administered by a central authority," and took over management for all Soviet zone forests larger than five hectares. Forest Management, the department of the Forest Service traditionally associated with industrial forestry, received new responsibilities in addition to its normal statistical duties. Foresters no longer set their district's harvests based on local site conditions and demand. Forest inventory specialists shipped raw data directly to the managers of the national economic plans who ignored forest structure to set maximum harvest levels based on reparations quotas.[94] Thus the Soviets controlled eastern Germany's forests sixteen years before the Party leadership collectivized the last private forest and farmland in the "Socialist Spring in the Countryside" in 1960.[95]

Soviet advisers in forestry and agriculture supervised Soviet zone management. These "experts" had little practical experience but were charged to "pass on 'the most progressive science in the world' to German doctors and engineers who have far better training and more knowledge," as Norman Naimark, the leading historian of the Soviet zone, noted.[96] Soviet forestry experts also ensured that no German "sabotaged" reparations harvests; they subjugated German foresters and bore Soviet power into the rural economy.

A common theme of Soviet zone propaganda posters admonished German workers, "Learn from Soviet Men to Learn Victory!" (*Von den Sowjetmenschen lernen heißt siegen lernen!*).

Soviet experts "Sovietized" eastern German forest management. The Soviets switched eastern German forestry to the Soviet forest-management calendar year to marry eastern German forest production to the rhythms of the Soviet economy. The old German forest management year of 1 October through 30 September "fit with the biological rhythms of nature," harvesting when the trees were dormant, the ground hard, and farm labor free to work in the woods.[97] Soviet forestry experts set a uniform quota for spruce for every state, including the lowland states of Mecklenburg and Brandenburg where the soils were too dry to grow much spruce. Thus the Mecklenburg government in 1946 was forced to buy 1 million cubic meters of spruce from Saxony to meet its reparations quotas to the Soviets. The Soviets divided harvest quotas into equal parts for each quarter to provide a constant flow of raw materials to Soviet factories. All trees marked for harvest had to be cut in the same quarter. As a result, foresters marked one-quarter of the annual veneer beech log harvest in July for harvest in the summer. But since there was low demand for veneer logs in the summer, this superb wood went as firewood. Soviet experts often delivered valuable saw and veneer logs to meet quotas for low-grade wood and pulp, even when they had export contracts in hand at high, U.S. dollar prices. Similarly, Soviet forestry experts forced Saxon foresters to harvest spruce in the summer, when it was prone to blue stain damage, to fill Soviet quotas. The Soviets, who re-exported this wood to Great Britain, rejected the spoiled lumber and ordered the Saxon forest districts fill their quotas immediately from other stands. Uneconomic summer harvests continued according to Soviet practice until reparations formally ended in 1954. Soviet controls and reparations led to further inefficiencies. Soviet military officials, personally held responsible for full delivery of reparations, were so nervous about short deliveries that they routinely over-harvested, sometimes selling overages on the black market.[98]

Direct Soviet control and terror assured maximum reparations deliveries. The imposition of Soviet "experts" over experienced German foresters also foreshadowed the creation of a new professional elite in the countryside loyal to the Party and the Soviets, one free of the feudal taint foresters had for many Marxist-Leninists. Soviet zone authorities purged the Forest Service. Young German forest scientists, such as Albert Richter, were sent to work under Soviet instruction in forest labor brigades, in effect as slave labor.[99] The Soviets expelled "many thousands of foresters," an East Prussian district forester noted, who feared that he and his colleagues would have to "abandon our green jackets for good" in the atmosphere of communist hostility toward professional foresters.[100] Many fled west. More than six hundred senior Prussian district foresters besieged British officials for junior positions in 1947, desperate for work at any level. Even though the Soviets kept the forms of National Socialist forest management, they purged the Forest Service and the professional and technical schools of most foresters trained before 1945, whether they were Nazis or not.

Soviet experts, therefore, directed a workforce dominated by unskilled personnel, itself forcing forest administration into a highly centralized structure.[101] This neutron bomb strategy—keeping National Socialist administrative forms while liquidating the people—hastened forest destruction.[102] Experienced foresters made up less than 10 percent of the Forest Service at every level between 1945 and 1949, depriving managers of critical knowledge of local site conditions and management history.[103] Foresters plan for regeneration and harvesting, but they also guard the forest reserves. The starving and cold population plundered the forest for fuel and food after the Forest Service purge.

The Party leadership created a new, socialist Forest Service from scratch. The first classes from the Tharandt and Eberswalde forestry schools graduated in 1950, just as experienced foresters, such as Albert Richter, returned to work.[104] The Party leadership used the forestry schools to remake the old hierarchy wholesale through its "cadre policy," installing technically competent and politically loyal professionals as a sort of "anti-elite elite."[105] The Party leadership used admissions to remake the Forest Service. The Party specifically excluded the sons of foresters, traditionally the most fertile ground for recruitment, relying instead on "sons of workers" and "suitability from a political perspective."[106] Political reliability and class origin, preferably of the urban proletariat, counted most heavily in admissions. The Party leadership was hostile to the cultural and social norms of the countryside, not just of the Junkers and "squires," but also of professional foresters.

The new forestry cadres would do more than extend the Party's reach into the countryside; the Party was grooming them to take over western German forest districts in anticipation of imminent reunification. Eberswalde and Tharandt graduated more than two hundred students in 1953, far more than the East German Forest Service could put to work. When faculty members questioned the need for such a large student body, Heinrich Rau, a senior Party official, spoke: "The SED regime [the East German state]," he declared, "is ready to take over West Germany's forests!"[107] A Social Democratic study warned that the cadres policy was "a sure indication of the Soviet zone government's plans for a reunified Germany."[108] The Party would purge West

German foresters after reunification just as they had purged foresters in the Soviet zone, but this time the Party would have loyal and competent foresters ready to take their place. The Party's anticipation of taking over the western zones was clear even to junior foresters. When a young forester complained in 1947 that part of his district had been unfairly allocated to the British zone, he was told not to worry: "we communists will be running all Germany soon enough."[109]

Still, "residual forms of agricultural professional ethics" limited the effects of the change. Foresters particularly resisted the new professional and political ethic, "socialist forestry," and the Party's "radical concept of compulsory modernization."[110] Foresters, and other workers and managers in the rural economy, took up the Party's extreme materialism unevenly. The Party was never able to stamp out entirely "bourgeois forest science," or traditional forest science and ecology.

Civilians in all four zones suffered from the occupiers' policies, and hunger and deprivation severely limited economic recovery. Machinery and tools were either missing or requisitioned, and the workforce weak from disease, malnourishment, and hunger. The bitterly cold 1946–47 European winter was the most severe ever recorded. The *Times*' German correspondent reported in early 1947, at the height of civilian misery: "That the British zone is now in a miserable state cannot be seriously disputed, though that it is much worse than anywhere else in Germany is open to doubt. The whole country is in the midst of a winter infinitely more bitter than the last, and this year 'the battle of winter' of Lord Montgomery's dramatic phrase is an undramatic reality of short rations, inadequate housing, and ever-present cold. 'Everything has gone except cold and hunger,' Germans say, and in the British zone, which is more industrialized than any other part of Germany, and where bombs and guns caused more destruction, the winter has intensified the bitterness that was already widespread in the autumn."[111]

The North Sea was two degrees below freezing; "There is no record of anything like it before," reported a British official. "Lightships off the east Norfolk coast are watching for the icefield said to be drifting westward across the North Sea" and even the Venetian lagoon far to the south was ice-bound. The Kiel Canal froze from the Baltic to the North Sea, as did the great network of canals over which most of western Germany's food flowed to the Rühr from grain elevators and warehouses on the North Sea. Wheat shortages forced western zone governments, already hard-pressed to meet even subsistence levels, to reduce civilians' daily ration from 1,550 to 1,420 calories, desperately importing vast quantities of wheat from America, cutting the bread rations of American and British soldiers, substituting potato for wheat flour to make bread to reach even this meager, starvation diet. As the London *Times* correspondent described, "In places where potatoes are not available it [the daily ration] will drop to about 1,190," well below starvation levels.[112]

Western Germany normally imported much of its food from eastern Germany. The Potsdam Agreement called for Germany to be administered as one economic unit, with food moving as normal from the Soviet zone to the West in exchange for reparations, coal and coke from the Ruhr Valley. Despite this, the Soviets halted food shipments to the western zones—to pressure the British and Americans to withdraw from Germany and to turn western Germans against the Allies, but also because Germany's food surplus was consumed in Soviet reparations, requisitions of food, and land reform chaos. Near-famine conditions raged throughout Europe in the harsh winter of 1946–47, but hit Germany particularly hard.

This cold drove the freezing Soviet zone population, already near famine and without coal, to take as much firewood from the forest in 1946–47 as the Soviets took in reparations. The cold and harsh conditions almost brought timber harvests in the western zones to a halt. But in the Soviet zone, there could be no letup in timber reparations harvests, no matter the toll on German woodsworkers. These harvests reached levels unimaginable even under the best of conditions. That the Soviets extracted such extraordinary volumes—despite ferocious cold and near-total shortages of food, clothing, shelter, and tools—was frightening testimony to the general suffering of forced laborers and political prisoners in the Soviet zone.

Forest labor was, above all, very hard work even in normal times. Most German men either were lost in the war or came home slowly and dispirited, suffering from a syndrome of hunger, illness, and uncertainty which plagued all workers. Equipment, tractors, and horses had disappeared under Soviet requisitions. Almost no motorized saws were available, only cheap handsaws of such poor quality that they flexed in half and stayed there. Brittle, underweight axes shattered upon use and were soon almost useless due to the lack of sharpening files. The malnourished and freezing men had no basic work clothing, protective gear, or boots, items as important as motor saws.[113] Safety equipment was nonexistent. There was never enough to eat and living conditions were lonely, primitive, and unwholesome. Only hard labor overcame these deficits.

The Soviets achieved their massive reparations harvests by terrorizing and exploiting foresters and forest workers.[114] The Soviets and the German Economic Commission discriminated against woodsworkers even though the physical demands and dangers of forest labor probably exceeded those of any other job except that of German miners toiling in the Soviet uranium mines,

"Wismut AG," on the Czechoslovak border.[115] Woodsworkers were paid one-half east mark per hour, far below the wages of industrial workers in comparable jobs. Woodsworkers did not get supplemental food rations to sustain them in their heavy labor, unlike coal miners, and they suffered terribly from cold and exposure.[116]

The Party finally gave woodsworkers a small raise after the Workers' Uprising of 16 June 1953 panicked the leadership, but forest wages still remained below those of industrial workers.[117] Forest labor relations in East Germany were designed to extract ever higher volumes of raw material. A piecework system, paying workers only for volume produced (despite equipment shortages or failures, the lack of food and clothing, and bad weather) was imposed shortly after the war, a supremely exploitative and dangerous wage scheme. The piecework system, as Marxist-Leninists must have known, was a long-standing grievance of workers and deeply resented feature of the most exploitative capitalist economies, yet Soviet planners nevertheless embraced it. Workers could earn more by overfilling their quotas, but only if the piecework rates were reasonable, constant, and fair. Instead, as workers overfilled their quotas, managers raised production hurdles and cut bonuses. As a contemporary reporter observed, "When workers beat their individual quotas to win a meager bonus as an '*Aktivist*' or '*Hennecke*' they only hurt themselves, as the Soviets then raise their quotas based on the increased production": working conditions more fitting to the early Industrial Revolution than to the postwar economy.[118]

Conditions in British zone forests were in many ways as harsh, although the workers had basic rights, better equipment, and supplemental food rations. The British army officers in charge of "Operation Woodpecker," the British government's short-lived forestry enterprise in the British zone, scaled back harvests during the severe 1946–47 winter, protecting British army as well as German civilian forest workers.[119] Forest work under such harsh conditions taxed British soldiers' strength, even though they enjoyed the benefits of full rations, good tools, boots, safety equipment, clean water, and warm beds.[120] Soldiers could work only one-month tours felling timber in the remote, cold reaches of the Harz Mountains where they lived in encampments "as lonely in many ways as any military post on the North-West Frontier."[121] How much harder, then, must it have been for conscripted and forced labor in the Soviet zone.

Workers in the Soviet zone forests labored without respite or release. The Soviets reduced the 1949 harvest quotas because of labor shortages aggravated by *Republikflucht,* but also because the Soviet Ministry for Internal Affairs closed the forced labor camps, most often former Nazi concentration camps, in January 1950 under international pressure.[122] "The scale of the human tragedy," observed the Social Democratic news service, "is immeasurable."[123] The Soviets delayed prisoners' release for more than two months while they improved camp conditions and fattened up inmates to moderate the pallor and habit of the concentration camp. Even so, 70 percent of the prisoners released to the West had tuberculosis, despite the Soviet claim that only 672 prisoners were infected. Before the camps finally closed, the Soviets loaded 15,038 political prisoners and POWs onto rail cars for shipment to the Soviet Union where Moscow Administrative Court sentences of twenty-five years' imprisonment in the gulag awaited them.[124]

Forced labor and concentration camps were central elements of and inseparable from Soviet reparations policy: intangible costs, like the suffering of POWs condemned to slave labor or forest destruction, were missing from reckonings of Soviet reparations. Even before Germany capitulated in April 1945, the Soviet military government set up thirteen NKVD (the Soviet secret police and father to the KGB) forced labor camps in the Soviet zone under the notorious Colonel-General Ivan, "Ivan the Terrible," Serov.[125] There is no way of knowing the total number of deaths in the camps, which Soviet zone observers and released inmates estimated at between 90,000 and 150,000, with another 30,000 to 40,000 people transported to Soviet forced labor camps.[126] Rainer Hildebrandt, founder of the Taskforce for Human Rights, estimated that 170,000 camp inmates died in Soviet zone forced labor camps and 36,000 were condemned to slave labor in the gulag.[127] The U.S. military government reported that more than half of the inmates the Soviets condemned to prison at the former Nazi concentration camp Buchenwald died from starvation alone.[128] The Soviets also impressed 50,000 skilled workers for Soviet research labs or to run German factories taken to the Soviet Union as reparations. Wismut SAG, the Soviet uranium mines on the Czech border, claimed more than 50,000 political prisoners and more than 100,000 forced labor workers. Many never returned.[129] The camps were a pool of basic labor, but also a tool of terror. A Social Democratic journalist questioned Walter Ulbricht in Berlin in May 1949 about the camps. Ulbricht freely admitted the existence of the concentration camps, a necessity to punish "warmongers." He threatened stunned Western reporters: "Warmongers include every Western journalist. Schumacher and Adenauer are the worst of the warmongers. When we catch these types, we lock them up. It is better to lock up the innocent in a concentration camp than to let warmongers such as these walk around free."[130]

Reparations fit Stalin's opportunistic nature and served, even if inefficiently, Soviet economic recovery and Soviet hegemony in Central Europe by destroying existing structure and creating conditions which favored central planning and one-party rule. Soviet propaganda — unwittingly bolstered by pessimistic Western journalists and writers — encouraged the myth that a Stalinist government could be democratic and in the tradition of European socialism, and that a postwar economic environment ruled by shortages and unchecked population growth required central planning. Thus, Stalin imposed land reform close on reparations' heels, the key part of the Soviets' postwar strategy of the "parliamentary road to socialism," of winning power through the ballot box by using the popular appeal of socialism and of agrarian reform. Land reform's assurances of order and equity complemented perfectly the poverty, hunger, and despair which reparations sowed. But Goethe's "primal conflict" of custom versus innovation remained the defining quality of the Marxist-Leninist war on the countryside, and the landscape itself would finally not bend to fit Marxist-Leninist axioms. Land reform, the Soviet strategy to liquidate class enemies in the countryside, to introduce Soviet-style collectivized agriculture, and to unite peasants with industrial workers, tested whether Marxist-Leninist modernity could overcome the countryside's traditional economic and ecological forms. The results of land reform's struggle between modernity and custom were revealed in eastern German farm and forest structure and in the well-being of farmers and forest workers.

The Bärenthoren demonstration forest in Sachsen-Anhalt where Freiherr von Kalitsch experimented with natural regeneration and uneven-aged mixed pine-hardwood stands on sandy soils after the First World War. (A. Nelson photograph)

OBERFORSTMEISTER UND PROFESSOR DER BOTANIK,
DIREKTOR DER FORSTAKADEMIE EBERSWALDE

Alfred Möller, father of the *Dauerwald* philosophy of radical ecological forest management.

Professor Richard Plochmann and Eugen Syrer of the University of Munich inspect the fine root structure of an oak seedling in a forest stand fenced to protect young hardwoods from deer browse. (A. Nelson photograph)

Modern Brandenburg pine stand clear-cut, plowed, disked, and raked, ready for replanting. The fire tower in the background stands guard against the ever-present threat of fire to pine planted outside its range. (A. Nelson photograph)

Cologne Cathedral rises out of the bomb waste on the west bank of the Rhine on 24 April 1945. Heavy Allied bombing during the war hit western German cities particularly hard. The ruins of the railroad station and Hohenzollern Bridge are to the east of the cathedral. (AP/ Wide World Photos)

Border posts change on the Oder-Neiße line in April 1945, three months before the Potsdam Conference, as Polish troops mark Poland's new western frontier. (Deutsches Historisches Museum)

A German family scavenges for bilberries along the cleared strip (the inner-German frontier) separating the Soviet and British zones in 1947. (Deutsches Historisches Museum)

German refugees from Soviet ethnic cleansing in the east wearily make their way through the central German landscape to safety in the west, 1945–46. (British Pathé)

City people scour the countryside for firewood and food after 1945. (Deutsches Historisches Museum)

German farmers in the French zone deliver their draft and farm horses to French officers for selection and shipment to France as war reparations in July 1946. (British Pathé)

A Soviet soldier grabs a Berlin woman's bicycle after a trade goes bad in the neighborhood around the Brandenburg Gate in 1945. (Keystone Pressedienst)

Workers wearing old army clothing and wooden shoes pause near a bombed building in Dresden in 1949. Shortages of work clothing, boots, and safety equipment caused great hardship, particularly for woodsworkers laboring in the extreme cold to meet Soviet reparations demands. (Corbis)

German workers demand land reform in June 1926. The communist land reform of September 1945 drew on land reform's traditional appeal to win political support. (Deutsches Historisches Museum)

"Junker Land to the Peasants!" Communist land reform propaganda pressed home a central theme of German politics: land reform and the expropriation of estates such as the manor house in the background. (Deutsches Historisches Museum)

"New peasants," mostly women, children, and old men, celebrate land reform with dancing in September 1945, a ritual often staged by land reform officials after the division of farms — here, the former estate Frauenmark. (Deutsches Historisches Museum)

The land reform functionary Quandt drives the first boundary stake into a field of the expropriated estate Gottin on 2 September 1945. (Keystone Pressedienst)

A new peasant family rests on the margins of their land reform farm carved out of the Kränzlin estate, probably in fall 1945. (Deutsches Historisches Museum)

"Day of the Solemnization of the Fields, 11 May 1947. City and Countryside—Hand in Hand!" Communist Party propaganda advancing the "Union of Workers and Peasants": the transformation of farmers into "workers" and farms and forests into "factories on the land." (Deutsches Historisches Museum)

"Our Forest Means: An Unconquerable Source of Health and Culture." This 1952 poster shows earnest workers planting pine seedlings to urge on Brandenburg foresters and woodsworkers to meet the First Five-Year Plan goal of replanting the more than 15 percent of the forest that was clear-cut to fill Soviet reparations and left barren. The poster's appeal to environmental and cultural values reflected the Party leaders' brief dalliance with radical ecology and *Dauerwald* in the early 1950s. (Deutsches Historisches Museum)

"To Learn from the Soviet People Means to Learn Victory. Peasants! Increase Yields through Application of Soviet Agricultural Technology." The transmission tower in the background recalled Lenin's 1920 speech to the Eighth Congress of Soviets: "Communism is Soviet power plus the electrification of the whole country." (Deutsches Historisches Museum)

Late in 1952 workers from the Stalinallee building project stand before a propaganda poster showing a Soviet expert teaching German workers "the world's most advanced science." Less than six months later they revolted in the June 1953 Workers' Uprising. (Getty)

Aerial photograph showing the diverse ownership and uses characteristic of the rural landscape in 1937 around Trebnitz, east of Berlin and near the Oder River. This distinctive pattern of diversity and stable hierarchies emerged from a century-long process of equilibration to ecological conditions and changing world markets. (Landesvermessungsamt Brandenburg)

Trebnitz fields and forest in 1953 showing the fracturing of farm and forest ownership after the Soviet's 1945 land reform. (Landesvermessungsamt Brandenburg)

Trebnitz field patterns in 1994 showing the results of forced collectivization and the complete reversal of land reform. This structure endures as Marxism-Leninism's most significant legacy in the rural landscape. (Landesvermessungsamt Brandenburg)

Woodsworkers fell mature spruce in the British zone. As in East Germany, modern tools were in short supply after the war. This crew used handsaws to cut down trees. (Corbis)

Ulbricht as Palinurus, the helmsman, rowing on the Baltic in a 1952 propaganda photograph widely reprinted in East Germany. (Keystone Pressedienst)

West German police protect the body of Helmut Griem from the East German People's Police, the VoPos. The VoPos shot Griem as he crossed the border to the West to meet his girlfriend, who kneels beside him (1 January 1950). (Getty)

A newsreel cameraman's telephoto lens captured VoPos in 1952 wearing Soviet-style uniforms introduced at the same time the Soviets fortified the inner-German border. (British Pathé)

Soviet tanks spread out into East Berlin on 20 June 1953 to suppress the Workers' Uprising. A tram run off its tracks blocks the street to traffic as a lone woman crosses a forbidding landscape of force and emptiness. This image foreshadowed the more famous photograph of a man's defiance of a Chinese Army tank in Tiananmen Square in June 1989 and Honecker's threat later that year to employ a "Chinese solution" to put down civilian protests. (AP/Wide World Photos)

*Republikflucht:* Refugees flowed into West German camps on the sandy Northern Lowlands on 1 April 1954 as many small farm and forest owners fled collectivization. (Getty)

Ballet dancers from Dresden's renowned Palucca Hochschule für Tanz were called out in September 1962 to help bring in the potato harvest, which was still failing in the chaotic wake of Ulbricht's hasty Socialist Spring in the Countryside in 1960. East Germany, much to Soviet Premier Khrushchev's irritation, had to import sugar beets and potatoes from Poland in the 1960s. (Getty)

Peasants taken from their farms to harvest trees rest by the roadside in spring 1963. Labor and livestock drafts on top of the chaos of forced collectivization badly hurt farm production and farm and forest workers' morale. (Getty)

# 4

## *"A Law Would Be Good": Land Reform*

*A certain* quantum *of power must always exist in the community, in some hands, and under some appellation.*
*—Edmund Burke,* Reflections on the Revolution in France

Germans could see the forest disintegrate under the burden of Soviet reparations almost before their eyes, as they walked in the woods or watched the thousands of trucks, railcars, and barges move logs and lumber to the east. Forest destruction raised existential fear, but most Germans realized that reparations were justified. Propagandists pressed home that cooperation with the Soviets—proving that the German people were "peace-loving" by "working through the conditions created by the fascists"—was the quickest path to ending reparations, to bringing the millions of POWs home, and to recovering the Oder-Neiße territories. So Stalin turned to land reform to prove Soviet populist credentials and to steer popular anger toward the British and Americans.[1] Land reform only seemed like the gentler aspect of Soviet policy. In reality, Stalin's land reform carried class warfare into the countryside. Stalin also made war on the countryside, demanding the remaking of farms and forests into "factories on the land" to fulfill *The Communist Manifesto*'s preconditions for socialism. Land reform—first to break the backs of the peasants and estate owners, followed by collectivization on the Soviet model—was also the essential first step to cementing Soviet power throughout Germany and to securing the fruits of Soviet victory.

The principal issue that united opposition against the Soviet occupation (even drawing some support from the Party's leadership) was Stalin's amputation of Germany's eastern provinces. Party propagandists argued that British and American failure to enact land reform forced the Soviets' hand, reasoning that the provinces could never be returned until all Germany was "democratic." The Party leadership could not passively accept their loss without sacrificing even more of the popular support the Party had lost due to reparations, and so they invoked land reform in the western zones as necessary for the return of the Oder-Neiße territories. The British and Americans were responsible for the loss of the eastern provinces, not the Soviets.

The Soviet zone press also charged that the failure of the western zone governments to enact Soviet-style land reform had caused the severe hunger in the western zones, threatening a humanitarian catastrophe unprecedented in European history.[2] When the British military government cut civilians' rations in July 1946—rations which already hovered at near-famine levels—the editors of *Neues Deutschland* promptly attacked the British government as *Volksfremde,* "enemies of the people," declaring that the "starvation of the German people is a deliberate Anglo-American policy."[3] More accurately, Soviet reparations and Stalin's land reform crippled Soviet zone farm production, which should have flowed to the historically food-deficit western zones under the Potsdam Agreement.[4] Food shortages in the Soviet zone also forced deep cuts in eastern German rations, shortages compounded grievously by Soviet requisitions for troop support and reparations. The *Neues Deutschland* editors attacked western German administrative and ownership structures: "The hunger crisis in the West is due principally to the fact that, with the exception of the German Communist Party, western German parties, politicians, and bureaucrats have failed to demand the thoroughgoing measures we have here in the East. The Western press admits that the care of the people's daily bread remains in the hands of the bourgeoisie—more precisely in the hands of Junkers, corrupt markets, Nazi farm administrators, and reactionary bureaucrats. The security of the workers' daily bread is, in the first instance, a *political* question" (emphasis in original).[5] Farm productivity flowed from central control and ideological conformity.

The German Communist Party leadership campaigned hard to prove that only the Party could win the return of the Oder-Neiße territories. As Max Fechner, a senior Party official, declared in a September 1946 article, "Clarity on the Eastern Question": "The Party shows the way to a true national solution and rejects any loss of German territory. The new borders [Stalin's border adjustments] are purely provisional."[6] Fechner blamed "fascist and reactionary demagogues in the West [who follow] the old Nazi aggressive solution of *Lebensräume* and revanchism" for forcing the Soviets to hold tight to the eastern provinces as long as western Germans demanded their return despite not implementing Soviet-style land reform.[7] The Soviets could not permit a unified Germany within its 1937 borders as long as independent farm and forest land persisted in the western zones; Fechner's "true national solution" demanded an ecological and economic revolution in the western German countryside matching the destruction in the Soviet zone and the liquidation of the Party's political enemies in the countryside. Germany would recover the Oder-Neiße territories only "when all German, West and East, internal political conditions are unmistakably democratic, antifascist, and antichauvinist" — when Germany was unified under Soviet control.

Fechner's defense of German rights to its eastern provinces, however, was too much for the Soviets. They would never give up the Polish and German lands they had taken either as spoils of war or as plums drawn from the Molotov-Ribbentrop treaty of 1939. Soviet diplomats asserted that the Allies had agreed at Potsdam to cede the Oder-Neiße territories to them and threw back protests against Soviet ethnic cleansing. As Soviet attacks on the U.S. presence in Europe mounted, Secretary of State James Byrnes, in his seminal Stuttgart speech on 6 September 1946, affirmed U.S. support for German self-determination and Germany's claims to its eastern provinces. Byrnes affirmed America's core interest in European freedom and declared America's role as a European power and leader of the Atlantic Alliance. The United States would not retreat from Europe as it had after the First World War, nor would it leave Germany to Soviet control.[8] Fechner, unhappily, found himself making arguments similar to those Secretary Byrnes had made only a week before his article appeared, that the Soviets had no right to Germany's eastern provinces.

Five days later the Central Committee denounced Fechner's defense of Germany's rights to the Oder-Neiße territories with Orwellian precision: "Recently the Party's position on the question of Germany's eastern boundaries was made clear and unambiguous [in Fechner's *Neues Deutschland* article]." The Central Committee then attacked "revanchist" critics [implicitly Fechner]: anyone who dared describe Stalin's border adjustments as "provisional."[9] Many Party members — and Fechner was fourth man in the hierarchy — wanted the eastern provinces returned for reasons of national pride to counter the political damage. The Party needed a bold program to protect itself, and land reform seemed a perfect policy: it cost nothing and destroyed class enemies at the same time.

Socialism enjoyed great appeal in Germany at the end of the war — both as the opposite of Nazism and as a humanist political philosophy. Marxist-

Leninist attacks on capitalism also echoed radical conservative language from the 1920s and 1930s, as well as *Dauerwald* foresters' attacks on industrial forestry. Many foresters throughout Germany still accepted the Nazi equation *Kahlschlag* = *Kapitalismus* (clear-cutting equals capitalism), that clear-cutting was a tool of "liberal high capitalism."[10] Since industrial forestry relies heavily on the tools of finance, cash-flow discounting, and statistical analysis, they were not far off the mark. Marxism, as the avowed enemy of liberalism and capitalism, might then better protect the forest than the "socialist" National Socialists. The postwar leader of *Dauerwald* foresters and Alfred Möller's acolyte, Hermann Krutzsch, abandoned western Germany in 1945 for the Soviet zone, leaving for what he saw as a more enlightened government open to radical ecology and organicism.

German foresters, with near unanimity, turned against industrial forestry and clear-cutting after the war. "Everywhere," noted Ryle, a senior British forest expert, "the cult of the pure [industrial] forest has entirely lost favor."[11] Ferdinand Beer, a leading West German forest scientist, issued a passionate call for an end to clear-cutting in "Clear-cut or Single Tree Harvest?" part of a wave of articles demanding an end to clear-cutting in eastern and western German forestry journals.[12] Early in 1947, the senior Soviet zone forest scientist and dean at Eberswalde, Kurt Hueck, called for a "new forest" of mixed species free of age classes: a close-to-nature forest, "an especially pressing task for forest science."[13] Most Germans, east and west, also desperately wanted a new Germany, free of National Socialism or any transcendental philosophy and war. The war and National Socialism, Henry Kissinger observed in 1956, had left a "legacy of disillusionment, if not disgust."[14] Many Germans distrusted abstract ideologies and grand schemes, longing for peace and a return to a life rooted in the *Heimat* (homeland) and nature.

Themes of lost natural harmony and structures ran through a popular book published in the war's closing days, Wilhelm Münker's *Judgment Day in the Forest: Forest Creatures Accuse,* in which natural hardwoods revolt against clear-cutting and industrial forestry.[15] Spruce, the epitome of industrial forestry, is brought before the Forest King, a majestic oak (*Waldkönig Eiche*). An honest, old-fashioned forester, *lieber Grünrock* ("dear green jacket"), read the charges: spruce was "the destroyer of peace in our woods."[16] He proved the damage caused to forest soils and to plant and animal life by pure conifer stands. The forester pleaded for a return to custom and tranquility: "Our very lives depend upon the restoration of a healthy forest."

The judge, Forest King Oak, ruled against spruce and industrial forestry but generously concluded: "Now we hardwoods offer up our sacrifice for the sake of the war effort. But the day will come when men can again freely create a true homeland [*Heimat*] for their children. You, spruce, will come out of this judgment better than you probably expected. We will make you welcome in our diverse, mixed forest." The judge's decision balanced economic and ecological values, rejecting *Dauerwald* as well as industrial forestry, reaching out to the middle landscape, to Goethe's balance between custom and progress. Land reform and Soviet promises of equity and democracy fit well with foresters' longing for peace and for a recovery of nature and stability.

Land reform, therefore, was the perfect device to win back the popular approval lost by reparations and Stalin's border adjustments. Yet the senior leadership of the German Communist Party hesitated, anticipating the damage land reform would wreak on the rural economy and food supply. At a 4 June 1945 meeting in Moscow, only a few weeks after Germany's capitulation, Stalin listened angrily as Anton Ackermann, the German Communist Party's chief theorist, advised that "the Party comrades do not recommend land reform just yet."[17] Ulbricht later admitted that most people, even communists, wanted only order and guarantees of work, not radical change or state ownership.[18] Furious at the Germans for their timidity, Stalin demanded immediate land reform and "liquidation of the Junker class" less than a month after the war's end.[19]

When Ackermann returned to Berlin on June 9 he carried with him a draft land reform law in Russian for a young German party member, Wolfgang Leonhard, to translate into German and prepare for publication.[20] Leonhard recalled the events of the Party leadership's meeting with Stalin in a 1998 interview:

> On the fourth of June 1945 suddenly Ulbricht disappeared, and you didn't know where and what. Only years later we heard that he went to Moscow and in Moscow on the same day was invited to the Soviet Politburo and Stalin personally. And Stalin gave him a new order. The order was immediately set up the Communist Party and help to create a Social Democratic Party, a Catholic Party, and a Liberal Party and set up then an antifascist Democratic United Front, and also fulfill this summer already, make all the preparations, maybe autumn, land reform.
>
> The confiscation of the feudal landowners and the division of the land in the hands of the peasants. So Ulbricht was there, Ackermann, much more intelligent, more capable, he wrote the programme. You needed a programmatic statement.[21]

When Leonhard asked if the Soviets would really share power with the bourgeois and socialist parties, Ulbricht retorted, "It's crystal clear: It must appear democratic, but we must have all the strings in our hands."[22]

Land reform promised to win popular approval for the Party and support for Soviet policies. It also balanced the harshness and opportunism of Soviet reparations with the promise of democracy and reform. As such, land reform was an ideal element in Stalin's "parliamentary road to socialism," which Nikita Khrushchev defined in *Pravda* as "winning a firm majority in parliament and then converting parliament from an organ of bourgeois democracy into an instrument of genuinely popular will."[23]

Land reform, "the Sovietization of central German agriculture," was the first major Communist Party initiative in eastern Germany after the war, an indication of the seriousness Stalin placed on controlling the countryside.[24] It "laid the socioeconomic foundations of the communist system" in eastern Germany. As Walter Ulbricht declared, "The elimination of large estate owners—followed by the expropriation of private firms, large banks, cartels, and syndicates—and the abolition of the power of the War Interest Groups is a prerequisite to the new constitution."[25] Hans Lemmel, Alfred Möller's successor as dean of the Eberswalde Forestry School, observed that land reform was "a communist agricultural revolution" made against the Party's traditional enemies.[26] Land reform's class warfare character was clear: a communist ally, the Social Democrat Otto Grotewohl and future premier of the East German republic, branded Junkers as "class enemies of freedom and historical poison."[27]

Land reform enjoyed great popular appeal across the political spectrum, similar to *Dauerwald*'s in timbre, and was a perennial theme of German political discourse.[28] The Weimar Republic's last chancellor (1930–32), Heinrich Brüning, defined agrarian reform for most Germans through his government's *Osthilfe* (eastern support) land reform.[29] The *Osthilfe* was a startling anomaly to his deflationary policies, costing more than $6 billion in current value, a staggering burden in the midst of the Great Depression and testimony to the seriousness of problems in Germany's east.[30] Reactionaries and Nazis attacked Brüning's conservative commissioner for the *Osthilfe*, the outspoken and passionate reformer Dr. Hans Schlange-Schöningen, as an "agrarian Bolshevist" for cutting farm supports, breaking up bankrupt large estates, and distributing the land to poor farmers and the unemployed.[31] Schlange-Schöningen worked to revitalize east Elbian farming, to reinforce the Prussian yeomanry on Germany's eastern marches, and to inoculate workers against Marxism through work on the land and ownership.[32] Opposition to the *Osthilfe* land reform program sparked the fall of the Brüning government, the last democratically elected government before Hitler.[33]

Schlange-Schöningen, despite his disfavor with the Nazis, survived the war, and the British military governor tapped him to direct British zone farms and forests. He emerged as the central reformer of postwar West German farm and forest policy and an insightful critic of Soviet zone land reform, aided by his experience as the Reichskommisar for the *Osthilfe* and impeccable anti-Nazi credentials. The British steadfastly defended him against U.S. and German criticism and from critics in the Soviet zone. Schlange-Schöningen's demand for an independent, mixed farm economy in the Soviet zone as well as in the western zones threatened the Party's legitimacy. The Party's central organ, *Neues Deutschland,* attacked Schlange-Schöningen with broadside titles such as "Schlange-Schöningen Must Go!" deploying rhetoric similar to the Nazis and German militarists. No longer an "agrarian Bolshevist," Schlange-Schöningen now emerged in communist propaganda as an "agrarian monopolist."[34]

Although the Party leadership drew fully on land reform's popular appeal, German communists' references to traditional land reform themes angered the Soviet military government. Colonel General V. S. Semyonov, a senior Soviet zone official, lectured his German comrades in August 1945 on the need to emphasize Friedrich Engels' work on Thomas Müntzer and the primary relevance of the Peasants' Revolt of 1524–25 and Karl Kautsky's orthodox Marxist writings to land reform, rather than pointing to Damaschke's romantic tract *Die Bodenreform.*[35] Land reform also served Lenin's 1920 *Bündnispolitik* (the Union of Workers and Peasants), strengthening the alliance of the peasants and workers by creating many new peasants on small farms, the only class of peasant, according to Lenin, which could build strong ties to workers.[36] Soviet zone land reform was built on Soviet collectivization, not on the conservative German political tradition.

The German communists published their demand for the "liquidation of the large estates of the Junkers, dukes, and princes, and the expropriation of their entire property and land" on 11 June 1945, seven days after their meeting with Stalin in Moscow.[37] Pieck introduced the slogan "*Junkerland in Bauernhand!*" — "Junker land into Peasants' Hands!" Wilhelm Pieck, chairman of the German Communist Party and future East German president, met his comrades' protests coolly, declaring that their worries about land reform were "nothing more than infamous slandering of the peasant."[38] Land reform was a temporary expedient and purely tactical: "In the course of future developments," Pieck soothed at a Party conference early in January 1946, "many possibilities will manifest themselves for undertaking corrections in this question."[39]

Pieck's "possibilities," many Germans sensed, were collectivization of all forest and farm land. The "Manifesto to the German People" was announced at the Party's First Congress in April 1946 (and the takeover of the eastern Social Democratic Party was celebrated), foreshadowing collectivization: "The Socialist Unity Party is the Party of an antifascist democratic parliamentarian republic that guarantees for the People every right of free expression of belief and conscience. [But] the Socialist Unity Party will not be satisfied with the realization of an antifascist democratic republic. Its goal is the socialist ordering of society."[40] Land reform would bridge a temporary antifascist democratic republic and a Marxist-Leninist socialist state, both chimera.

The Party ran its only free election to select delegates to the Land Reform Commission after three months of planning and of courting the "bloc parties" to reap the anticipated popular support.[41] Voters chose 52,000 delegates to the Land Reform Commission. The delegates' tasks were to oversee land reform and safeguard citizens' rights. More than half the delegates elected were political independents, the "target," in Walter Ulbricht's words, of land reform policy.[42] The Party's position was "particularly precarious in the agricultural economy and rural society": land reform would "anchor" small peasants, Lenin's favored class of farmers, to the Party.[43] The strategy worked at first; Party membership soared six times in rural Mecklenburg, from 3,200 members to 19,500 in October 1945 and to 32,000 members by year-end.[44] Wilhelm Pieck highlighted land reform's usefulness in August 1947: "I'd like to emphasize that land reform created a great number of supporters in the villages, 300,000–400,000 people. The Soviet Occupying Power and the Party of the Working Class gave German peasants their land, a good preparation for the "Union of Workers and Peasants" [the *Bündnispolitik*]. It's also good for German-Soviet friendship, since 300,000–400,000 of our supporters got their farmland from us."[45] Pieck understandably confused the German Communist Party with the Soviets, "us." Pieck, in a later telephone conversation with the Russian adviser to the Soviet commander in chief, stressed land reform's political role in "neutralizing the bourgeois parties" and asked for faster division of forestland to strengthen the peasants' loyalty to the Soviet Union and the Party.[46] Forestland, since large landowners held it disproportionately and because the division of forests did not threaten the food supply, was a favored source for land reform handouts.

The Party finally launched land reform on 4 September 1945 after pro forma peasants' assemblies. The first Land Reform Directive declared: "Land reform must end Junker rule in the villages and liquidate the Junker feudal estates which have always been a bastion of reaction and fascism in our land and a main source of aggression and wars of conquest against other peoples."[47] Nearly identical directives followed promptly for the rest of the Soviet zone.

The noncommunist bloc parties joined the German Communist Party in a declaration of support on 13 September 1945, with only the Liberals and the conservative Christian Democrats asking for compensation for expropriated farm and forest landowners. The Christian Democrats also warned that collectivization loomed, a fear on many minds.[48] Ulbricht promptly fired the Liberal and Christian Democratic leaders and hounded them from the Soviet zone, recalling later, "They were struck down by the workers and therefore fled to the power regime of monopoly capital and landowners [western Germany]."[49]

Soviet and German Communist Party "functionaries" completed land reform in only six weeks, taking without compensation more than 3.3 million hectares of farm and forest land—all Soviet zone farm and forest land over one hundred hectares, more than one-third of the total.[50] The Party leadership kept 1 million hectares of farmland as "the People's Own Estates" (*Volkseigenengüter*, VEG), handing out the balance to farm and industrial workers and refugees. When Allied foreign ministers of the Four Powers (the Soviet Union, the United States, Great Britain, and now France) met on 14 April 1947 to form a joint land reform policy as agreed at the Potsdam Conference, Soviet zone land reform was already a year and a half old and irreversible.[51]

Although land reform emphasized the expropriation of war criminals' land, Nazis owned less than 5 percent of farm and forest land in the Soviet zone.[52] "Expropriation," an East German forest historian noted, had "very little to do with who was a Nazi and who was not."[53] Many non-Nazi small landowners or former local government officials lost their farms and forest for criticizing the Soviet regime.[54] Anton Hilbert, a senior Soviet zone land reform official, noted in 1946, "The only way to escape being labeled as a war criminal or Nazi is to have been in a concentration camp or a Hitler prison—anyone who protests against the madness, in today's food crisis, of tinkering with soil and land are arrested."[55] The Nazis had seized the farm and property of the Mendelssohn-Bartholdy family, descendants of composer Felix Mendelssohn, because they had Jewish ancestors. No sooner had the family resettled on their farm after the war than the Red Army evicted them again.[56] Carl-Hans Graf von Hardenberg lost his house and farm to the Nazis for his role in the 20 July 1944 plot against Hitler, yet lost them again to the Soviets in September 1945. But if Nazis were not at the core of land reform, then neither were the Marxist-Leninists' arch-bogeymen, the Junkers.

Junkers made up less than 30 percent of land reform victims. The average farm expropriated was two hundred hectares, an efficient size given the region's dry, sandy soils and hardly the mammoth estates ruled by autocratic, absentee landlords attacked in the Soviet zone press.[57] Ninety percent of the expropriated farms were less than five hundred hectares, still an appropriate size for northern German farms.[58] The majority of land reform's victims were modest farmers.

The Junkers were the strawman used by the Soviets and German communists to justify class war against the complex rural society. There were, in fact, few Junkers in the Soviet zone. Soviet ethnic cleansing and seizure of the core of Prussia east of the Oder-Neiße line—and the Allies' dissolution of Prussia at Potsdam—had already broken Junker power months before land reform. Stalin's border adjustments had effectively expropriated the estates of landowners holding 2.4 million hectares east of the Oder-Neiße line and 11,000 estates over 100 hectares. "Junker land into Soviet and Polish hands!" would have been a more accurate slogan than "Junker land into peasants' hands!"

As the Soviets expropriated farm and forest, they systematically destroyed property registers, maps, and estate records, liquidating the institutional memory of traditional rural society under a decree "providing for the complete destruction of all records of previous land ownership."[59] Destruction of legal and cultural records also made land reform irreversible in the event of German reunification, which nearly everyone assumed was imminent. Land reform created a socialist tabula rasa in the countryside. The editors of *Neues Deutschland* reported in August 1946: "Old land registers, titles deeds, and other documents of those large Junker estates that land reform divided up were recently destroyed as the last evidence of the power of the old feudal overlords. The unencumbered transfer of the large Junker estates to new peasants has now ushered in a new age."[60] As surveyors ran new transects, they obliterated existing boundary markers, cultural symbols which had sparked the conservation movement a century before. Tractors tore apart hedgerows dividing ancient fields while axes felled alleys of pleached linden and poplar.

The Soviets attacked architecture, landscape, and memorials to liquidate memory, custom, and culture. They destroyed more than 10,000 old manor houses, barns, and stables in 1947 as "symbols of a feudal age."[61] Farmers who tried to join management of their land within cooperatives, often with the former estate owner as manager, were punished. The Party leadership could not permit the form of the old estates to survive lest their former owners claim them after reunification nor could they allow the rhythms and patterns of the old landscape to endure. The Party leadership ruled, "any attempt at collective or communal management is sabotage of land reform!"[62] And they were right. Land reform's goal was reordering the rural society, economy, and landscape to fit Marxist-Leninist norms, not efficiency or equity. The Soviets judged land reform's success by its political effect and never planned for land reform structures to endure more than a year or two.

The Soviets and German land reform functionaries doled out small allotments averaging seven to eight hectares to new peasants and small farmers, "too small for the light soils of eastern Germany": in the words of a new peasant: "too large to die on and too small to live on."[63] New peasants paid roughly DM 200–290 per hectare for their land (US$430 in current value), based on the price of one to one-and-a-half tons of rye, or one year's harvest.[64] Ten percent of the price was due by the end of 1945, the balance in goods, services, or cash within ten to twenty years. Later East German historians qualified this as a recovery of overhead and transfer costs: to legitimize forced collectivization in spring 1960 and to defend against claims to actual ownership. The new land reform farms were classified as personal property and inheritable. Yet title and property rights were murky. The farms were classified as "work-property" that could not be sold, leased, mortgaged, or pledged as security.[65] The only transfer permitted, apart from inheritance, was a return to the land account, the *Bodenfonds*.[66] The new peasants' farms and forest had the quality neither of a capital asset nor of personal property, nor were they yet part the People's property, the VEG.

Despite the Party's hopes, a popular movement did not spark land reform. Ernst Goldenbaum, chairman of the (communist) Democratic German Peasant Party, claimed: "the independent collective action of the workers and peasants with the help of the democratic bloc parties and the mass organizations initiated land reform, not the government."[67] Land reform was popular, and the bloc parties (after the Liberal and Christian Democratic leaders were expelled) did support it. But if land reform was a "revolution in the countryside," as communist propaganda insisted, then it was a revolution from above.

American correspondents touring the Soviet zone countryside in late 1945 witnessed "much doubt and complaint," although the new peasants they saw seemed content.[68] But this contentment did not survive the first harvest year, and Marxist-Leninist theory never caught on. Few Germans, in either the West or the East, cared for ideology after Hitler and the war. Only the most active Party cadres and followers felt loyalty to the Party—probably not more than 10 percent of the total population.[69] By the end of the first postwar winter of hardship, shortages, and Soviet reparations, most farmers, new and old, were increasingly critical of the Party.[70]

Most new peasants got just bare land for their mortgages, without barns or houses, tractors and machinery, much less seed, fertilizer, or fuel. Only 16,000 of the 209,000 new peasants got houses, and only 58,000 had living quarters; the rest often lived in unheated stables and outbuildings.[71] The new peasants were not only inexperienced but often strangers in their districts, resented by old farmers, enduring, as Keats lamented, like Ruth amid the alien corn. "New peasants," complained Schwerin's vice president, were "cast into a hostile social environment without any economic foundation."[72] Older, experienced farmers, who normally would have helped their neighbors, shunned the new peasants whose presence carried the aroma of expropriation. Farm expropriations naturally worried existing farmers, as did the Party's class warfare rhetoric and demonizing of modest farmers as "kulaks."

Farmers brought in the 1945 harvest with difficulty and far short of normal yields. And then, work on the land reform parcels stalled as many new peasants began flocking to traditional cooperatives or quit their farms.[73] With collectivization on farmers' minds, few either bothered or knew enough to prepare their fields for winter sowings of oil seed or winter wheat.[74] There were no fall cover crops to enrich the soil and there would be no spring harvest. Worse lay in store as Soviet soldiers spread throughout the countryside; free of Western observation, they requisitioned food, seized equipment, and drove off the few surviving farm animals.[75]

Farm administrators by mid-1946 began commandeering old farmers' carefully husbanded seed, fertilizer, and fuel and expropriating draft animals, plows, and harrows to give to the new peasants.[76] Forced collectivization loomed as a possibility to correct the growing political embarrassment as old farmers and traditional cooperatives were thriving even as the socialist collectives failed. In late 1947 the Social Democratic Party's news service reported: "Now that the land reform project is widely acknowledged as collapsed, old established farmers must ever more follow the same path as new peasants—*Kolchoz!* [socialist collective]. That's the prospect for farmers" (emphasis in original).[77] Memories of Stalin's forced collectivization between 1929 and 1933, and the murder of millions of Ukrainian peasants, must have heightened eastern German farmers' anxiety; every successful independent farmer in the Soviet zone had to fear that someday the Party might brand him or her as "kulak" or "Junker," no matter how modest their farm or birth.

The land reform regulations and procedures drawn up by the elected Land Reform Commission promised "the most democratic principles" and that land reform victims, even though their farms were taken without payment, would be treated fairly and humanely.[78] Land reform administrators arbitrarily cast farmers off their families' land and out of their homes, underscoring the "class antagonisms" Marxist-Leninists felt for farm and forest owners.[79] Gerhard Grüneberg, Erich Honecker's deputy in building the Berlin Wall in 1961 and the hard-line minister of agriculture and forestry in the 1970s, declared: "Implementation and securing of the democratic land reform took hard class warfare. The presence of Soviet class brothers in the uniform of the Soviet army prevented the counterreaction from escalating to open violence."[80] Popular opposition surged as Red Army soldiers brutalized the rural population and Party functionaries abused and ignored land reform regulations. Evictions, as Naimark commented, "were not infrequently accompanied by rampages by Soviet soldiers, first when they entered the local agricultural regions in April and May of 1945, then again in September 1945 when the Soviets took the initiative—along with German authorities—in carrying out far-reaching land reforms."[81] This meant suppression with force of landowners' defense of their farms and forest.

High-ranking land reform administrators and even some communists joined with local officials from village to state level to protest land reform's brutality, some resigning in frustration.[82] The most effective protesters, such as Thuringia's vice president, Dr. Kolter, were arrested, sent to former Nazi concentration camps, or murdered.[83] Anton Hilbert, a socialist who left western Germany to support agrarian reform in the Soviet zone, witnessed crimes close to the Nazis' *Krystallnacht* pogrom against German Jews in November 1938, "so that one could almost believe they were conceived in the same brain."[84] Schlange-Schöningen visited Hilbert in Thuringia in May 1946 to observe Soviet land reform firsthand. He wrote: "A new style of Nazi power rules in communist clothing. It's not a question of land reform but liquidation of the intelligentsia, just as in Russia. In two years' time, today's Russian zone showcase will be a land of absolute hunger." Indeed, soon after land reform the Soviet zone population suffered near starvation.[85]

Grains and sugar beet yields plummeted 30 percent and potato production slid 22 percent in 1946, the first full year after land reform. Within a year the Soviet zone population suffered from near famine. East Germany was the only European country still rationing food in 1954.[86] Food shortages and rationing lasted until May 1958, and meat and butter rationing returned in 1961.[87] Even so, fixed prices, control over the food supply, and persistent shortages were food rationing in all but name. The East German people were never truly free of rationing until the Wall fell in November 1989.

Food shortages brought land reform's discontent from the countryside to the city. As Carl J. Friedrich and Henry Kissinger observed, "Perhaps no single other factor contributed so much to the Soviet zone population's discontent with the régime as this failure to provide adequate food."[88] The Party leadership fed East Berlin during the Berlin Crisis of 1948–49 and the Soviet blockade of West Berlin only by requisitioning food from the countryside, which reduced even the rich Magdeburg Börde farm region (with fertile loess soils) to tighter rationing and bread shortages.[89] A British correspondent reported: "The raiding and robbing of the fields for food is a regular occurrence. Some of the resources of the district are being directed to maintaining the supplies the Russians are sending to Berlin. Resentment is rising. It expressed itself at the weekend in a strike of the workers of the Schäffers and Budenberg machine works as a protest against the lack of meat and fats. Thirty arrests are stated to have been made. The Russians are searching the baggage of all travelers to Berlin and confiscating food and other goods."[90]

Because of hunger, and workers' need to steal time from work to search for food, industrial output fell to between 50 and 70 percent of prewar levels. The Soviet military government's paper reported that a survey of fifty plants showed 15–20 percent of workers absent "on false medical certificates, collecting food in the countryside." The paper's editors sternly lectured factory and business managers to improve the workers' living conditions and feed them, as if the exhortation to greater effort itself gave the managers the resources to feed their workers.[91]

The rural population (new peasants and old farmers alike) was uneasy with land reform's arbitrary implementation and the functionaries' flouting of their own regulations. The Party leadership responded with thinly veiled scorn. Walter Ulbricht visited anxious farmers in the Bitterfeld region north of Leipzig, reporting: "A farmer fretted to us about land reform, 'But we have no law for this, here there is no law.' 'A law?' I asked him, 'If the farmers here decide to confiscate the estate owner's land, that is their democratic right.' The farmers responded, 'Yes, you are surely right, but a law would be good.' We calmed them: 'All right, fine, if you want a law, we will make a law so that you can carry it out in a completely orderly fashion.' "[92] The farmers, correctly seeking a more secure sense of their tenure, questioned the dignity and seriousness of land reform's title transfer—How should they hold their land then, particularly under this harsh and arbitrary regime? Ulbricht dismissed the farmers' concerns as trivial, yet the farmers asked the most important question—What quality did ownership have in Ulbricht's new Germany? Not just equity and human decency, but simple productivity depended upon the answer. If ownership was imperfect and rights fragile and temporary, then farmers had little incentive to invest in the long-term health of the soil or infrastructure, or to limit their harvest in the forest.

Six months after land reform began, Edwin Hörnle reported to the Soviet military government on serious problems, including the "incorrect" partitioning of forestland, stating, "many new peasants still have no lodging, no outbuildings, most still don't have their own farmhouse. But since time is so critical, the peasants' own initiative will solve the problem, along with help from all sections of the people."[93] Soviet officials warned Hörnle not to discuss his report in public.

Hörnle's phrase "peasants' own initiative" meant loosing new peasants on the forest to cut construction lumber.[94] The Party leadership launched the "New Peasants Program" in 1946 to promote this initiative but did not offer financing or building materials. This left many new peasants in the position of having to make bricks not only without straw, but also without clay or mud.[95] Ulbricht, frustrated with the failure of the New Peasants Program to improve housing—and with farmers' mounting dissatisfaction—scolded: "Too much is decreed, too little carried out with real energy!"[96] Ulbricht refused to invest in the rural economy, so conditions improved slowly and the New Peasants Program failed to stem the hemorrhaging of the rural population to the cities or, more ominously, to western Germany.

New peasants fell upon the forest, prodded by their insecure sense of ownership, shortages, and hunger following land reform. The Party also reversed forest enclosure, reopening the forest to livestock grazing, fallen wood gathering, and litter raking.[97] The withdrawal of forest guards made the effects of illegal cutting for firewood and grazing of livestock even worse.[98] A major study of the Soviet zone forest concluded: "The granting of land reform forests deceived the peasants that the Party intended to support the new peasants' holdings. Then, the Party forbade foresters to instruct the peasants in management of their forests as Soviet experts also forbade state foresters to intervene in the peasants' destruction.[99] Thus no one could expect peasants to follow sound management practices. Dire need drove refugees, 'new peasants,' and small farmers to exploit their forests to get the basic resources needed for survival."[100]

The Party seized the former Reich forest, 55 percent of the total forest area, 1.6 million hectares of the best quality and best-stocked forest even before the Party launched land reform in September 1945.[101] Soviet land reform expropriations after September 1945 brought a further 1 million hectares of forest, a third of the total, under the Party's control, paradoxically (for a land reform program) cutting private ownership from 45 to 32 percent. Then the Party leadership kept more expropriated forestland than they took of farmland as *Volkseigenewald,* or "People's Own Forest."[102] The Party leadership distributed to new peasants less than half the forest from the "Land Account" (*Bodenfonds*) in average parcels of one hectare, an even more unsustainable structure than the land reform farms.[103] The original preponderance of Reich forest (55 percent of the total) led to the state's control of over two-thirds of the zone's forest by April 1946.

Experience confirmed foresters' caution about land reform. They saw in the 1930s the chaos that arose from the *Osthilfe*'s splitting up of forest ownership. A district forester warned: "We must preserve the many varied forms of forest ownership, particularly private ownership. We must direct all our resources and planning to prevent forest partitioning."[104] A postwar land reform administrator echoed this warning: "It is particularly important that land reform not take forestland. Small peasant forests have always been the problem child of rational forest economics. Forests have broad ecological functions which small forest management can't serve. Private forest cooperatives arose in the previous centuries from the peasants' understanding that small peasant forest holdings are wasteful. Of course many new peasants, who lack coal, just clear-cut their forest and don't bother to replant."[105] Kurt Mantel, a leading West German forest historian and policy scientist, warned of the dangers of land reform for forestry: destruction and abandonment.[106] Once land reform smashed apart the large forest management units, the small, isolated allotments degenerated into an unregulated commons that the new owners despoiled. New peasants reasonably feared collectivization was imminent, so they took their value while they could.[107]

East German statisticians placed almost equal blame for forest destruction in the postwar years on peasants' firewood harvests as they did on Soviet reparations. The claim deserves attention, even if it doesn't excuse overall Soviet policy. There is no question that the people desperately needed firewood, normally 20 percent of the harvest. The Soviet zone was cut off from its regular sources of coal from the Ruhr and former German Silesia. Party officials described Berliners' struggle as a people under siege: "Men and women scared out of the rubble the last scrap of wood, they destroyed barricades, buildings, barriers, scaffolding, etc., for fuel." Poachers cut 140,000 cubic meters in Berlin's city forests on top of the 500,000 cubic meter harvest organized officially. "Every free particle of wood," commented the editors of a leading western German journal, "evaporates like a drop of water on a hot stove."[108] *Neues Deutschland*'s editors in an article "Not Only for the Stove . . ." put the best face on the steady destruction of the revered 4,500-hectare Grunewald forest along the Havel, stressing the need for wood to repair Berlin's bridges. A senior Party official further noted in late December 1946: "Wood is our savior in our time of need. We sacrifice forests and parks to the saw and axe and bemoan the massive firewood cutting—even though a tree can be cut for firewood and regrow in fifty years while coal needs many centuries."[109] By mid-1946, one year after the end of the war, there was no more wood to take. "The Tiergarten and Humboldthain are wastelands, the Grunewald is 'thinned out,' as the professionals say. Berliners need 1.2 million cubic meters of firewood to survive the winter, and there is no more wood within Berlin's walls."[110]

Political considerations and the Party leadership's desire to maintain the people's favor contributed to the Party leadership's tolerance of peasants' "nonsocialist" (outside the Plan) destruction. As late as 1958, the Party leadership still hesitated to limit the firewood cut on private forests, which ran at two times annual growth, citing the state's decision to cut off coal supplies from the

West: "We must not be under the impression that we can command the forest owners what to harvest when there is no coal."[111] In addition, the Party leadership expected the peasants to collectivize voluntarily. There was little sense in antagonizing peasants and increasing the already strong opposition to collectivization when in less than two years all remaining private farm and forest land would be forcibly collectivized.

There are no accurate records of harvests on private forests, only incomplete figures reported long after the fact to the U.N.'s Food and Agriculture Organization (FAO). The records of firewood cutting which do exist suggest that the harvest may in fact have been close to normal. City people depended more on coal for heating and cooking and thus probably took heavy volumes from city forests.[112] Families in the countryside, however, depended less on coal for heating and cooking and thus were probably less likely to increase their firewood harvests when coal imports stopped. The great forests in the countryside probably did not see large increases in firewood harvesting despite Party propagandists' claims.

Still, the Party leadership and the Soviets had to allow civilians to cut firewood: many people living in the cities did not have sufficient coal or charcoal to warm their meager suppers much less heat their bomb-blasted flats, which often lacked running water or indoor plumbing. But there was a limit to how much the Soviets would allow firewood cutting to siphon away from reparations. Official Soviet zone statistics charge peasants with cutting more than 10 million cubic meters in 1947, as much as Soviet reparations harvests. Given the extreme hardship the state had in harvesting a like amount for Soviet reparations, it is doubtful civilians could have taken an equal volume from an area four times smaller. The most likely answer is that much of the harvest recorded as firewood went to the Soviet Union as timber reparations: the Party leadership hid the full measure of reparations' damage by exaggerating peasants' firewood harvests.

It is likely that firewood harvests remained somewhat above 20 percent of the total (and were concentrated on the 600,000 hectares of independent forest), still a devastating concentration of demand. Party statisticians allocated to peasant harvests an additional 6.3 million cubic meters (not far from annual growth) — this volume probably went for Soviet reparations. This brings the total volume of Soviet reparations to more than 100 million cubic meters for the nine years of reparations from 1946 to 1954, a total value (in 1950 prices) of more than $1 billion ($9.3 billion in 2003 value).[113] It was unlikely they would sacrifice a most valuable form of cash reparations — timber — merely to warm the homes of their former German enemies. If the Soviets were willing to let land reform drive civilians to near famine to achieve

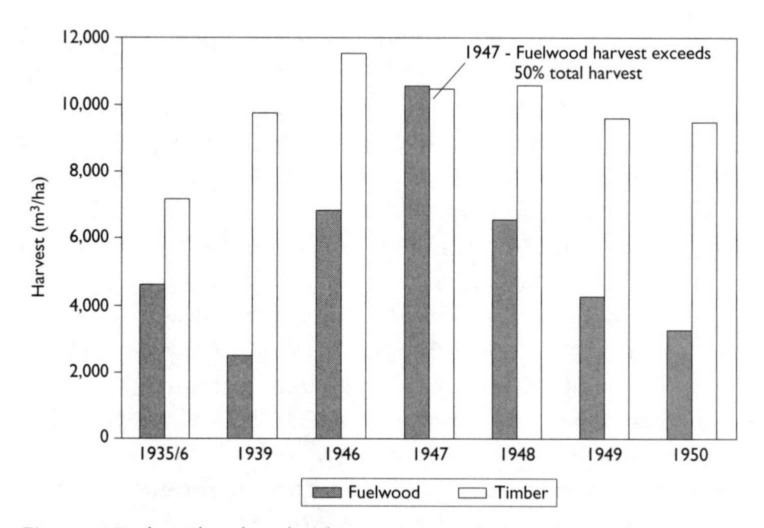


Figure 5. Fuelwood and timber harvests, 1935/6–1950. Data gathered from Wünsche and Schikora, "Der Waldfonds der DDR," 75; Stolper and Roskamp, *Structure of the East German economy;* Haden-Guest, Wright, and Teclaff, *World geography of forest resources,* 289.

their political goals, it is doubtful they would share timber reparations, a primary postwar Soviet goal.

Instead of catering to new peasants and farmers, the Party impressed them to cut and haul timber for Soviet reparations.[114] Soviet requisitions of farm livestock cut the horse and cattle population between 65 and 70 percent, and many of the farm horses that remained were frequently taken for skidding logs out of the forest and returned exhausted to their owners.[115] Sheep stocking fell 40 percent. Pig counts fell 20 percent, but only because they were not as easily driven as cattle and horses.[116] Farmers soon learned they had no more independence than their draught animals: "A young man, known in the district as a good communist, told the Party officials boldly, 'I'll decide who comes on my land and soil [*auf meinem Grund und Boden*].' The mayor retorted, 'Mein Lieber, you [using the intimate, and in this context condescending, *du* form] are only a *Kolchose.* We decide what you must do and what is permitted. You have your fields to farm, and you will plant and harvest what we tell you to' " (emphasis in original).[117]

From the first postwar days, Norman Naimark commented, "collectivization was on everyone's mind."[118] The new peasants' farms were simply too small to farm successfully on the light soils and they had scant resources, no credit, and almost no infrastructure. The Soviets denied new peasants extension help. Hans Lemmel, Alfred Möller's successor at Eberswalde, forecast land reform's failure and ultimate collectivization: "It remains to be seen how long the artificial, mostly too small, capital poor and especially extremely poorly equipped new landowners can survive and the consequences the reform will have on the people's food supply."[119] Knowing land reform's sure effect on farm production and that all agricultural experts warned against land reform, the Party leadership had to have collectivization in mind from the beginning. Fritz Lange, a member of the German Communist Party Control Commission, recalled, "The land reform for us was above all a political problem and with it [the need] to destroy the strongest underpinnings of the reaction: the large landowners and the manors."[120] The Party's emphasis on political goals deeply disappointed the idealistic Anton Hilbert, who was drawn to Thuringia by the promise of agrarian reform: "Are we not conscious that the elimination of larger farms must eventually be reversed through imposition of pure socialist economic forms? Land reform's rationale is purely political, and has nothing to do with factual arguments."[121] The Party's political goal was control of the hostile countryside and collectivization, voluntary if possible. Get the administrative forms, ownership correct, the orthodox Marxist-Leninist cadres were thinking, and success would follow.

The Party campaigned from the start to move farmers into socialist collectives.[122] Propaganda promised technical training to overcome the "backwardness" of peasant life through education "in the greatest school of all, the socialist collective."[123] The Party leadership withheld extension help and refused credit to independent farmers to steer them into the socialist collectives. The traditional cooperatives into which many smaller farmers sought shelter not only subverted Lenin's *Bündnispolitik*, but also looked suspiciously like old wine in new bottles. So a Thuringian cooperative was told abruptly that its members would get no seed in spring 1946.[124] The communists would not allow any challenge to their political power or allow the cooperatives' successes to thwart the destruction of their class enemies.

Harsh production quotas were designed to force struggling farmers to collectivize. The Party raised the delivery quotas of larger farmers (above those set for comparable farmers in Russia) to drive them into the socialist collectives that were languishing from the lack of experienced farmers.[125] These quotas were clearly intended to break the farmer holding more than twenty hectares, particularly medium-sized farmers holding over fifty hectares, *Großbauern* or "kulaks." Delivery failures meant losing one's land or arrest and sentencing to one of the many former Nazi concentration camps.[126] Because the Party could not distinguish between "saboteurs and profiteers" and honest farmers, the Party punished uniformly for short deliveries.[127] Fear of collectivization, arrest, and the concentration camps led farmers to advertise in newspapers for seed and produce to fill their quotas.[128]

Farmers continued to stream into traditional cooperatives despite the pressure to conform — or they abandoned their land to flee west — anything rather than join socialist collectives. By the end of 1949 more than 22 percent of new peasants had abandoned their small plots to escape the inflexible delivery quotas and debt. Sixty thousand farmers, a further 30 percent, abandoned their farms by March 1952.[129] This trend worsened and by 1950 the state had to take back 20 percent of the new farms. Fear of collectivization and distrust of the Party drove abandonment, as a senior East German forester wrote in an internal report to the State Planning Commission in 1958.[130] Acknowledging the serious threat *Republikflucht* and abandonment posed, in 1946 the Party called a Land Reform Conference in Berlin to introduce an urgent "*Aktion*" to settle politically reliable, "active antifascists" in the "most reactionary districts." Independent farm families were evicted from their houses and farms, forced to turn over everything they owned to loyal "*Aktivists*."[131] The "active antifascists," however, were even less successful at farming than the new peasants, and the politically reliable activists soon were also fleeing the republic.

The Party branded farmers who resisted collectivization as "reactionary peasants," enemies of progress hobbled by a feudal attachment to "nondemocratic, old production relations."[132] In fall 1945 a Soviet judge condemned a new peasant and Party member charged with "sabotage of land reform" to three years' imprisonment in a former Nazi concentration camp.[133] Special Soviet courts condemned seventeen- and eighteen-year-old sons of recalcitrant farmers to forced labor camps in the Soviet Union, often on the pretext that the boys were former Hitler Youth.[134] Army and Party police routinely picked up students, many youth leaders in one of the "bourgeois bloc parties," off the street or arrested them at home in front of their parents.[135] Soviet judges condemned these young men and women to prison for "political reeducation" or to forced labor camps in the Soviet Union.

Angry at the new peasants' refusal to accept voluntary collectivization, the Party leadership intensified its struggle against "enemy elements sabotaging land reform." In a five-hour speech to senior Party cadres on 14 April 1948, Ulbricht demanded that farm production increase to "industrial levels," foreshadowing Industrial Production Methods and the remaking of farm and forest as "factories on the land."[136] At the September 1948 Party conference where he introduced the "New Course," Ulbricht called for intensified "class warfare tactics" in the countryside and faster collectivization of private farmland, ominously branding reluctant farmers as "rich peasants and kulaks."[137]

The Party leadership responded to *Republikflucht* just as it reacted to the crisis in the rural economy: it decreed that women must have more than two children. Premier Otto Grotewohl presented the "Law for the Protection of Mothers and Children and for Women's Rights" (27 September 1950) to the People's Chamber (the *Volkskammer*), declaring; "The two-children habit is the practice of a dying population." The law proclaimed "the duty of the progressive Soviet zone family to bear enough offspring to provide adequate manpower" for the state.[138] Not only farms and forests were to be remade as "factories on the land"—Industrial Production Methods would also direct human reproduction as part of the Party's uniform template for all economic and biological activity.

Land reform changed the countryside's physical and social structures. Now, collectivization would create an indissoluble alliance between workers and peasants and erase the distinction between rural and industrial labor, Lenin's *Bündnisdoktrin.* Edwin Hörnle, head of the Soviet zone Department of Agriculture and Forestry, laid out the two-stage process of communist land reform clearly: "Land reform has changed the village's social structure. Now the socialist collectives [VdgB] will change the village's spiritual structure."[139] Liquidating large landowners through class warfare changed the social structure of villages; the spiritual structure would change when the peasant entered socialist collectives, voluntarily liquidating themselves as the peasant class to become rural analogues of industrial workers and thus advance in the Marxist-Leninist class structure.

The Party cadres took over forest management with the founding of the East German republic in October 1949. Along with increased state control and afforestation, replanting of the vast Soviet clear-cuts began with the First Two-Year Plan (1949–50).[140] This new policy of investment reflected economic reality and the First Two-Year Plan's priorities of political and economic control, calling for afforestation of 40,000 hectares each year.[141] Yet the Plan itself called for clear-cutting at least 50,000 hectares of forest each year, and more than 400,000 hectares of forest were still unplanted in 1949. Not only did the Plan not attack the huge backlog of unplanted forest, it did not even replant the annual harvest clear-cut. This, finally, was too much. Foresters overfilled the 1949 afforestation goal by 20 percent, coming close to replanting the 1949 clear-cuts, and by June 1950 foresters had replanted 75,264 hectares, a remarkable achievement in light of the slow start.[142] The forest recovered its 1945 unplanted forest area by August 1953 at the unavoidable expense of severe skewing of the age-class structure toward the youngest age classes.[143] The normal area of unplanted clear-cuts in an industrial forest, 1 percent of the total forest area, was not reached, however, until the late 1960s. The lag in bringing clear-cut land back into production contributed to the characteristic dominance of the youngest age classes of East German forests which remains a distinctive quality to this day.

Thus, the farm and forest landscapes were poised precariously between collapse and collectivization at the end of 1949. The rural landscape's social and ecological structure was inherently unsustainable, made more so by the loss of the eastern provinces in 1945. Although the Party leadership tried over the next forty years to recover stability, resilience, and productivity in the rural economy, the natural landscape never escaped from the original burdens of Soviet occupation or from the imposition of Marxist-Leninist ideology.

These tenuous initial conditions are read most clearly in the forested landscape, a metaphor in forest ecology for the damage Soviet occupation policies wreaked on political, economic, and cultural life. Growth in the Soviet zone forest plunged from 3.5 cubic meters per hectare in the 1930s to 1.8 cubic meters in 1949. Vigor and resilience also plummeted as stand structure deteriorated. Salvage of storm- and insect-damaged trees surged over 25 percent, a waste from which East German forestry never was free. Stocking—forest reserves and capital—plunged over 50 percent between 1945 and 1949 to 60 cubic meters per hectare. The forest's basic structure, its age-class distribution, lurched, skewing toward the youngest age classes as the Soviets failed to replant their vast reparations clear-cuts. Future forest managers in East Germany never were able to recover a more stable stand structure but instead chased the quality and value curve downward throughout the forty years of East German history. Thus land reform and reparations created, through political, economic, and ecological innovations, a forest of greatly reduced diversity and complexity with highly restricted flows and an inherently unstable structure. The damage was so deep, and the initial conditions so determinant, that neither the East German polity, nor the East German landscape, ever recovered.

As German foresters surveyed the vast areas of unplanted clear-cuts in the early 1950s, they grimly noted the highgraded, thin stands left in the wake of Soviet reparations and firewood cutting. Afforestation was the obvious solution: to replant the vast clear-cuts and bring the forest quickly back into production. Yet the best course of action might have been—to do nothing. But certainly not to spend vast resources replanting the artificial, uniform pine and spruce plantations that the Soviets had cleared for them.[144] In the empty expanses of clear-cut forest there was an opportunity to complete the ecological revolution of the early nineteenth century, for a close-to-nature forest to emerge from ruin in the same way the native forest recovers naturally from fire. Left alone, the unplanted clear-cuts would have reseeded naturally, the first step toward a forest close-to-nature, an early-succession forest of limited economic value, perhaps, but one ultimately of greatest ecological and economic vigor and stability. Sixty years later—today—we would have had a new, diverse forest instead of exhausted plantations still on the brink of ecological and economic collapse. Such an experiment, letting a new forest renew itself without a central plan, was anathema to the new East German leadership, governed as they were by Marxist-Leninist economic science, extreme risk aversion, and a fetish for control.

A close-to-nature forest was no more acceptable to the Party leadership than a free Germany with a liberal market economy, and a close-to-nature forest was unthinkable without liberal political institutions and individual rights first. The resolution to East Germany's crisis in forest ecology was political, similar to Willy Brandt's prescription for the divided Germanys—"Let grow together what belongs together," or democratic reunification. Dr. G. Schröder, a senior East German policy planner, linked the economic and ecological crisis in the forest with reunification, writing with unintentional irony in 1957: "In reality, a peacefully reunited Germany could easily solve the long- and short-term problems of sustainable forestry which plague a divided Germany. Only reunification can resolve the economic and ecological problems arising from the division of the German economy and the heavier harvesting in East Germany."[145] His interest in political reunification stems from his liberal orientation, his respect for diversity and appreciation of the value of natural ecological and economic structures evolved over time. Schröder also appreciated the need for greater tolerance for risk and for individual foresters to let go of the illusion of control which underpins European industrial forestry. A liberal political system and market economy was, as Schröder intuited, an essential precondition to ecological stability and productivity.

So, at the very beginning of East Germany in October 1949, as the German communists took over formal control from the Soviets, ecological problems were formidable but resolvable through the experience and structures at hand. Foresters were ready, with a unanimity rare before May 1945, to continue the restoration of a diverse, complex forest. Most foresters rejected radical *Dauerwald* or extreme industrial forestry. A close-to-nature forest, however, was far from an idyllic preindustrial, virgin forest, or *Urwald*. Close-to-nature forestry's mandate was far more ambitious: the ecological restoration of a forest never documented or studied, the forest that would regenerate naturally once clear-cutting ended and the native forest ecology restored itself. East Germany's bosses imposed industrial forestry and central planning on a forestry profession reaching out to complex structures, just as they imposed the shortage economy and their dictatorship on the East German people. Their means reflected their harsh materialism and orthodoxy: they purged the Forest

Service of liberal and conservative foresters, harassed and persecuted "bourgeois" foresters and scholars, and threatened their children, all tactics of the post-Stalinist terror, to enforce uniformity, control, and power.

East German foresters looked forward to the fifties with hope as Soviet control receded. East Germany, in addition to inheriting a relatively sound, yet still artificial, forest structure, also inherited *Dauerwald* and the tension between close-to-nature and industrial forestry—part of the longer-term struggle between custom and innovation that Goethe and Mendelssohn illuminated. Von Kalitsch's assistant, Hermann Krutzsch, brought the dogmatic, religious variant of *Dauerwald* philosophy to communist East Germany, wielding *Dauerwald*'s anticapitalist, romantic rhetoric. Krutzsch, this apolitical and romantic autodidact, succeeded against all expectations and the Party stumbled into a brief, disastrous experiment with radical ecologism in the early 1950s, the Menz Resolution (*Menzer Beschluß*), an extreme variant of *Dauerwald* whose irrationalism exceeded the scope even of National Socialist "Green" environmentalism.

The Party leadership hoped through the 1950s for popular acceptance of socialism and for the legitimacy it never won at the polls. The Party tried to coerce independent farm and forest owners to collectivize voluntarily, to modernize the rural landscape as an expression of popular will. Land reform so distorted the rural economy that peasants should have had no alternative other than to modernize and join in solidarity with the worker. Instead, they found traditional, complex structures to thrive on the land (the farm and forest cooperatives) or abandoned the republic altogether.

As peasants and small farmers fled East Germany in the 1950s the Party leadership abandoned the outward forms of democracy and forcibly collectivized farm and forest land through the "Socialist Spring in the Countryside" in early 1960, ending land reform and signaling the Party's abandonment of democratic forms. The Socialist Spring in the Countryside foreshadowed the Wall in August 1961 as well as the ongoing collapse of economic and ecological diversity in the countryside. The Party reduced farm and forest land to simple structures by the end of the 1950s, the natural landscape reflecting in its structure and system qualities the poverty and brittleness of the East German republic as a whole.

# 5

# *The Landscape's "Socialist Transformation" and Flight from the Countryside (1949–1961)*

*Nature will be reported. All things are engaged in writing their history. The air is full of sounds, the sky of tokens, the ground is all memoranda and signatures, and every object covered over with hints, which speak to the intelligent.*
—*Ralph Waldo Emerson,* Representative Men

Land reform and Soviet reparations fomented an ecological and economic revolution in the rural landscape into the late 1940s. Near famine, economic waste, and punishing reduction of diversity and structure knocked East Germany behind the western part of the country, reversing their 1936 positions. The cost to the forest was staggering—fourteen years of growth taken from the forest in only four years and the near destruction of the East German farm and forest economies.

German foresters were helpless during the years of direct Soviet occupation to protect the forest from reparations and land reform. The Soviets first purged experienced forest ecologists and silviculturists in 1945, working them at hard labor cutting trees for reparations under inexperienced Soviet "experts" who treated them as inferiors. Now, as the end of formal Soviet occupation neared and the new East German government took shape, many foresters looked forward to recovering in the new decade of the 1950s the ecological and economic riches squandered since the end of the war.

The German Communist Party leadership, reconstituted in April 1946 as the "Socialist Unity Party" (*Sozialistische Einheitspartei Deutschlands,* the SED), announced the formation of the East German state in October 1949. The rural economy and landscape were emerging from the chaos of land reform and reparations. Since reunification no longer seemed imminent, the East German state would have to survive as an independent polity and conserve its population and resources or collapse.

As East German foresters took responsibility for forest management after October 1949 waves of articles on close-to-nature forestry infused with excitement again flooded eastern and western German journals. It looked to many German foresters in late 1949 as if the new East German government would follow pragmatic policies and conservationist, close-to-nature forestry. The Party leadership did formally adopt, not just close-to-nature forestry, but radical environmentalist *Dauerwald* forestry in 1952. Indeed, for the next two decades of East German history, while Walter Ulbricht ran policy and government, innovation and statist revolution defined East German political and economic life.

Hermann Krutzsch, von Kalitsch's assistant at Bärenthoren in the 1930s, settled in the Soviet zone to proselytize for *Dauerwald* in its homeland, the dry pine forests of the North German Plain.[1] Krutzsch bore the sacerdotal, romantic elements of green National Socialism into East Germany, lacing forestry journals with ribbons of *völkisch* jeremiads and exhortations for *Dauerwald.* These articles culminated in 1949 in two special issues of *Forst- und Holzwirtschaft:* "Optimal Stocking Forestry [*Dauerwald*]—The Demand of the Hour!"[2] The authors attacked both National Socialist forestry and Soviet forest reparations with remarkable candor and demanded *Dauerwald* forest management. As in 1922 and 1934, *Dauerwald*'s moment seemed nigh, particularly since reparations, which mandated clear-cutting, would be ending in 1954.[3]

Johannes Blanckmeister, the preeminent postwar forest silviculturist and follower of Rudolf Steiner, downplayed *Dauerwald*'s philosophical, irrational timbre to emphasize its promise of greatly increased productivity once clear-cutting stopped.[4] Blanckmeister's new name for *Dauerwald,* "Optimal Stocking Forestry," had a hard edge to it and sounded "economic" and scientific, unlike *Dauerwald*'s rather wooly-headed romantic definition: "permanent forest." Blanckmeister defined *Dauerwald*/Optimal Stocking Forestry as a "synthesis of biology and practical technique, of nature and economics," a system which harnessed the power of nature to increase labor productivity: heady thoughts for Walter Ulbricht and his colleagues.[5] One of Blanckmeister's colleagues adopted Marxist-Leninist rhetoric to correct romantic "ideologues" who demanded a return to the *Urwald* (the primeval forest); romantic thought, he lectured, violated Marxist-Leninist dogma and dialectical materialism. Industrial forestry was wrong not because plantations were unnatural or artificial, but because they were "capitalist wood farms." Another prominent forester offered *Dauerwald*/Optimal Stocking Forestry as the socialist alternative to capitalist exploitation, a formula calculated to exploit the Party leadership's credulity.[6]

East and West German foresters gathered in West Germany in 1950 for the last time to draft a national program for close-to-nature forestry. Karl Dannecker, a senior West German forest scientist and founder of the *Dauerwald*-inspired "Working Group for Close-to-Nature Forestry," presided with his partner Hermann Krutzsch, leading a large East German contingent.[7] Albert Richter, East Germany's most influential forest manager, questioned *Dauerwald*/Optimal Stocking Forestry's neglect of forest management.[8] Richter believed that close-to-nature forest management, not *Dauerwald*, placed resolution of economic and ecological goals at the heart of German forestry, a task *Dauerwald* was ill suited to tackle.[9] *Dauerwald* theory missed the social and economic contexts of forestry, particularly the immense pressure on the forest from the almost unlimited demand for wood in postwar Europe. Nevertheless, twenty-one senior West German foresters signed onto Dannecker's *Dauerwald* platform.[10]

Krutzsch backed Dannecker, complaining that some of his colleagues had used the conference to justify East German harvests. When Krutzsch returned to the East, the Party leadership disciplined foresters who had failed to present a "socialist front." Further contact with West German foresters was forbidden.[11] In his conference report Dannecker later commented wistfully, "For a long time we have not heard from our East German colleagues."[12] But the *Dauerwald* debate had not yet gone completely underground in East Germany, and *Dauerwald* supporters still had one last, hollow victory to win in East Germany, the remarkable "Menz Resolution" which formally abolished clear-cutting.

Despite persecution of foresters for not defending the republic, Blanckmeister and Krutzsch won a surprising victory for *Dauerwald* at the Menz Conference (*Menzer Tagung*) of 14–15 June 1951. Krutzsch led a pilgrimage to Bärenthoren, *Dauerwald*'s birthplace on the sandy lowlands of Kreis Zerbst in Saxon-Anhalt, just as Alfred Möller had done at the Dessau Forestry Conference in 1922.[13] All 110 East German forest scientists and practitioners approved new forest management regulations for the People's Forest (*Volkswald*), forbidding clear-cutting and age class management, although some foresters grumbled that the Menz Resolution rendered forest management and traditional practice "*Tabu.*"[14] Blanckmeister's manifesto called for hardwoods to increase from 20 percent to 80 percent of the forest (a close-to-nature forest), a radical reversal of the existing structure.[15] The Menz Resolution, however, did not apply to the 30 percent of forests still in private hands and cooperatives.[16] *Dauerwald* triumphed at the Menz Conference as it had at the 1922 Forestry Conference at Dessau and with the National Socialists' creation of the Reich Forest Service in 1934.

The Menz Resolution and *Dauerwald*/Optimal Stocking Forestry took effect with the harsh First Five-Year Plan (1951–55). Walter Ulbricht, the Party leader, parroted *Dauerwald* rhetoric: the People's Forest "must have an optimal stocking of living wood" by 1956 and harvests from now on would take less than growth.[17] But the First Five-Year Plan, perhaps relying on Blanckmeister's assurances that *Dauerwald*/Optimal Stocking Forestry would release stored productivity, set a contradictory goal of supplying "the People's economy with constantly increasing quantities of wood."[18] The chaos of Soviet reparations harvests returned. Work crews fell upon the forest to meet aggressive Plan targets, and harvest levels did not even approach growth until the early 1960s.[19]

Blanckmeister's principal goal, and the goal of most German foresters, was to restore a diverse, close-to-nature forest of native hardwoods and to end clear-cutting.[20] Yet *Dauerwald* was a catastrophically ill-suited footing for a national forest policy: its orientation was romantic and vague, giving Plan bureaucrats full scope to rationalize maximum, annually increasing harvests. Even though the long-term economic benefits of *Dauerwald* were not proven, no one doubted that clear-cutting, *Dauerwald*'s antithesis, yielded far greater short-term returns. The Party leadership chose both extremes: to reduce clear-cutting but keep harvest levels at high levels. In practice, this meant highgrading: taking the best trees and leaving the worst. The already battered and fragile older age classes shrank before the storm.[21] A leading West German forest scientist saw that even though the Nazis used *Dauerwald* rhetoric to hide cuts in stocking, at least National Socialist harvests never exceeded 150 percent of growth. Soviet reparations harvests in contrast took four years of growth in one year. East German harvests in the 1950s continued at a comparable pace, even though *Dauerwald* was official policy.[22]

The Party leadership, struggling to recover from the ecological and economic devastation of Soviet reparations, grasped at Blanckmeister's promise that *Dauerwald*/Optimal Stocking Forestry would deliver maximum economic and ecological benefits at the same time. The Party leadership was a sure target for *Dauerwald*/Optimal Stocking Forestry's abstractions. After all, Marxism contained religious and scientific strains similar to *Dauerwald,* the "mythological, prophetic, and utopian sense of Marxism" as well as a "scientistic" belief in the power of reductionist science and human reason to solve problems.[23] In following both *Dauerwald* and industrial forestry at the same time, the Party leadership repeated a contradiction inherent in Marxism.

In the end, when he had to choose between religious and scientistic Marxism, Ulbricht—already coming under the spell of cybernetics—chose science and innovation over nature and custom. When dismayed foresters protested that the severe harvests violated the Menz Resolution and East German forestry legislation, Ulbricht scolded them, lecturing that their criticisms were "either not productive or purely theoretical."[24] Plan quotas took clear precedence over conservation.[25] Ulbricht was determined to pull maximum, annually increasing volumes from the forest regardless as to forest health or the advice of scientists. And although formal Soviet reparations ended in 1954, massive volumes, now labeled as "exports" and valued at low, 1944 fixed prices, still flowed to Russia, reparations in all but name. Thirty-seven trains with more than one thousand railcars loaded with timber crossed out of the rail head at Frankfurt am Oder to the east in June 1954, not a far different picture from that from the height of forest destruction in 1948.[26]

As a result of management chaos, forest production crashed and state and private forests failed to meet their 1953 Plan goals.[27] Forest structure deteriorated but, more seriously, peasants infuriatingly overfilled their firewood harvest quotas by 14 percent when more valuable sawlog production lagged at 73 percent of the Plan quota. Peasant firewood harvests often camouflaged valuable sawtimber harvests smuggled into West Germany, diverted into "own use" outside the Plan.[28] The Party leadership blamed "poor Plan discipline" while starving forest management of funding. The fault lay in the Party's fixed-price and quota regime and the simultaneous plunge into radical environmentalism and extreme industrial forest management.

The ultimate cause of *Dauerwald*'s failure was brutally simple, as the chief of East German forest management explained: "One cubic meter of wood from clear-cutting was cheaper than one cubic meter from thinning."[29] Heavy cutting for Soviet reparations continued unabated. In the first five years of the East German republic stocking in the most valuable age classes over sixty years fell 32 percent.[30] Harvests continued at extreme levels despite blaming firewood cutting—"peasant abuse"—for the deterioration in forest structure even though the People's Forest took up 70 percent of total forest area in 1949 and the state effectively controlled most forest in any event.[31]

The dynamics of trial and rejection of *Dauerwald* from the Menz Resolution in 1951 to the return of industrial forest management at the Second Forest Conference in 1956 mimic the green National Socialists' embrace of

*Dauerwald* in 1934 and its rejection in the First Four-Year Plan as the Nazis readied for war. East German historians blamed *Dauerwald*'s defeat under National Socialism on "capitalist relations, under which forest plundering could not be stopped."[32] Yet the National Socialists continued to restore the natural forest and foresters debated *Dauerwald* freely throughout the Second World War.[33] Even given "capitalist relations," National Socialist forest managers managed far more ecologically, even under the extreme pressure of the Second World War and imminent defeat.

West German foresters were working through similar problems; they struggled to reconcile economic needs with the goal of restoring a close-to-nature forest. The roots of the conflict between nature and economics, between custom and innovation, lay in the early nineteenth-century simplification of the forest—Cotta's great ecological revolution which transformed the European landscape. Despite early profits, economic returns had been slowing since the late nineteenth century, about the time the second rotation matured, increasing the urgency of Cotta's plan to return to a more complex forest structure. Cotta's original plan called for a gradual recovery of diversity and complexity and the restoration of mixed hardwoods in uneven-aged stands; the question was not if a forest close-to-nature should be restored, but when. German foresters, East and West, understood this, and Blanckmeister managed to convince the Party leadership. But once Ulbricht and his colleagues saw that close-to-nature forestry meant the Party role was no longer primary, they jettisoned *Dauerwald* at once.

Restoration of forest health and vigor meant finding space between *Dauerwald*'s radical ecologism and nature worship and industrial forest management's "scientism" and extreme materialism. Eilhard Wiedemann, a senior West German forest scientist, reached back to the first days of plantation forestry, to the work of Wilhelm Pfeil, the founder of the Royal Forest Academy at Eberswalde and protégé of Wilhelm von Humboldt.[34] Pfeil had argued for the equal importance of ecological and economic factors, countering G. L. Hartig's "Eight General Rules" which guided even-aged industrial forest management.[35]

The next iteration of close-to-nature forestry, the flexible and ecologically oriented "Iron Law of Site Conditions" (*Eiserne Gesetz des Örtlichen* [EGÖ]), evolved from Eilhard Wiedemann's work in collaboration with Albert Richter, the last instance when East and West German foresters worked together, even if at a distance.[36] The Iron Law of Site Conditions devolved decision making down to the local forester familiar with ecological, cultural, and historical contexts of the forest. It evolved as the third way between extreme *Dauerwald* and Marxist-Leninist industrial forestry and was the most important advance in silviculture in the postwar period.

The Party leadership now looked to Richter to rescue the forest from the consequences of the Menz Resolution: the managerial chaos of *Dauerwald* and the physical destruction of maximum harvest. Richter's sustained yield management, free of *Dauerwald*'s irrational proscriptions, tolerated clear-cutting to meet society's need for wood.[37] Nevertheless, the Party leadership waited until the Second Central Forest Conference in Leipzig (10–12 February 1956) to reinstate industrial forestry (age class management) in language which must have worried foresters, particularly Blanckmeister and Richter.

The conference report called for increased technological innovation and "the highest possible mechanization" to reach maximum, annually increasing production and increased labor productivity with lower costs: familiar Marxist-Leninist prescriptions notable for a persistent failure to recommend investment.[38] The conference's focus on technological innovation mirrored the Party leadership's growing fascination with the administrative sciences and cybernetics.[39] Thus Richter's new silvicultural guidelines, based on Pfeil and Wiedemann's Iron Law of Site Conditions passing decision-making authority down to foresters in the field, "were seldom observed in practice," as East Germany's leading silviculturist observed.[40]

The Party leadership rejected ecological management sotto voce, still dipping into *Dauerwald* rhetoric as late as 1958 to court public opinion, when a senior Party economist wrote: "It is the merit of the 'First Workers' and Peasants' State on German Soil' that men of the East German republic rescued *Dauerwald*, this classic forest management method. Our new forest shall be one of mixed species and uneven age classes!"[41] The Party leadership waited ten years, until 10 October 1961 after building the Berlin Wall, to revoke the Menz Resolution and *Dauerwald*/Optimal Stocking Forestry.[42] Love of the forest and the natural landscape still resonated deeply in East German culture and the Party leadership could not abandon such a powerful myth casually.

The Party leadership turned its back on *Dauerwald*/Optimal Stocking Forestry because it conflicted with maximum production. They would do the same with Richter's core principle of sustained yield, "perverting" what the director of East German Forest Management praised as Richter's "genius." The Soviets inflated inventory data to justify massive reparations harvests. Now the Party changed the meanings of words to camouflage the harshness of its systematic assault on the rural landscape. Sustained yield equaled Plan quota: "The annual sustained yield harvest is to be understood as the production quota which itself defines the sustained yield harvest."[43] Party theorists redefined social and ecological goals to fit Marxist-Leninist ideology: "Sustained yield is in our conditions the striving toward a continuous maximum raw wood production."[44] Party theorists similarly redefined social and cultural uses of the forest to mean "providing a constant supply to the People's economy of raw wood, bark and naval stores."[45] Forest science itself now meant "dialectical and historical materialism," subordinating biological and ecological science to an irrational political ideology.[46]

The State Planning Commission gave forestry its production quota based on the demands of the Plan. The Party leadership, however, clung to ecological rhetoric even as it pressed extreme harvests, manipulating inventory data to hide the destruction. G. Laßmann, a constant champion of Marxist-Leninist orthodoxy, claimed the fall in People's Forest harvest below growth in 1956 as a vindication of the Party's forest policy.[47] Laßmann, however, inflated his growth figures grossly, using 4.24 cubic meters per hectare as growth when it had fallen to 2 cubic meters per hectare, almost half West German forest growth.[48] The 1953 harvest took as much as twice the forest growth. Stocking fell 18 percent between 1949 and 1955 despite Ulbricht's call in the First Five-Year Plan to increase stocking by 94 percent to 160 cubic meters per hectare by 1956.[49] Ulbricht not only missed his goal but missed it by a huge margin.

The Party perfected its control over the forest in the 1950s with the creation of 104 State Forest Districts. The new, politically reliable Marxist-Leninist Forest Service used the State Forest Districts to manage state and private forest, the most important administrative development in the 1950s.[50] State Forest Districts had the status of *Volkseigene Betriebe* (VEB) and followed the demands of the national economic plans.[51] State Forest Districts managers were ordered to "create conditions for socialist management" and to harvest constantly increasing supplies of wood through systems analysis and centralized controls: cybernetics.[52] Harvests were set by the ministry as inflexible quotas which required clear-cutting.[53] The director in charge of a State Forest District had to coordinate his management with the Party organization and Party union, just as Party officials kept watch over professors and students in the forestry schools.

Many of the most valuable reforms of National Socialist forestry were unwound. Individual State Forest Districts reported to the Ministry for Agriculture and Forestry through the *Bezirk* structure after July 1952, a structure which strengthened the cadres in the countryside but weakened ecological management.[54] The Party again changed forest districts' boundaries to match political boundaries (erasing the National Socialists' ecological boundaries of the 1930s and 1940s) to increase central control and political oversight. This foreshadowed the dissolution of the five states at the Second Party Conference at the end of July 1952 when the "last remnants of federalism, self-government and state tradition were dissolved" and People's Forest harvests were brought into line with the First Five-Year Plan.[55]

Forestry lost the independence it had won under the Nazis in 1934 and again became part of a "unified agricultural structure." The chief of the East German Forest Service declared, "It is not to be tolerated that forestry problems be handled apart from the overall development of agriculture" as the forest moved under the control of the Agriculture Ministry, headed by the notorious and harsh Gerhard Grüneberg in the 1970s.[56] The People's Chamber passed a new Nature Protection Law (*Naturschutzgesetz*) on 4 August 1954 to replace National Socialist environmental legislation.[57] Nature preserves and environmental protection were now clearly subordinated to the demands of the national economy. At the same time the categories "forest policy" and "forest history" disappeared from the indexes of the principal forestry journal *Forst und Jagd*.

The regulation "Status of the State Forest Districts" (10 December 1954) introduced socialist accounting procedure balancing revenues with costs, the *Ausgleich*, to "overcome cameralist thinking."[58] This change had as severe an effect on forest management as production quotas and central planning.[59] Inevitably, "When costs had to be covered, the price followed," a leading East German forest policy scholar later recalled.[60] Under a system where harvest levels increased annually and prices were set to match costs, there was no trigger to stop or slow exploitation.

Simple bookkeeping and the important tasks of managerial accounting perplexed the central planners, who tinkered with accounting reform throughout the 1950s and 1960s, looking for an optimal combination of information, control, and confirmation of the laws of "Marxist economic science"—qualities not to be found in any reporting system. The Party's economic planners adapted Soviet control systems, such as the gross (*Brutto*) accounting system, to fix prices and enforce managers' tasks, particularly meeting Plan quotas.[61] The gross accounts system encouraged waste because purchased materials were included in calculating prices, giving managers an incentive to boost content. Interplant and interindustry flows, even intraplant flows, were included in calculating prices.[62] Forest managers made heavier use of fertilizer, machinery, labor—even senselessly harvesting sawlogs for pulp—all to build gross production. The pernicious influence of gross accounts spread to afforestation. The Plan rewarded managers with premiums for raw area replanted and volume rather than for the survival of seedlings, directly leading to huge losses and inefficiencies in this most critical task of plantation forestry: replanting clear-cuts.[63] East German statistics, "blown up" by the gross accounts system's double counting of flows, were totally unreliable.[64]

Management of the People's Forest, therefore, was settled early in the decade, for good or ill. Independent forest and farm owners, however, faced a decade of turmoil and uncertainty. Many of the small, independent forest owners also owned small farms and resisted both the Party's blandishments and pressure to enter socialist collectives and to place their thin hectares of fractured forest under the control of the State Forest Service. The reluctance of farmers and new peasants to accept collectivization voluntarily and their increasing activism and dissatisfaction in the mid-1950s frustrated and alarmed the Party leadership.

As peasant unrest mounted, the State Planning Commission instructed Ignaz Kienitz to prepare a secret report on the forest and farm economy and peasant attitudes. Kienitz posed the study's central question: "Do peasant forest owners trust our government's farm and forest policies?"[65] Many of the responses from interviews with a broad cross section of peasants were so negative that he counted only the opinions of "peasants loyal to the Workers' and Peasants' State," although he screened "*Staatsbejahung und Lippenbekentniss*" ("yesing" the state and lip service). His study documented widespread dissatisfaction and deeply rooted mistrust of the state even among the "loyal" minority sampled.

Peasants trusted neither the new land tenure structure nor the capability of the state to manage the rural economy. As Kienitz observed, "All forms of peasant mistrust are grounded in the fear that they will lose their forests."[66] The socialist land tenure structure lacked legitimacy, a function of its genesis in expropriation and Marxist-Leninist rhetoric's hostility toward private farm owners and "bourgeois" foresters. Pre–land reform owners had seen the brutal expropriation of farms of more than one hundred hectares in 1945 and had no illusions about the Party's ultimate intentions. New peasants saw their forests as a gift from the state, fully subject to rescission and not as legal property.[67] Kienitz reported that "in no event" would most peasants choose socialist management, preferring either their own management or to join a traditional cooperative.[68] The Party leadership would never get farmers to collectivize themselves; the Party would have to impose socialism on the rural landscape.

Furious at the peasants' ingratitude and impatient to collectivize private forests, the Party leadership picked up again its land reform complaint against small landowners: they lagged in replanting clear-cut stands and were unworthy stewards. Failure to invest in afforestation (replanting clear-cuts) was a serious charge. As the state increased its control over private forests, afforestation did indeed improve. People's Forest afforestation rates in the early 1950s were more than 16 times greater than those for private forests, dropping to 3.5 times by 1955 as State Forest District oversight and state aid increased. Still, private forests came out of land reform in poor condition compared to the People's Forest, with stocking levels only 80 percent of the anemic levels in the

People's Forest.[69] The poor quality of private forests in the early 1950s, however, follows mostly from the fact that the state took the best quality forestland in 1945 and Soviet reparations harvests took a large proportion of the private forest harvest cut between 1945 and 1949.[70]

Private forest afforestation accounted for almost 25 percent of all afforestation on 30 percent of the area between 1955 and 1958, a not unreasonable share. The State Forest Service cleared its afforestation backlog from Soviet reparations only in 1959; private forests followed closely in 1960.[71] Private forestland, particularly once the bulk of it was in traditional cooperatives, was no worse managed than State Forests when one considers its poorer initial quality, owners' lack of resources, and the state's withholding of extension advice.

The best-managed forests, however, were neither the better quality People's Forest nor independent forest, but the traditional forest cooperatives (the *Waldgenossenschaften).* Successful independent management outside the socialist collectives was not what the Party leadership wanted from land reform. For them, the form of management mattered more than results. As Kienitz revealed in his discussion of conditions in Bezirk Erfurt: "The relatively large proportion of old communal forests are in most instances well managed. Their transition into socialist collectives is, however, very difficult. A consciousness-raising education effort could take decades if an economic transformation must be reached voluntarily. [Members of forest cooperatives also] have the impression that proceeds from increased harvests and prices do not have to be reserved for afforestation but are treated as a net profit."

The heavy regulations governing afforestation, and the Party's onerous fees and high interest rates, slowed forest owners' replanting of clear-cuts.[72] Kienitz complained that no government department seemed to know who was in charge. The State Forest Districts oversaw private forests' Plan duties and had "responsibility to increase the wood production of cooperative and peasant forests," but the counties (*Kreis*) also had formal responsibility for overseeing private forest quotas until 1958.[73] The State Forest Districts were responsible for providing planting stock but had been excluded from afforestation planning and prohibited from "interfering" in peasant forest management since 1945.[74]

The Party first issued new forest management regulations at the start of 1951, raising peasants' taxes between 500 and 700 percent, almost unbearable pressure on private forest owners to collectivize.[75] High fees for afforestation were "always a sore point," saddling smaller owners with costs many times their annual revenue.[76] Farmers in Bezirk Dresden could pay only 20 percent of their afforestation fees in early 1957 despite the Party's arrest as "saboteurs" of farmers who failed to pay their taxes and fees. Peasant owners in the large lowland Bezirk Frankfurt am Oder refused all payment in 1957, instead sending in a petition protesting the burdensome fees and taxes. The protest movement almost stopped afforestation and precipitated the Party's decision to collectivize forestry in advance of agriculture.

Kienitz summarized these problems, describing the managerial chaos: "In 1956 the *Bezirk* and county councils of the Peasant Unions [VdgB] did nothing. The responsible District Commissions in 1957 had no instructions. The results were firewood harvests outside the Plan, forest destruction, and a failure of peasants to meet economic goals. Problems with payment to woods workers demanded resolution. The state failed to pay DM 600,000 in wages for compulsory work on the People's Forest in 1956 [and was slow in paying for wood delivered by peasants]."[77]

In the face of the Party's managerial failures and in fear of imminent collectivization, the peasants rejected socialist management, streaming into efficient and democratic traditional forest cooperatives, the *Waldgenossenschaften*. Eighty-eight percent of peasant owners joined forest private cooperatives, shunning the socialist collectives. Few peasants chose voluntary collectivization.[78] It seemed, to the horror of the cadres, that the peasants might succeed in communal management—but without the Party! The increasing militancy of the cooperatives' members made this resounding defeat for the Party's program of voluntary collectivization even more stinging.

Private forest owners preferred to keep their timber harvest for themselves, to barter within the socialist economy for scarce resources or services, or to smuggle wood across the frontier for hard currency. Such "own use" and "diversion from the socialist economy" rankled more than mere poor management. A healthy private market persisted through the 1950s with prices well above the state's. Peasants took almost half the 1956 sawlog harvest for firewood, "lost" production that cost the economy annually DM 29 million.[79] Sawlog deliveries to the state languished at 70 to 80 percent of private owners' Plan quotas and were often taken from inferior thinnings. The state lost four to five cubic meters per hectare for an annual loss of DM 65.2 million.[80] Kienitz observed, with "thirty to forty years of state management one could achieve 600,000 cubic meters more annually than current yields," the equivalent of ten years of harvests.[81] The Party leadership could not steer production on private forests through quotas and fixed prices, or even through coercion, as long as the East-West frontier remained partly open and free markets in West Germany beckoned.

Forests were a significant native resource in a small country that was poor in other natural resources. As private forest owners held back wood from the "socialist economy," they denied the state more than one-third of the East German forest. Party economic experts predicted that coal reserves would be exhausted by 2000, noting that the forest offered a renewable "internal reserve of energy and organic raw material" critical to the national economy.[82] Therefore, Kienitz demanded that forestry create "all the preconditions and all management policies necessary for maximum wood production."[83] Private forests had to be collectivized so that forestry as an industry could meet its national quotas and support autarky. The inexorable logic of central planning and autarky demanded full state ownership and industrial forest management.

Class warfare was never far below the surface. Kienitz complained of "an intensification of certain capitalist tendencies" and a "general retreat to capitalist forms" in the private forest economy. Kienitz answered the study's initial question, "Do the peasant forest owners trust the agricultural policy of our government?" in the negative when he complained "many peasants presume to *demand* "free" markets" (emphasis in original).[84] Peasants not only rejected the socialist collectives, they actively embraced the Party's avowed enemies: a liberal market economy and traditional forest cooperatives. When Kienitz complained that the forest cooperatives were "not successful economically," he meant that their harvests "disappeared" into the private economy and that private owners did not meet their Plan quotas.[85] Kienitz concluded his analysis with an ingenuous call to Marxist-Leninist ideology: "The social-economic structure of peasant forest ownership explains everything. The problem is with the large farmer class [*Großbäuerliche Element*] that will have nothing to do with socialist management and the otherwise well-managed peasant owner who concerns himself only with his forest. New regulations at the very least must bring precisely this group of peasants to an accounting. Owners must also set their harvests according to stocking and accept higher production quotas."[86]

The Soviets designed land reform to prepare for a landscape of large-area collective farms, the "Union of the Workers and Peasants" (the *Bündnispolitik*), and to cement the Party's sole power in the countryside. The tortuous, drawn-out process of collectivization in the Soviet zone in the 1950s revealed the Party leadership's intent to use land reform to win popular approval: ultimate collectivization was never in question. Kienitz identified the failure to collectivize immediately in 1945 as the fundamental "mistake of the land reform. Above all, we should have secured the main principles of the land reform—collectivization and large area forest management, land reform's unalterable goals."[87] As long as the Party leadership thought it could draw political advantage from land reform, however, it delayed collectivization.

Kienitz's arguments for prompt collectivization were unassailable from a managerial and ecological viewpoint. But until the end of the decade the Party leadership still hoped for voluntary collectivization, for political as well as ideological reasons, and feared the political consequences of reversing the land reform overtly, from the top down. But Kienitz did not stress the central importance of Marxist-Leninist administrative forms sufficiently, only the ecological and economic benefits of large-area management.

And so, the Party's guardians of orthodoxy attacked Kienitz. The form of management itself, not efficiency or equity, was all that interested the Party leadership. Many farmers did join with their neighbors and community in cooperative management, but the wrong form. They overwhelmingly chose the traditional and democratic old forest cooperatives, shunning the Party's socialist collectives. And then, to the Party leadership's fury, they succeeded.

Kienitz challenged the Party leadership's strategy of pressing for voluntary collectivization against an ever-more-recalcitrant peasantry. The heat of the opposition raised against him and the reality of imminent forced collectivization revealed how the Party leadership still wanted popular approval in the rural landscape, its denial of evolving reality, and its willful self-delusion. Kienitz's crime was to tell the truth: the peasants did not trust the "First Workers' and Peasants' State" and the Party would have to force socialism upon them.

Kienitz's heresy and undermining of socialism earned him a prompt attack by Laßmann, an orthodox Marxist-Leninist:

> [Kienitz's plan] completely contradicts the principle of voluntary transition to socialist large-area production methods. [This transition] can not be accomplished through regulation. The initial phase of the law-governed development stage from capitalism to communism can neither be bypassed nor evaded. Such a suggestion contradicts the "Union of the Workers and Active Peasants" (the *Bündnispolitik*) and would find no support among members of forest collectives or individual owners.
>
> Altogether his recommendation proposes a "third way" to the transition to forest mass production. This fundamentally conflicts with the desires and interests of private forest owners and will not lead to socialism. On the contrary, it would seriously harm the socialist transformation in agriculture and forestry and imperil the convergence of the worker and peasant classes, benefiting only the enemies of socialism.[88]

Written just before the Wall went up, Laßmann's polemic throbs with anxiety. Laßmann and the Party leadership demanded a "socialist transformation" of the rural landscape; traditional forest cooperative societies not only took resources out of the Party's control, they also perpetuated old ideas and structures in the countryside and thus were intolerable. What had been the point of land reform and the wholesale destruction of land records, manor houses, and forest memorials, and the tearing apart of ranks of old hedgerows and allée of pleached linden and beech if the Party still lost control of the countryside? Laßmann was right in one sense. Forced collectivization and the Socialist Spring in the Countryside did alienate the rural population and spurred torrents of *Republikflucht*, ultimately forcing Walter Ulbricht and Nikita Khrushchev to seal the last remaining portal to the West in August 1961 with the Berlin Wall, proving the failure of the regime's political campaign in the countryside and vindicating Kienitz.

Informal politics may have sparked Party bureaucrats' attacks on Kienitz as much as ideology, for Kienitz indirectly attacked the State Planning Commission and State Forest Service management. The first step toward full, formal collectivization came on 1 January 1958 when State Forest Districts ordered many private forest owners to bring their forestland within the socialist farm collectives (LPGs), cutting private ownership from 30 to 24 percent.[89] Government surveyors again obliterated boundaries on a vast scale in a reprise of land reform's cleansing of memory. The signposts and cairns of traditional rural life faded even further.[90] Later in the year only 14 percent of the East German private forest was still individually managed.[91] If these forests were not well managed, the responsibility lay with the Party's planning bureaucrats. So Kienitz's real target was not individual landowners — only a fraction of the forestland was under their control — but the Party's incompetent and dirigiste forest and farm bureaucrats. Kienitz's analysis, startlingly clear and prescriptive despite the necessary overburden of Marxist-Leninist pieties, aroused remarkable resistance to what otherwise was a sensible and even inevitable plan. Once again, foresters learned from Kienitz's punishment not to challenge Marxist-Leninist ideology or Party political policies or to debate even in terms most respectful of orthodoxy.

Thus Party leadership completed collectivization of most private forestland in 1958, eighteen months before the Socialist Spring in the Countryside forcibly collectivized all remaining independent farmland.[92] The Party leadership had two conflicting goals for independent farm and forest land outside the "socialist economy" in the 1950s. One was political: to secure the Party's power in the countryside. The other was practical and immediate: to repair land reform's damage to the farm and forest economy and increase rural production in an economy still burdened with food rationing and shortages. The Party leadership resolved these two goals through this "second land reform," liquidating private ownership and abandoning for good their hopes that the rural population would ever voluntarily accept socialism.[93]

Forced collectivization reflected political trends throughout the Soviet bloc, except for Poland and Yugoslavia, between 1957 and 1965.[94] Farm collectivization followed between fall 1959 and spring 1960 in the Socialist Spring in the Countryside, the Party cadres waging "hard class warfare against the resistance of reactionary forces in the villages," as the author of East Germany's leading work on economic geography described this final expropriation.[95] Gerhard Grüneberg, agriculture minister in the 1970s, characterized collectivization: "In the spring of 1960 the voluntary decisions of individual peasants to enter socialist farm collectives [LPGs] were realized and socialist production methods were victorious in the countryside. Workers and peasants together created a firm foundation of socialism in the countryside. Those were also years full of class warfare in which enemy action hindered the realization of socialism. But through patient work many reluctant peasants were won over to the socialist farm collectives."[96] Party propagandists echoed the myths of land reform, asserting that the Party "orchestrated the will of the peasants through intense political, ideological, and organizational work in the countryside to convince peasants of the wisdom of the Party's policies and bring all agricultural production into socialist collectives."[97] But, as with land reform, the Socialist Spring in the Countryside was a revolution from above which served the Party leadership's political and ideological purposes alone.

Emigration, the key and only index of the Party's success in the absence of elections, had moderated in response to relaxed economic policies introduced after the Workers' Uprising in June 1953 (the "Days of Western and Fascist Provocation"), and the economy slowly improved with the currency reform of 1957, the lifting of maintenance costs for Soviet troops, and the end of formal rationing.[98] Flight to West Germany moderated as the economy improved, so the emboldened Party leadership scrapped the Second Five-Year Plan of 1956–60 two years early, in late 1958, to follow the more ambitious and ideologically infused First Seven-Year Plan of 1958–65.[99]

Global economic and social environments were favorable as the Party leadership shaped the First Seven-Year Plan as the great "Golden Age of Economic Growth" (1950–73), a period of prosperity unmatched in history, lifted all economies, even the sclerotic Soviet bloc economies.[100] But East German growth still underperformed badly relative to the West and fell behind despite sporadic absolute growth.[101] The Party leadership's performance as managers of the national economy was abysmal: East Germany did not recover its 1936 level of industrial production until 1953. Although its electricity production doubled between 1936 and 1955, West German electricity production rose almost four times in the same period.[102] Consumer goods production did not recover its 1936 levels until 1955, as food production "apparently" recovered prewar production levels.[103] Yet the Party leadership had to reimpose butter, milk, and egg rationing in 1961; bread and meat supplies were irregular and the quality and quantity of consumer goods deteriorated in the 1950s.[104] Food shortages continually plagued the East German people and rationing in some form was a permanent fixture of East German life.

Erich Apel, Ulbricht's economics czar, in 1959 called for "a great leap forward" to develop socialism rapidly, responding to the Second Berlin Crisis (1958–61) and the need to jumpstart East Germany's stalled economy.[105] East Germany's "great historic task" was "to catch up with and surpass" the West German economy by the end of 1961 and prove the superiority of socialism. This would win the cold war battle of ideas and secure the Party's legitimacy.[106] Ulbricht's and Apel's ignorance of the relative power of the West German economy is staggering—they must have drawn confidence from the "advantages of socialism," in the power of "Marxist-Leninist administrative science," and in the technological achievements of the Soviet bloc, beliefs reinforced by the Soviet launch of Sputnik 2 in early 1958.

Apel's and Ulbricht's deadline for overtaking West Germany by 1961 passed, yet Ulbricht's confidence in the power of socialist management science and in cybernetics was unshakable. Science and information technology would surely propel the East German economy ahead of the West's—even ahead of the Soviet economy. The East German government abandoned the goal of overtaking West Germany only after Ulbricht's overthrow at the Eighth Party Congress in 1971 and after Erich Honecker's policy of dissociation from West Germany.

As State Forest Service control over private forests solidified, Schröder predicted a "new, socialist era in forestry," asserting that "the essential preconditions for the success of socialist forest management have been laid down" with the establishment of the State Forest Districts.[107] Just as the Party leadership increased its control in the rural landscape in the late 1950s, Ulbricht and the Party elites also increased their control over all aspects of life, principally through the aggressive First Seven-Year Plan (1958–65). Foresters felt the plan's dirigisme and control through the companion program of "Socialist Reconstruction."

Socialist Reconstruction, the First Seven-Year Plan's main program, cut investment and development of basic industries to concentrate where East Germany had "natural advantages to create the material and technical basis for the victory of socialism.[108] A Ministry of Agriculture and Forestry official defined Socialist Reconstruction as a "scientific method which is immune to resource limits" and biological constraints. Socialist Reconstruction would deliver constantly increasing forest and farm production without increased employment or investment and free of biological constraints.[109] These startlingly amateurish statements were nothing new; the call for increased labor productivity while investment declined — derision of resource constraints or biological limits — were core features of all East German national economic plans.

Investment in particular, whether building forest capital through harvesting less than growth or investing in new plant and equipment, was anathema. Because they enjoyed the advantages of Marxist economic science unfettered by the profit motive or capitalist relations, planning alone would yield increased productivity. Apel acknowledged that there were "two ways to increase productivity, better management and increased investment." He dismissed investment, citing Ulbricht's dictum that productivity gains from increased investment were "in most cases an illusion."[110] More could be done with existing resources by increasing "the productivity of existing plant and equipment with more rational management."[111] Scientific management and detailed technical planning alone would fuel productivity growth.

Socialist Reconstruction further centralized decision making, liquidating small farms and forest as "irrational hangovers from capitalism."[112] State Forest District management instructions now were explicitly Marxist-Leninist, detailing the "productive power of labor" as central to "increased productivity and, therefore, absolute increases in productivity" despite the relative unimportance of labor costs in forestry and the importance of biological constraints.[113] Forest productivity would increase without additional investment and despite biological constraints.

Yet the conundrum of increasing production while preserving the asset persisted at the end of the decade. Alexis Scamoni, East Germany's leading forest scientist, and Gerhard Schröder proposed planting fast-growing, short-rotation hardwoods, particularly poplar (*Populus spp.*), to replace spruce pulp and relieve demand pressure on the forest. "Better," advised Schröder, "to spend the money at home" on fast-growing hardwoods than to import the Siberian pulp which began to flow as the First Seven-Year Plan integrated imports and exports within the concurrent Soviet Seven-Year Plan.[114] To relieve East German raw wood shortages, pulp imports from the Soviet bloc started in 1956 when the Soviet Union recognized East Germany as a sovereign state and one year after East Germany entered the Warsaw Pact.[115] The shift to imports also reflected a limited surrender of the autarkic principles established in the Second Five-Year Plan of 1956–60.[116]

Soviet pulp imports were a mixed blessing. Transport problems and delays of even a few weeks caused by a slow thaw of the Siberian ports in April forced cutting higher-value domestic sawlogs to replace delayed Siberian pulp.[117] Close to a third of the East German raw wood consumption was filled by Soviet pulp, but the Poles demanded U.S. dollars to settle freight charges, and the East Germans had to make substantial investments in Siberian pulp plants. But integration into the East bloc outweighed what was rational. East German analysts credited Soviet imports for the partial forest recovery in this period by "allowing harvest levels to drop," as if harvest levels had a dynamic independent of the Party.[118] Harvest levels did drop significantly, and slowly stocking did recover.[119] But these were distinctions without a difference as long as near-universal clear-cutting and even-aged management continued. It did nothing to reverse the long-term decline in vigor or even to correct the distorted age class distribution.

Ideology and self-delusion took over forest management as cold war tensions heightened in the late 1950s and Ulbricht pressed for a settlement of Berlin's boundaries and for a peace treaty. Apel and Ulbricht disdained such analytic tools as managerial and cost accounting as "bourgeois tools of the businessman," unnecessary when the Party had the heady brew of the most advanced science and technology. "Profit" emerged as the bogeyman, even though profit is an artificial—but useful—accounting abstraction if soberly calculated. Apel stated: "Scientific management and technology, not the profit motive, will drive the socialist economy. Workers need not fear technology since their jobs are guaranteed."[120] What workers really had to fear was the wholesale shifting of amateurish economic targets and ideological goals onto their shoulders, a burden to be carried without help from the state. Foresters also had reason to dread Socialist Reconstruction, as it targeted both private forests and People's Forest as ideal candidates for a "revolution in management." Ideology and a religious belief in the "advantages of socialism" undergirded Socialist Reconstruction, but Ulbricht and his new cadres of technical experts also were coming under the spell of cybernetics and systems theory, making an unholy mix of religious and scientistic Marxism.

Experienced foresters, even some educated in the new, socialist forest academies, resisted Socialist Reconstruction's stark materialism and short-term focus. This opposition, and Ulbricht's insecurity following the 1956 Hungarian Revolution, triggered another purge of the Forest Service, a "cadre triage" (*Kaderauswahl*) purging politically unreliable foresters, many from Eberswalde and sympathetic to the Hungarian rebels' cause.[121] Eberswalde foresters were traditionally more innovative in forest ecology and involved themselves less in East German politics. More Eberswalde professors and their children, following the course of many other East German professionals, fled to the West than from Tharandt. Professor Hildebrandt, Albert Richter's assistant, escaped to West Germany in 1957, soon followed by Richter's two sons. This heightened the Party's natural suspicion of foresters and ignited intense repression as the Party searched for "spies."[122] The Party, under the pretext of foiling "economic espionage," collected evidence against senior foresters and punished any they deemed politically unreliable.[123] The Party disciplined Albert Richter yet again for his sons' flight from the republic. The rounds of purges, petty humiliations, and spy hunts signaled the end of independent ecological research.[124] But first, in 1959–60, a final muted round in the debate between close-to-nature forestry and industrial forest management took place, the last time East German foresters could criticize the Party's forest policy.

Committed Marxists faced off against each other this time, thanks to Gerhard Schröder's (director of the Forestry Department [*Forstabteilung*]) integrity. Perhaps only a Party member such as Schröder could have taken on Socialist Reconstruction so vigorously. Schröder presented a long-term plan to restore a forest close-to-nature at the Fifth World Forest Conference in 1960, extending his running debate with the orthodox G. Laßmann between 1959 and 1960 in the forestry journal *Forst und Jagd*.[125] The debate turned on Recknagle's 1913 identification of the central dividing point in German forestry, whether forestry's focus should be on short-term returns, implying clear-cutting and industrial forest management, or on long-term productivity, implying natural regeneration and close-to-nature forestry.

Schröder championed the long term: "The main goal of forestry has always been to build a diverse forest through early conversion of poorly growing pine and spruce stands [to create a close-to-nature forest]."[126] Forest management's key objective should be to increase forest capital, to maximize "the sustainable volume of usable timber. This is the deciding question for increasing workers' productivity."[127] Schröder, contradicting the humorless Ulbricht, recommended capital investment over increased workers' productivity, yet another heresy.

Laßmann in turn demanded industrial methods, complaining that forestry had the lowest productivity growth rate of any industry, shrewdly parroting Ulbricht's "second way" of increasing productivity without increased investment.[128] The debate was settled late in 1960 when the Party dismissed Schröder from his post and sent him to State Forest District Steinhagen in Mecklenburg, far from Berlin. No director or senior official would ever again contradict the Party, and industrial forest management, followed by Gerhard Grüneberg's Industrial Production Methods, reigned until 1989 and reunification.

The unambiguously titled order the "Unity of Management" at the same time decreed that all East German forests were industrial ("production") forests managed under a strict age class regime and regenerated though clear-cutting. The main task for the next thirty years would be "achieving maximum raw wood production though rapid increases in worker productivity."[129] To accomplish this, the Party once more reorganized the State Forest Districts under a central Forestry Department to increase central control and to enforce Plan tasks. A bureaucrat in the Forestry Department wrote: "State Forest District reconstruction plans must make full use of all the advantages of our social order to achieve rapid increases in productivity. They must commit to meet and over-fill the production quotas of the Seven-Year Plan. Thus we contribute to the maintenance and securing of peace through an increase in the speed of our building of socialism and as an example for both Germanys."[130] Optimal stocking, a focus on long-term investment in forest capital, was irrelevant next to the short-term goals and to proving the superiority of "socialist forestry" over the "bourgeois forestry" of West Germany.

Ulbricht, an autodidact fascinated with technology, listened to the loyal forestry cadres' advice in 1959 and agreed to cut the 1965 harvest in the final year of the First Seven-Year Plan to 80 percent of 1958 levels to stabilize stocking by 1985.[131] But harvests still remained high relative to growth despite pulp imports from the Soviet Union, so high that "even moderate thinning operations meant worsening of the age class structure."[132] There was no alternative to continued heavy harvesting in the older age classes until chip-board and oriented strand-board production came on line to relieve pressure on larger dimension timber. So despite the moderating advice of foresters, the First Seven-Year Plan failed to address the dominance of the youngest age classes.[133]

The increasing prominence of ideology, whether building the Union of Peasants and Workers or demonstrating the superiority of socialism, and Marxist-Leninist class relations theory in forest management plans showed how abstractions dominated management.[134] Socialist Reconstruction explicitly meant "continued radical ideological change and implementation of truly socialist common labor in all areas," effectively institutionalizing the economic crisis and destabilization begun by reparations and land reform.[135] Economic instability and permanent revolution had become drugs to the Party leadership, who needed a shortage economy and permanent crisis to legitimize its authoritarian political system.

The Party leadership's harsh shift to authoritarianism and dirigisme in the First Seven-Year Plan and Socialist Reconstruction—and forced collectivization—sent the economy into a slump. As the first full year of Socialist Reconstruction ended in 1959, food shortages returned. By 1960 *Republikflucht* climbed again to almost 200,000, reaching its apogee of more than 200,000 in the first seven months of 1961.[136] Between 1949 and 1961, 3 million refugees from a population of only 17–18 million fled East Germany, many from the rural landscape. Emigration deepened the economic crisis caused by the Socialist Spring in the Countryside in 1960 and forced collectivization, the harsh and failed First Seven-Year Plan, and Socialist Reconstruction.

The First Seven-Year Plan collapsed early in 1961 and was formally annulled at the Party's Sixth Central Committee Conference in early 1963.[137] Bruno Leuschner, an Ulbricht loyalist, director of the State Planning Commission since 1952, and member of the Presidium since 1954, admitted six months before the Plan target date that East Germany had failed in its "main economic task" of overtaking West Germany.[138] The economy in a shambles and hemorrhaging workers, the Party leadership decided early in 1961 to seal the border with West Germany; construction on the Wall began on 13 August 1961.

The Wall marked the final resolution of East Germany's major incongruities and "consolidated socialism."[139] A stubborn private economy had survived Soviet reparations and even regained some strength in the 1950s. Significant, even if diminishing, qualities of personal freedom also continued into the 1950s. But this could not continue in a state that claimed a monopoly on power and truth. The Party crushed ecological and social diversity and forced East German society to conform to reductionist Marxist-Leninist norms. Foresters after 1961 could no longer safely challenge Marxist-Leninist forest economics. The Wall finally gave the Party carte blanche and doomed East Germans to a generation of shortages and dictatorship and the forest to accelerating forest decline.

By 1959 socialist forest management had cycled through naked exploitation under Soviet reparations, radical *Dauerwald,* "Optimal Stocking Forestry" from 1951 to 1954, and Richter's rational "Site Appropriate Forestry" in the mid-1950s, to return finally to pure extractive values with Ulbricht's Socialist Reconstruction, a roundtrip to the regime of Soviet reparations. As Ulbricht began to look more to the scientific and technical elites in the 1960s for management advice rather than to the orthodox cadres, forest management improved. The next decade saw the high-water mark of scientific management and cybernetics in East Germany, a version of scientistic Marxism or "cybermarxism." For a brief moment toward the end of the 1960s it looked as if cybernetics might supplant orthodox Marxism-Leninism. Forestry, even given the dominance of the Plan, briefly benefited from the rise of Ulbricht's technocrats, but the raising of even a subordinate god to Marx and Lenin triggered reaction: Ulbricht's fall and the end of cybermarxism.

# 6

## *The Landscape Transformed (1960–1961)*

*It is clear that the forces of diversity are at work inside the Communist camp, despite all the iron disciplines of regimentation and all the iron dogmatisms of ideology. Marx is proven wrong once again: for it is the closed Communist societies, not the free and open societies, which carry within themselves the seeds of internal disintegration.*

*The disarray of the Communist empire has been heightened by the gross inefficiency of their economies. For a closed society is not open to ideas of progress—and a police state finds that it cannot command the grain to grow.*
*—John F. Kennedy, State of the Union address, 1963*

The East German landscape evolved into more than an arena for class warfare in the 1950s; it became the frontline of the cold war. As the sharp shocks of reparations, land reform, and forced collectivization passed, the industrial forest slowly recovered and the fractured farm structure reknit either under the shelter of traditional cooperatives or, through "hard class warfare," within socialist collectives. The guiding principles of the 1950s command economy—for central control, Plan discipline, and maximum production—remained intact. Despite rhetoric of reform and ecology, neither the biological nor the economic structures emerging from Ulbricht's chaotic decade of experiments were close-to-nature or productive. Impoverished system qualities continued not only to blight economic and ecological structures into the so-called Period of Reforms of the 1960s, they defined them and made the Soviet bloc distinctive and "second world."

The 1950s also had witnessed the final division of Central Europe which the Soviets began at the Potsdam Conference (17 July–2 August 1945). This division soon hardened with the perfection of a formal East-West frontier girdled by deep reaches of border fortifications, forest cantonments, and military exclusion zones. As the East German economy faltered and food production crashed under the dead hand of Ulbricht's command economy, the people of the rural landscape flowed in ever greater numbers into West Germany. *Republikflucht,* flight from the East German republic, and the growing economic disparity with West Germany threatened the survival of the East German republic.

The first step to regaining the political and economic momentum lost since 1945 was to close the republic's boundaries through the Wall to stave off imminent collapse. Then cybermarxism, Ulbricht's cybernetic New Economic System (NÖS), could restore economic health within the republic's closed border confines. The Party's rhetoric of reform—slogans of modernity and cybernetics, sustainable forest management and conservation—could not disguise the poverty and sheer waste of the economic and ecological processes they obscured. The telltales of decline and brittle, artificial structures and the discrepancy between rhetoric and reality were most easily seen in the natural landscape, the staging ground for Soviet armies poised on the East-West frontier as well as for the Party's ongoing war against the countryside. As political borders hardened, the Western nations of the Atlantic Alliance raised their guard.

In the late 1950s United States Air Force pilots started making high-altitude reconnaissance flights from fields in southwest Europe in new Lockheed U-2s. They flew high above the headwaters of the Main River, crossing into the East over the Leipzig Basin and into the skies above the North German Lowlands. Photo interpreters poured over the endless reels of aerial photography shot over East Germany, scanning the layers of radar, thermal, and optical data burnt onto the film emulsion, peeling them back to uncover the thirty Soviet armored and mechanized infantry divisions laagered beneath the forest canopy.

Reflections from the Northern Lowland forests also lit the film with the luminosity and color temperatures characteristic of long-needled, loose pine, easily distinguished from spruce's density and warmer heat signatures in the Southern Uplands. Analysts could mark, as they scoured the forest canopy for signs of the armies lurking beneath, the slow reforestation of Soviet clear-cuts. The naked eye, even from the U-2's extraordinary cruising altitude of twenty-two kilometers, could see the sudden emergence of collective farms' large fields within the chaotic, patchwork pattern of smaller private farms. The uniform, reduced structures of the farm and forest landscape emerging from land reform and forced collectivization mirrored the brutal simplicity and power of the Soviet armies below.

Observers needed, however, neither the U-2 pilot's vantage point at the edge of the atmosphere nor sophisticated remote sensing tools to mark the East-West frontier, the Soviet bloc's 1,386-kilometer-long and almost 6-kilometer-deep system of fortifications dividing East and West Germany. It dominated Central European geography, seeming "almost to be a feature of Europe's physical geography like the English Channel or the Alps."[1] The frontier system's sheer physical mass gave it the character of a small nation, at 8,000 square kilometers not much smaller than Connecticut.[2] The East-West frontier, perfected with the Berlin Wall in 1961, complemented the inner rural landscape of simplified farm and forest, revealing in binary clarity the emergence of a homogeneous and static natural and political environment east of the Elbe River in the 1950s.

Boundaries and frontiers are of central importance to ecosystem management and, along with issues of legitimacy, were defining issues of the cold war. The Allies had drawn the demarcation line between eastern and western Germany in 1944 not knowing that it would metastasize into the most heavily defended border in the world and mark the Soviet point of contact with the West. The Soviets demanded more political and economic control over their zone of occupation than a porous border allowed; they wanted insulation to hide the scale of Soviet reparations and brutality. The Soviets also thought opportunistically and suspected British and American motives; a closed border and a police state better camouflaged their troop strengths and deployments and improved their control over people, resources, and ideas. From the beginning, even when reunification seemed imminent, the Soviets pressed for closed frontiers.

The Allied Control Commission agreed in a collegial spirit to a Soviet request on 30 June 1946 to prohibit Germans from crossing the frontier in either direction.[3] In the immediate postwar years people moved more to the East, drawn by stories of better conditions. The Soviets doubled their border guard strength in mid-1949 as the flow of people, undeterred by border check points, turned back to the West; a senior western German border policeman declared the border was "now so effectively sealed that illegal crossings both ways had stopped almost completely."[4] The Soviet zone darkened as cross-border flows of information, people, and goods slowed to a halt. Lethal border fortifications were complete along the entire length of the East-West German frontier by the end of 1952, surrounding West Berlin's outer perimeter but not yet separating the Soviet zone from West Berlin, at a staggering initial cost of $2.5 million per kilometer.[5] The People's Police (the VoPos) had killed 1,826 people fleeing the republic and arrested escaping citizens in unknown numbers in the prohibited zone behind the frontier by the time the Wall fell in November 1989.

Despite this, young people and professionals, weary of the constant propaganda and uncertainty, risked death or arrest and sentencing to a Soviet forced labor camp to flee, with almost 200,000 streaming out in *Republikflucht* in 1950. By the end of 1953 more than one million East Germans had decamped for the West since 1949, drawn as much by the promise of independence as by political ideology or material benefits. After the Soviets and East Germans crushed the Workers' Uprising in June 1953, *Republikflucht* alone threatened the Party. The Berlin Wall sealed off the only open portal in August 1961, saving the East German republic for another 28 years.

West Germany's open society and better life lured East Germans westward. But West Germany's assertion of legitimacy over all Germany and its growing economic power and identification with the "capitalist camp" must have worried the invidious Party leadership just as much as *Republikflucht*. Konrad Adenauer's conservative West German government signed the "Germany Treaty" in May 1952, taking back full sovereignty from Britain, the United States, and France and committing any future reunited Germany to the Atlantic Alliance. Thus refugees fleeing westward added to the strength and wealth of the republic's mortal enemy, one directing all its growing diplomatic and economic power to isolate and dissolve the East German republic.[6] The Soviets and East Germans used their control over the frontier and inner-German flows—particularly in Berlin—to show the West German people that Adenauer's pro-Western policies threatened reunification, and to stop the West German parliament from ratifying the Germany Treaty. The frontier, backed by the Soviet armies encamped under the cover of forest, was the Soviet bloc leaders' best weapon either to control the political terms of reunification or to prevent it.

Stalin reacted angrily to Adenauer's Germany Treaty and to Britain's and America's earlier rejection of the "Stalin Note" of 10 March 1952 in which he proposed a neutral, reunified Germany. Recognizing that reunification under his terms was increasingly unlikely, Stalin raised the price of West Germany's alliance with Britain and America. He instructed the East German Party leadership in early April 1952 at a near-midnight Kremlin meeting: "You must organize your own state."[7] Wilhelm Pieck, East German president, recalled Stalin's chilling comment that the first line against the Americans was "Germans (Stasi [secret police]), behind [it] Soviet soldiers."[8] East Germany was the Soviet's frontline state in the West. Therefore, on the same day Adenauer signed the Germany Treaty binding West Germany to the Atlantic Alliance, the Stasi closed and strengthened the fortifications on the East-West German border and around West Berlin's outer perimeter.[9] When Ulbricht promised in a news conference on 15 June 1961, just before the Berlin Wall went up, "No one intends to build a Wall," the Wall between East and West Germany had already been up and killing East Germans for nine years.

The frontier fortifications which followed from Stalin's order to "organize your own state" in spring 1952 caught Westerners by surprise; they looked on in disbelief as construction troops backed by VoPos ploughed a two-meter-wide strip twenty-five meters east of the border, "either in preparation for building some obstruction or to mark the border unmistakably," as a startled British journalist noted.[10] The scale of the operation puzzled Western observers. By early June it seemed a "hedging and ditching" operation, the ploughed strip now clearly "the area in which police may shoot on sight," with a five-hundred-meter-wide minefield and warning zone beyond.[11] The VoPos uprooted everything in the five-kilometer prohibited zone, as Drew Middleton reported in the *New York Times*, "with ruthless energy. Bridges have been blown, roads blocked, towns and villages evacuated to insure that in this zone nothing will live or prosper save the ubiquitous troopers of the East German People's Police (VoPos) and Russian soldiery." Stasi officers cleared "unreliable" citizens, letting them take only ten kilograms of household belongings and clothing and only one hundred east marks from their bank accounts — the balance of their property and savings was forfeit to the state. They were forced to sign affidavits attesting that they had moved voluntarily, freely giving their homes and farms to reliable Party members. Now the East Germans referred to the demarcation line as a "state border," and the Party dropped references to sector boundaries and thus to Four Power rights throughout Germany.[12]

Everyone could see the physical dimensions of the emerging Wall system, now complete save for the border between the Soviet sector of Berlin and West Berlin: how long and deep it was and how many soldiers had their submachine guns pointed against their countrymen behind it. We may never know how many people died trying to breach it, nor did anyone ever know its meaning until November 1989; did it separate two sovereign countries, or was it a temporary line sundering what belonged together? Many thought the May 1952 militarization of the East-West border and the Stasi takeover of the frontier signaled a permanent border, that the Wall system marked a permanent division not only of Germany but also of Europe. The sheer ugliness and brutality of the Wall suppressed hopes for reunification in even the most patriotic German and gave rise to a false sense of a climax stage in international politics and myths of emerging convergence between the Soviet bloc command economies and Western democracies.

Soviet appropriation of the East-West frontier as the leading edge of its own security frontier emerged further in October 1953 when long columns of VoPos—swinging their arms forward and up across their bodies, their chins thrust forward and up to the sky—goose-stepped up the Unter den Linden wearing new, Soviet-style uniforms, their baggy, olive-drab trousers bloused in the tops of high, Russian peasant boots.[13] The symbolism of power was not lost on the East German people. The Party "defends the border" with deadly force, as Ulbricht in an open 1961 letter lectured draftees who questioned shooting fellow Germans: "It is perfectly legal when necessary to shoot Germans who represent imperialism." The editors of the *Free German Youth* magazine scolded "those young people for lack of character and advised them to look to Herr Ulbricht as an example."[14] Ulbricht's example was playing out along the inner-German frontier where the VoPos had already shot and killed hundreds of East Germans on his orders.

Soviet interest in the East-West border moderated in the mid-1950s following Stalin's death on 6 March 1953 and Khrushchev's new government and reforms. Khrushchev wanted better relations and trade with West Germany and to reduce tensions in Central Europe so he could focus on domestic problems, above all on Soviet agriculture. So he invited Konrad Adenauer to Moscow in June 1955 to discuss opening diplomatic relations, using German POWs still held in Soviet forced labor camps as a lure. Adenauer then agreed to open diplomatic relations with the Soviet Union in September 1953, relaxing the Hallstein Doctrine, which precluded relations with countries that recognized East Germany.[15] A little more than nine thousand POWs, all that remained of the many hundreds of thousands—possibly millions—in camps at the end of the war, came home.

Adenauer tried to preserve Germany's rights to the Oder-Neiße territories and the Hallstein Doctrine's claim that West Germany alone represented the German people. Yet in a major broadcast three days after the signing, Moscow Radio announced that Germany's frontiers had been settled for good at Potsdam and that the "east German government was the proper government of east Germany."[16] The same day, the Soviet government announced the arrival in Moscow of East German Prime Minister Otto Grotewohl "for negotiations on 'questions of interest to all sides.'"[17] Alarmed at the rapprochement between their patron and their principal adversary, he had hustled to Moscow for reassurance and to repair Adenauer's damage to East German prestige.

Walter Ulbricht feared that Soviet courting of West Germany might lead to the sacrifice of East Germany to Soviet strategic and economic interests. He demanded a treaty (more than the agreement Adenauer signed) to restate Soviet support for East Germany. The resulting Moscow Treaty (20 September 1955) conferred on East Germany a sovereignty that differed only in detail from the unilateral Soviet declaration of 26 March 1954.[18] The Soviets also delegated to East Germany control over "all except allied traffic to and from Berlin."[19] This puzzled Western observers: East German VoPos already controlled the traffic between West Germany and West Berlin, and East Germany already had formal sovereignty within the Soviet bloc. Still, Ulbricht craved Soviet emphasis of East German legitimacy, and (even though his strategy was reactive to Adenauer's and Khrushchev's) he stressed that his Moscow Treaty was the "antithesis rather than the counterpart" of Adenauer's agreement.[20]

Ulbricht's insecurity and Soviet policy revealed the inner truth of East Germany's diminishing power relative to West Germany and, in absolute terms, to the Soviets. The Soviets granted East German sovereignty only after they had locked it within the Warsaw Pact Treaty, reasserted their absolute right to intervene in East German affairs, and held onto their Four Power rights in all Germany.[21] Although they dissolved the post of Soviet high commissioner in Germany, his powers and responsibilities were transferred to the Soviet ambassador in East Berlin.[22] The Soviets clearly were in no hurry to give up their special, Four Power rights or their veto power over German reunification, or to surrender complete control of the Berlin and East-West frontiers.

The East Germans used the control over the borders, which they did have, aggressively, as Western observers feared: "Herr Ulbricht, speaking at Moscow yesterday, made some dark remarks on this subject [the free flow of traffic between West Germany and Berlin] which might qualify for threats of a squeeze on west Berlin," holding "this vulnerable city as the hostage with which to pry recognition from west Germany."[23] East German border troops held two American congressmen and their families at pistol point for the better part of a day after VoPos noticed a radio in their jeep—"Spies!"[24] The Soviet commandant in Berlin rejected American protests, retorting that East Berlin was "no longer an occupied sector" but the capital of a sovereign German Democratic Republic.[25] Finally, as the Four Power agreement provided, the Soviets pried the Americans free of the VoPos and into the custody of the Soviet commander, who courteously received them and sent them back to West Berlin. Such nasty East German behavior may have aggravated the Soviets as much as the British and Americans.

Walter Ulbricht also used the Moscow Treaty to warn East German civilians that they had no choice but to accept the Party's permanence, saying: "after the signing of the Moscow Treaty, no one in the German Democratic Republic

[East Germany] can say any longer that things may change."[26] The forbidding and gloomy physical geography of the frontier and the sour and surly VoPos, even more than Soviet support for the Party's legitimacy, gave meaning to Ulbricht's threats and his claims for East German exceptionalism.

Yet the Party leadership only postponed facing their looming strategic dilemma: whether to obey the Soviets or to strike out independently. Despite Ulbricht's trumpeting of Soviet support, their interests were diverging as West Germany became more powerful and the Soviets more interested in détente. A fundamental problem of East German policy came to the fore: how to sustain legitimacy as the popular appeal of Marxist-Leninist ideology, never very strong, waned. East Germany had weak natural boundaries: the republic was geographically transitory and without historical precedent as an independent entity. Its existence was a fluke of Four Power politics and reliant upon Soviet grace. Political and economic dynamics forced the role of satrapy upon the East German Party, a role which chafed.

The Wall and the militarized frontier made manifest this vulnerability and dependence. The *Times* of London presented the Party leadership's strategic dilemma: "And yet a satellite along ordinary lines is exactly what east Germany can never become. Its international status is provisional, and when it comes to the point, for example when the unified command was set up at Warsaw in May, this has to be recognized. The present value to Russia of east Germany is mainly political. It offers a bargaining counter with the west when questions of European security have to be discussed."[27] Ulbricht's strongest point of leverage both with the Western Allies and with the Soviets balanced on the East-West frontier and on Berlin, the complex intersection of Four Power and all German interests.

Another frontier besides the East-West borders plagued West German relations with the Soviets and dimmed the Party's popular appeal—the fate of Germany's provinces east of the Oder-Neiße line. But the Soviets did not want to, nor could they, return Germany's land, unraveling Stalin's complicated border adjustments, ethnic cleansings, and land grabs since the 1939 Molotov-Ribbentrop treaty. The Party leadership at first blamed the failure of the Western Powers to enact Soviet-style land reform for the loss of the Oder-Neiße territories. But as land reform faded and then vanished in the wake of the Socialist Spring in the Countryside, the Party leadership tried to make the eastern provinces, and their peoples and culture, disappear. Thus, the Party renamed two Berlin railway stations late in 1950: the Stettin Station, named for the capital of lost eastern Pomerania (Szezecin after 1945), as the "Station of Polish Friendship," and the Silesia Station, also lost to Stalin's border adjustments in 1945, as the "Station of Soviet Friendship."[28] The humorless Party leadership missed the irony and latent insult to German opinion. After public protest the Party retreated and the Stettin Station emerged as the "North Station" in December 1950 and the Silesia Station as "East Station" in May 1951.[29] Both the Party leadership's clumsy renaming of these two stations and the public protest and ridicule show how important the issue of the Oder-Neiße territories remained.

The Party guarded the frontiers between memory and politics which went up internally in East Germany as ruthlessly as they defended the physical border with the West. But the borders between remembrance and ideology were harder to police. A whole region of Germany might be lost, but the people's subconscious still itched with remnant memory. As with the renaming of the Silesia and Stettin stations, the Party leadership increasingly turned to alternate realities. Rather than explain why East German economic performance lagged behind West Germany's, the Party leadership asserted the "advantages of socialism."[30] A senior member of the Politburo explained: "I've always been suspicious of comparisons with capitalists' achievements. We should really think over politically and ideologically the type and manner of comparison with the capitalist states. Maybe it is better to assume from the outset the advantages of the socialist GDR [East Germany] over the leading capitalist industrial states. I have in mind not only the advantages of the socialist mode of production vis-à-vis the capitalist regimes."[31] "Real, existing socialism" was not far behind. The Party leadership crossed over the boundary between reality and fantasy, asserting the advantages of socialism just as they preached conservation and *Dauerwald* while clear-cutting continued unhindered.

Communist "cultural advisers," also on the offensive to obliterate memory, combed through the shelves and stock of libraries, bookstores, and publishers' warehouses, removing and destroying novels, guidebooks, and atlases which described the geography and culture of the lost Oder-Neiße territories, as well as books which failed to "meet the social demands of the workers." A librarian looking out from his office window into East Berlin could see his East German readers making their way across the Potsdammerplatz to his library in the West. He scrounged for newspapers and magazines to lend his hungry Soviet zone readers, particularly begging Western journalists and diplomats for Sears, Roebuck catalogues, explaining, "Our customers like to know about things beyond the curtain."

Martin Ochs, in a front-page article in the *New York Times,* wrote that the communist effort "dwarfed the Hitler book burning episode of 1933 and employed a strict Soviet-style system of literary appraisal," a purge which consumed nine million volumes, "almost the entire old stock of light reading and parts of scientific, politico-economic, historical and children's books" and shut down independent publishers.[32] The purge extended to Marx and Engels: Party officials ordered the Party's publishing house to drop the original editions of Marx and Engels in favor of the 1882 Russian edition that Marx revised to state that a world revolution could start in backward Russia as well as in industrial Germany.[33] The East German people were often hungry and cold in the hard winters after the war. But the cultural famine of Marxist-Leninist pieties and the death of imagination, memory, and longing—even for simple things of the past—blighted their lives as well.

The Party leadership also decided to root out the stubborn strains of conservation and *Dauerwald* philosophy in the Forest Service. It zeroed in on Eberswalde, the home of Alfred Möller and Alfred Dengler, whose classic *Silviculture under Ecological Principles* remains in print in its seventh edition.[34] Eberswalde's faculty historically had enjoyed fruitful collaboration with owners of large Prussian estates, and Eberswalde retained a "mild feudal character," a mark of the faculty's respect for tradition and its independence from the state. Eberswalde faculty also had a long-standing connection to Humboldt University in Berlin and the broader scientific community. Albert Richter, a revered professor, observed, "This independence and certain worldliness definitely endured in the minds of Eberswalde students into the 1960s," despite the 1945–46 and 1956 purges.[35]

The Party leadership, at the same time it completed the Socialist Spring in the Countryside early in 1960, targeted foresters in the field, with agents collecting evidence under the pretext of foiling "economic espionage." Stasi agents and informers searched for "spies" among the faculty while Party officials browbeat scientists and scholars working on forest ecology and history. The campaign culminated in the 10 July 1962 show trial of Eberswalde's Dean W. Erteld and Vice-Dean Alexis Scamoni. Erteld was charged with "consorting with *Republikflüchtlingen*," men who had fled East Germany, and with criticizing East German foreign policy to a West German forester.[36] Both men were further accused of tolerating an "ideologically unreliable faculty" and of failing to give their children and students "proper socialist educations." Albert Richter, himself purged in 1945, termed this "a methodical act of infamy and intrigue." The Party now held senior administrators and scholars personally responsible not only for the political reliability of faculty and students, but also for their residence in East Germany.

The judges dismissed Erteld, stripped him of his pension and doctorate, and sent him down to work in the field.[37] The persecution of a respected senior professor, dean, and National Prize winner deeply shocked foresters. Scamoni and junior faculty also were punished as the Party systematically cleansed

Eberswalde. High Party officials then harangued the Eberswalde faculty, accusing them of being "reactionaries" for failing to educate their children as good socialists. They demanded a complete revision of the institute's curricula to deemphasize science and silviculture in favor of Marxist-Leninist economic science. The Party leadership finally stripped Eberswalde of its status as a *Fakultät* (Faculty) and cut its historic link to Humboldt University.

The Forest Service purge in the early 1960s reflected the intensified central control and assertion of ideology inherent in Ulbricht's 1960s reform program, the New Economic System, or NÖS. The Party attacked foresters' "forestry-only" perspective and their allegiance to "bourgeois forest economics," that is, conservationist, close-to-nature forestry.[38] Foresters' rejection of Marxism-Leninism and refusal to ignore the destruction wreaked by Soviet reparations angered the Party leadership. Their criticism of Industrial Production Methods inevitably shone unwelcome light on Soviet reparations, the primary cause of the forest's distorted age class structure. The chief of the Forest Service scolded:

> In addition to our enemies' rabble-rousing exhortations, foresters hold many false opinions about the role and meaning of forestry. In particular, they overemphasize forestry's value to the economy, that it is the "navel" of the People's Economy. [Such foresters] tackle the People's Economy's fundamental problems with a "forestry-only-perspective."
>
> Such opinions mostly surface when particular measures in favor of industry are proposed and decided upon, measures whose implementation causes difficulties for the State Forest Districts and demand special effort. Foresters also believe that the Party and government should have conserved all forestry production after the war for the development of forestry [instead of to cover reparations].[39]

Everything was political to the Party bureaucrats and cadres. G. Laßmann, a senior Party cadre, accused *Dauerwald* foresters of sabotage: "Class enemies often stirred enmity and doubt between workers, administrators, and forest scientists with the purpose of building opposition to the Party's and government's measures. For the same purposes opponents also discovered a 'theory' of so-called ruin and destruction of our Republic's forests. All foresters are still today not free of this delusion."[40] Perhaps conservationist foresters' pleas for a close-to-nature forest policy were, after all, as political as they were ecological in timbre.

Foresters refused to accept the Party's extreme materialism. The chief of the Forest Service lashed out at forest scientists and ecologists in a 1960 article lambasting the editors of the leading forestry journal, *Forst und Jagd* (Forestry and Wildlife), for publishing too much forest ecology research and not enough practical articles to help foresters meet their Plan quotas.[41] Science served only the Plan. Thus the Party changed *Forst und Jagd*'s title to *Socialist Forestry* in January 1962. "What We Plant Today We Will Harvest under Communism!" heralded the caption of its inaugural cover showing a heroic socialist forester directing the view of two young girls over a field of uniform seedlings to a mature stand of pure pine beyond, a visual metaphor for the transition from socialism to the communist future.[42]

Farmers, too, longed for the comfort of traditional forms and custom, which were not only familiar and suited to local environment but far more productive and stable than socialist farming and forestry. Ulbricht's emphasis on ideology and the stress on continuous innovation and increased mass production naturally worried farmers and foresters, as he demanded, "building socialism in the countryside in a great revolutionary upheaval."[43] No one apart from the Party leadership wanted revolutionary change—most wanted stability and enough to eat.

Independent farmers who resisted collectivization suffered continuous harassment and their children prejudice in their schooling, always behind the children of workers or collective farmers. The London *Times*' correspondent reported widespread unrest: "Farmers in the Magdeburg district have demanded that collective farms be restored to private enterprise and have claimed the right to own tractors, which are hired out by the State. There has also been an admission of brawling between farmers and Party members in agricultural districts along the Polish border," the conflict heated by the productivity of these still independent, formerly German farms across the Oder River.[44]

Farmers and peasants clung to tradition and custom—suspecting the Party's continuous revolution and innovation—and the loss of the Oder-Neiße territories still chafed. A Party propaganda pamphlet of early 1960, "Single Farmer Arnold and His Relationship to Socialism," scolded independent farmers for hanging onto "bourgeois concepts of freedom" rather than joining the collective farms' "cooperative order and democracy."[45] Persistent structures, old boundaries, and memory obsessed the Party leadership, who pressed the modernist principles of Socialist Reconstruction, "the most rational organization of production on the basis of the most advanced science and technology and the full use of the creative initiative of our workers. But that is possible in agriculture only with large socialist cooperatives, not with many small, fragmented individual farms. There one always has the old boundaries."[46] The Party leadership was determined to liquidate boundaries in people's minds as well as in the rural landscape, desiring a uniform landscape of massive collectivized farms, one forest of even-aged pine and spruce plantations, and one class of human being in the countryside, a class indistinguishable from industrial workers.

The Party launched its final collectivization drive in spring 1960, the Socialist Spring in the Countryside, completed on 14 April 1960. Cadres and "agitators" poured into the countryside to browbeat farmers, a fight the editors of *Neues Deutschland* praised as "hard class warfare" to complete land reform.[47] Reports of tractors triumphantly plowing across age-old boundaries, glowing accounts of liberated peasants finally free to tear apart old hedgerows and knock over boundary markers filled the Party press. The London *Times*' correspondent reported the final liquidation of remnant small farms and homesteads in East Germany:

> Party officials, delegates of Communist "mass organizations," members of the police and of the judiciary, are let loose on the countryside, and if a farmer has the reputation of being the best in the area—which often means that he provides the backbone for the local resistance to collectivization—scores of them concentrate on him.
>
> They start by describing the bright prospects of a Socialist future, continue with threats, and end with accusations. Loud-speaker vehicles are stationed outside his house, and for hours on end deafening appeals are made to him to hand his land over to the cooperative, and if he displays reluctance he is denounced as an enemy of the State, a criminal enemy of the new Socialist order, a saboteur of peace. During the night floodlights are directed at his windows.
>
> The "persuader teams" work on him in shifts and do not hesitate to move into his bedroom and prevent him from sleeping: and the "brain-washing" process continues until it finally leads to the required result.
>
> The Communist Press, naturally from a different angle, proudly proclaims the measures used, and how many "persuaders" had to work on farmer X from Y village until he decided to join the cooperative "voluntarily," and how this was the signal for all the other farmers to follow suit.
>
> The war of nerves does not end with the farmer's submission; he has to sign a document stating that he acted of his own free will. Communist officials may even insist on his joining in the celebrations held to mark the socialization of his village.
>
> Refugees describe many tragedies among farmers—suicides and imprisonment of those caught while trying to escape. Much land in east Germany has been worked by the same family for generations. "Rather than work as a slave on my own land, I prefer to work as a paid hand in western Germany," one refugee said. "I could not prevent them from robbing my land, but they shall not get my work in addition."

> The reason for the drive must be political. Herr Ulbricht knows that the farmers and peasants, including a good many who owe their land to the Communist land reform of 1945, form an obstinately conservative element in the population, and is determined to destroy it at the roots.[48]

Conscious of international opinion, Ulbricht always denied that the Party had forcibly collectivized independent farms. In 1965, in his first international trip as head of state, he gave Egyptian President Gamal Abdel-Nasser a rambling lecture on the Party's achievements in agriculture. Ulbricht droned: "We haven't nationalized farms in the German Democratic Republic, and I suspect we won't in the next five to ten years. Why not? We want the peasants to choose collectivization themselves. The land is the peasants' property for us. They bring their land with them into the collective farm. According to our experience, the development of collective cooperation is the best path for our country, for the steady forward development of socialism. There was real class warfare in our countryside. There were large farmers [*Großbauern*] who owned up to 99 hectares who kept and farmed their property even after the land reform. We have nothing against farmers, but created a new power economically."[49] Of course, farms under 100 hectares were not "large" by any yardstick, as the original 1945 land reform legislation recognized.

The "new power" was the collective, already dominant in the rural landscape long before 1965. Farmers fled under the Party's pressure, the flight peaking over the long Easter weekend of 2 April 1960. The exodus recalled the flood of refugees just before the Workers' Uprising in June 1953.[50] VoPos ran dragnets through Berlin's trains, arresting and taking hundreds of refugees off to prison and forced labor camps for "reeducation."[51] Almost 200,000 escaped in 1960 despite the intensified watch on them, and another 207,026 fled in the first seven months of 1961. These numbers were below the peak flow of 331,390 refugees in 1953, the year of the Workers' Uprising, but only thanks to the Stasi's increased surveillance and the East-West frontier's formidable new fortifications. The 1960–61 refugee flow, mostly through the narrow, intensely watched Berlin boundaries, was even more remarkable for the vastly greater danger compared to 1953. The bleeding of people threatened the republic's survival. A senior Soviet diplomat wryly noted that "soon there will be nobody left in East Germany except for Ulbricht and his mistress."[52]

Just as in the 1946 season following land reform, farm production crashed after the Socialist Spring in the Countryside. The spring sowing of maize stalled. By May 1960 some districts were at 2 percent of their quota, while others were at only 44 percent. Khrushchev, already irritated with Ulbricht's decision to force collectivization against his advice, suspected passive resistance to his pet project of large-area maize cultivation.[53] Ulbricht, staggering under the loss of farmers to the West after the Socialist Spring in the Countryside, even appealed to West German farmers to come work the fields of East Germany.[54] But West German farmers were no more willing to live under "socialist production relations" than were their countrymen in the East: almost no one answered Ulbricht's call.

The faltering rural economy forced the Party leadership to raise food prices, triggering still more *Republikflucht.* Milk, meat, and fats were scarcer than usual, prompting the editors of *Neues Deutschland* to demand that collective farmers work "at least up to the standards of their capitalist past." Even Party loyalists complained of serious food shortages, questioning the Party's competence to manage the economy. One fifty-year-old railroad worker, a communist beaten up by the Nazis in 1933, openly grumbled to a British reporter: "How would the English people feel if they were still rationed today and had less to eat than during the war? Pretty fed up, I imagine, and that's just how many people feel here just now. Politics for ordinary people is a question of a full stomach; if we had good food here, and plenty of choice, then they would carry whatever policies they pleased and Herr Ulbricht would be fêted by the people as even Hitler was!"[55]

Peasants did not work as hard within the collectives as they did for themselves, or even on their small, private garden allotments within the collectives, and farmers still resisted industrial methods and specialization. Authors from the Walter Ulbricht School of Political and Legal Science reported: "Peasants have not yet given up their old ways of life and thought."[56] Ulbricht anticipated that the form of the socialist collective itself would inspire peasants to harder work. But peasants failed to accept their responsibilities either to history or to deconstruct the new East German labor law—Walter Ulbricht scolded collective farm workers for taking the eight-hour day too literally.[57] Under the slogan "production efforts in preparation for the peace treaty," the Party leadership increased farm and forest work norms in 1961, one year after the Socialist Spring in the Countryside, and sent VoPos and Party officials into the countryside to round up "work-shy elements" for forced labor in the thin fields of the collective farms and forests.[58]

The state newspapers, indulging in socialist "self-criticism," lamented the deteriorating food situation and farmers' continued resistance to collectivization, reporting that thirty "rebellious" members of a collective farm had walked out of a district meeting after Party functionaries refused to break up the collective farms. Prosecutors charged one of the farmers, a Party member sitting on the village council, with "defending Social Democratic ideas" and expelled him from the Party. *Neues Deutschland* reported that hungry workers in a plant near Berlin, "incited by a West Berlin intelligence agency," wrote a letter to Ulbricht blaming the food shortages on over-hasty collectivization, asking that the responsible officials be punished. Food supply and farmers' morale both plummeted after collectivization, as they had in the aftermath of land reform.

The opening in the Iron Curtain in Berlin kept the people in a state of uncertainty as *Republikflucht* threatened the East German republic's survival—the two Germanys would have to either come together or separate more permanently. A British correspondent noted how uncertainty and constant shortages of fats, sugars, fresh vegetables, and fruits—the grayness and meanness of everyday life, what the people called the lack of the "thousand little things"—blighted East Germans' lives:

> The workers are weary of the norms and the new labour code; the farmers of the enforced collectivization which has replaced the earlier more gradual approach; the intellectuals try to isolate themselves in their studies and their research, but constantly find that they must pay tribute to communist orthodoxy; the doctors, in spite of the privileges they have been granted in the past few months, are concerned about their children's education.
>
> The zonal boundary stretching from the North Sea to the Czechoslovak frontier is now so vigilantly guarded that few manage to escape through the barbed wire entanglements, the tall watch towers, the armed patrols with police dogs, unless they live in the border areas and know intimately the lie of the land. Eighty-five per cent of all refugees now come through east Berlin—on foot or by the elevated or underground railways which link the outlying eastern suburbs with the western sectors. They avoid carrying conspicuous parcels or suitcases which might arouse the suspicion of the People's Police. Families came secretly, with a few concealed valuables.
>
> The east German Government would like nothing better than to stop this running sore of Berlin, and is constantly putting pressure on the Russians to do something about it. If there were a real Iron Curtain, without this gaping hole, dividing off the country, the population might perhaps settle down to life under communism.
>
> In the past 12 years more than 2,500,000 have abandoned their shops, their farms, and their homes; and the Federal [West German] Government fears that peoples of non-German stock, Slavs, Mongols, and even Chinese, may be brought in from the east to take their place, thus gradually altering the character of the country and perpetuating its division.[59]

Half the refugees streaming across the sector boundaries in Berlin in the early months of 1961 were under twenty-five years old, a humiliation for a regime which prided itself on its appeal to the youth. This embarrassed Ulbricht's most trusted lieutenant, Eric Honecker, the first chairman of the

Party's youth organization, the Free German Youth (FDJ), and its boss until 1955. The Wall, of course, was the permanent solution to the hemorrhaging of the communist youth elite. In the meantime, the Politburo cut the FDJ's top age from twenty-five to twenty-two in March 1961, lowering at a stroke the percentage of FDJ youth in the refugee flood.[60]

Nikita Khrushchev, alarmed at the population drain on the Soviets' front-line state (since 1949 more than three million East Germans had vanished in *Republikflucht*), decided on the Berlin Wall early in 1961 to stave off an East German collapse.[61] Working from East German sources, Douglas Selvage suggests that Khrushchev, not unwillingly, agreed to the Wall on 26 July 1961 under Ulbricht's urging.[62] Either way, Ulbricht could not have acted without Soviet approval. Erich Honecker supervised the building of the Wall, assisted by the future agriculture secretary and driver of Industrial Production Methods, Gerhard Grüneberg.

The Berlin Wall, even though it came close to the midpoint of the cold war and more than a decade after the frontier closed, came to symbolize Soviet bloc brutality, as Goronwy Rees described in 1964:

> The Wall is a kind of masterpiece of the squalid, the cruel, and the hideous, the most naked assertion one could find anywhere that life was not intended to be anything except nasty, brutish and short. It is quite incredibly ugly, being built out of a kind of porous concrete brick that is altogether greyer, darker, deader, less responsive to light and shade than any material has any right to be. It is crowned by a *cheval de frise* of rusty barbed wire, erected in such a way that it is effective, not for keeping people out, but only for keeping people in, as if they were cattle.
>
> To its ugliness, The Wall adds a maniacal ingenuity. Buildings present bare, blocked faces, so that no one may escape through them; others have been destroyed, trees cut down, areas evacuated, so as to ensure that, on the Eastern side, nothing shall obstruct a free field of fire on The Wall. Firing points, observation posts, batteries of searchlights, loudspeakers, are mounted on roof tops, in windows or embrasures, and elaborately linked with each other, so that any movement towards The Wall can immediately be observed and brought to a halt. These precautions are directed as much against the watchers as the watched, and every guard comes himself under the observation and into the field of fire of other guards.
>
> The porous grey concrete of The Wall is stained, on its Western side, by irregular white patches which mottle it like the symptoms of a disease; they seep through from the other side, which is white-washed so that any shadow of movement will show up against it. Beyond The Wall is a line of movable anti-tank obstacles; beyond them again, barriers painted red and white like customs posts. Between these barriers and The Wall lies the *Todeszone*,

> or *Vernichtungszone,* the zone of death or destruction, in which the guards have orders to liquidate immediately anyone who does not halt after a warning shot.
>
> The Wall is something new, it's smart. It gives a genuine *frisson.* ("I love that Wall," said Mr. Khrushchev.) And here also one can mount the observation platforms which allow one to look over The Wall and gaze down the long, grey, infinitely dreary vistas of deserted and decaying streets leading into East Berlin; at any moment one expects to see the spy coming in from the cold.[63]

Ulbricht wanted more from the Berlin Wall than the Soviets were willing to grant. He wanted the Soviets to demand a peace treaty ending the Second World War, to settle Berlin's status as East Germany's capital, and to reinforce the Party's legitimacy. Khrushchev wanted to stabilize the Soviet's frontline state and then to force Ulbricht to repair the farm economy crippled by the Socialist Spring in the Countryside. Khrushchev answered Ulbricht's naggings in a 26 February 1962 Moscow meeting: "On 13 August [the day of the building of the Berlin Wall], we achieved the maximum of what was possible. One must consider the situation realistically." He urged Ulbricht to focus on improving the East German economy, particularly the farm economy, instead of worrying about international politics.[64] Alexei Adzhubei, Khrushchev's son-in-law and editor-in-chief of *Izvestiya,* attacked Ulbricht at a meeting of East bloc journalists in the same month: "We got the impression that Ulbricht is unable to deal with the fundamental question: how to achieve results in agriculture—they should work on it. Phrases cannot replace potatoes, which East Germany does not have."[65]

Ulbricht had shortchanged agriculture to devote scarce resources to pursue his fetish of "overtaking West Germany" and even to compete with the Soviet Union. Alexei Kosygin, first deputy chairman of the Soviet Council of Ministers, complained: "In East Germany there are accommodations, city centers, etc., that are not planned for the Soviet Union until 1970. Comrade Khrushchev said that he is upset that little is being invested in agriculture. If a decision [has to be made], whether city centers are to be built or investments made in agriculture, then the latter. In general, agriculture is the sore point of all the people's democracies."[66] Anemic farm production, Ulbricht's stubbornness in forcing collectivization, and his meddling in international politics widened the rift between the Soviet and East German elites.

Ulbricht met with Khrushchev again in Moscow one month later, where "Comrade Khrushchev referred to the difficulties in agriculture and asked whether it is true that East Germany bought potatoes from Poland," particularly pointed because of Ulbricht's angry criticism of the Poles' failure to collectivize. "Comrade Kosygin interjected that East Germany is importing sugar and before, it was exporting it. A long conversation evolved over the development of agricultural machinery."[67] Sugar beets (*Beta vulgaris*) had been a staple of European agriculture since Napoleon introduced them as a substitute for British West Indies cane sugar. Sugar beets grew well in the northern European climate and yielded high-quality sugar. The pulp residue made nutritious animal feed, a staple of European dairy husbandry. The cake by-product, high in nitrogen, made an excellent fertilizer and soil conditioner. East Germany's failure to be self-sufficient in sugar ate up valuable foreign exchange reserves, and not only deprived the East German people of a simple luxury, sugar, but also hobbled milk and butter production. Supplies of high-quality animal feed were always short, and people felt keenly the lack of fats in their diets.

The Socialist Spring in the Countryside in April 1960 and the closing of the Berlin sector boundaries in August 1961 gave Ulbricht the structures and control he needed for his great program of innovation in the 1960s, the NÖS (the New Economic System), and his parallel reform of Marxist-Leninist ideology through cybernetics. Soviet bloc elites recognized that their ideological pull and their economic and military power were eroding. The best chance to recover the fervor and power of earlier days, maybe the last chance, was to harness the great postwar scientific and technical revolution, the Information Technology Revolution (IT Revolution) to perfect the command economy, renew Marxism-Leninism, and restore vigor and productivity to East Germany's farms and forests. The Wall bought time to prove that communists could learn from the mistakes of the past and take advantage of modern computer and administrative science, demonstrating socialism's superiority to capitalism and advancing Marxist-Leninist ideology at once. The success of reform in the 1960s would depend on the cadres' openness to change and whether a command economy could be manipulated to perform as well as liberal market economies through artificial "levers" and controls. Ultimately success—in forest and farm policy even more immediately than in the industrial economy—depended on releasing control, opening the economy to outside flows and risk, and allowing ecological and political structures to emerge from the bottom up.

# 7

## *Cybermarxism and Innovation (1961–1971)*

*To innovate is not to reform.*
*—Edmund Burke, "Letter to a Noble Lord"*

Soviet bloc leaders faced promising economic and political environments in 1961, the midpoint of the "Great Golden Age of Economic Growth" (1950–73). It was an age of productivity unmatched in history despite economists' predictions of a long Malthusian depression after the Second World War.[1] The Information Technology Revolution (the IT Revolution), together with advances in agriculture, transportation, and finance, emerged as the engine of this extraordinary growth. Soviet bloc planners also held powerful advantages after Stalin's death in 1953 and Nikita Khrushchev's reforms: central planning, superb scientific and technical schools, demonstrated prowess in space, formidable nuclear and conventional forces, and the daunting claim to embody history. The Soviets, confident after their victory in the Second World War, exuded the aroma of the church militant as they entered the 1960s.

Two separate images of Khrushchev fused in the Western mind into one image of implacable Soviet will: Khrushchev's 1956 challenge to Western diplomats—"History is on our side. We will bury you!"—and his japing at the U.N. in October 1960, waving and banging his shoe on his desk in answer to a delegate's charge that the Soviets held East Europe captive.[2] Western political and business leaders feared that the West might not be competitive, particularly as the Soviets, with Walter Ulbricht of East Germany in the vanguard, vowed to harness the IT Revolution and cybernetics to outstrip the West.[3]

The Soviet bloc's stunning technological achievements—Yuri Gagarin's April 1961 space flight and the testing of the world's most destructive nuclear bomb six months later—seemed to confirm their leaders' choice of technology as the catalyst for renewal.[4] As Khrushchev said in 1962, "The Soviet nation will make itself even more communistic by feeding on more and yet more education and technology."[5] Cybernetics would propel the Soviet bloc economies forward to overtake the West and renew waning Marxist-Leninist ideology. This would secure the Party's future, win victory over capitalism, and move humankind closer to communism.

Digital computers and systems analysis seemed tailor-made for Marxism's strain of scientism—in Jacques Barzun's definition: "the fallacy of believing that the method of science must be used on all forms of experience and, given time, will settle every issue."[6] Intellectuals and managerial elites throughout the Soviet bloc looked to cybernetic theory to make their economies work better and to revive Marxist-Leninist ideology. Cybernetics formed the cornerstone of Ulbricht's reform program of the 1960s, the New Economic System (NÖS), and of the Party elites' image of the natural landscape as a hybrid of machine and organism whose development, like that of society, was determined by inexorable law and human reason.

Soviet bloc elites had closely followed the extraordinary contribution systems analysis and computer technology had made to Western growth and productivity. Robert McNamara brought to the Pentagon as secretary of defense in 1961 his Whiz Kids and sophisticated planning and budgeting systems from the Harvard Business School, the U.S. Air Force, and the Ford Motor Company. He turned his prodigious energy and unparalleled experience toward reforming the military and planning the war in Indochina. McNamara's skills in management science may not have been more advanced than those of his counterparts in Gosplan, the Soviet State Planning Committee, or in the East German Council of Ministers, but McNamara did have the benefit of U.S. business managers' creativity and innovation and their appetite for risk—advantages outside his control but critical to U.S. power. The power of U.S. business was revealed in the IT Revolution's signature firm, International Business Machines (IBM) and its iconic Model 360 mainframe computer and operating system.

Soviet bloc elites looked enviously on IBM's Model 360 as a metaphor for an ideal socialist economy. The Model 360 was as much an advance in how managers thought of their tasks as it was a technological, manufacturing, and marketing triumph. Risking five billion dollars (at a time when its annual revenues hovered around two billion dollars), IBM "bet the firm" in a widely publicized campaign, fielding the Model 360 in 1964 in a brilliant feat of innovation.

The Model 360 was pure cybernetics. Its operating system, OS/360, lay at the core of the innovation, launching the precedence of programming over hardware. The OS/360 operating system centralized tasks through a revolutionary "shielded supervisory system" which reserved core operations and decision making to the supervisor program, a rough cybernetic variant of Lenin's "democratic socialism." The Model 360 and OS/360 catalyzed the U.S. economy, and as a unit they were the single most successful innovation ever.

Khrushchev called for a pause to review "the entire planning system" in 1962, citing the lagging performance of Soviet bloc economies vis-à-vis the West.[7] The publication of Yevesy Liberman's 9 September 1962 *Pravda* article "Plan, Profit, Premium" had already given the Soviet Communist Party's imprimatur to cybernetics, the central component of Ulbricht's economic reform plans of the 1960s. *Time* magazine featured Liberman's portrait on its 12 February 1965 cover, a mark of the global fascination with the scientific-technological revolution and hopes for Soviet reform. Liberman, however, was not the intellectual force behind Soviet reforms. Senior Soviet economists led by the academician Vasilii Sergeevich Nemchinov put Liberman forward as their public face. Erich Apel, chief of the East German planning apparat, adapted Nemchinov's work for the NÖS programs of economic reform and the restructuring of the farm and forest economies.

Decentralization, "levers to simulate markets," and cybernetics characterized Soviet and East German reforms of the mid-1960s. Nemchinov's program was consciously cybernetic: to "divide problems, hence decisions, into big ones and small ones. The big ones will be determined by the top of the hierarchy, as before. The small ones will be decided by the lower rungs, possibly by the lowest—the enterprise. But the sum of all decisions, has, in some way, to equal the big ones," or perhaps to equal more than the sum of the parts.[8] Cybernetics and early variants of chaos theory captivated the Ulbricht clique, offering a model of the East German economy as a cybernetic system capable of self-organization.[9] The reformed, cybernetic Plan would operate much like IBM's OS/360, delegating most decisions to the lowest level possible while reserving important decisions to the center and harmonizing all activity.

But the IT Revolution, already in full flower of creative destruction in the late 1950s, posed lethal challenges to the foundations of Marxist-Leninist ideology: to the iconic importance of the worker, to class struggle, and to Marx's "economic science." Marxist-Leninists might refuse to see risk takers' and investors' worth, but this much was painful and manifest: managers and technical experts, not the worker, now stood at the center of economic activity, critical human assets in waging the cold war. What were the guardians of Marxism-Leninism to do when modernity overtook and exposed its central myths: the worker and class struggle? One could change the myth, advancing a new class of professionals to the front pew, or keep the myth and accommodate reality on the margins. The East German Party leadership chose both. In forestry and farming this meant a brief return to traditional forest science and conservation as technical specialists for a time were moved ahead of the cadres, while harvests briefly fell to sustained yield levels.

Cold war competition made it critical to value correctly the relative worth of worker to technocrat. The Golden Age had bestowed unequal blessings on the West and increased its relative power despite perceptions of raw Soviet muscle and will. The growing economic divide in productivity between East and West finally focused the attention of Soviet bloc elites on modernizing Marxist-Leninist *Praxis*. Soviet economists and intellectuals had been questioning the command economy since the 1950s as the West's economic growth exploded and the Soviet growth lagged. After all, the Soviet economy in the previous fifty years had evolved "from one run by Genghis Khan with a telegraph to one run by Kafka with an abacus."[10]

Even though the Soviet bloc economies were declining relative to the West, life had improved in the 1950s. Yet greater prosperity and peace perversely weakened the Party's power, which still depended on coercion. Fading memories of the Terror and the Second World War reinforced the sense that Marxism-Leninism was no longer relevant, particularly to the rising scientific and technical elites. Absent change, the Party leadership feared "a gigantic rationing and apportioning system" might emerge in an ideological vacuum and stall communism's forward impetus.[11] How were they to keep even themselves, the Leninist "conscious and organized vanguard of the proletariat," politically engaged when the Plan had finally conquered the worst shortages? More than in any other Soviet bloc country, this erosion of ideology threatened the East German elites' "basis for societal *solidarity* and for the *legitimacy* of their rule" (emphasis in original).[12] To meet this challenge to their power, Khrushchev launched "a great campaign against bourgeois ideology" throughout the Soviet bloc, deploying the IT Revolution as a weapon against relative decline and as a new source of legitimacy.

Then, from the opposite end of the world, a purer and more rigorous vision of communism emerged from Mao's China in the 1960s and directly challenged the Soviet bloc elites' ideological bona fides. These factors—increasing personal economic security, lower levels of coercion, economic challenges from the West, and the threat from China—combined to gnaw at Soviet bloc leaders' confidence in their ideology and their security. A new revolution, this one grounded in information rather than class, might rescue them. So Ulbricht turned to cybernetics to ground his reforms of the 1960s.

Erich Apel, head of the State Planning Commission, designed the new cybernetic reform program, the NÖS. An odd choice, Apel was responsible for the failed First Seven-Year Plan (1958–65), which collapsed early in 1961 under the weight of central control. Apel, however, was a scientistic Marxist who favored East German economic independence from the Soviet Union, as did his mentor Anton Ackermann, the Party's postwar chief ideologist who advocated a "special German road to socialism."[13]

Khrushchev saw the Soviet bloc's weak, chronically lagging farm production as his Achilles' heel, a troubling weakness given Soviet prowess in science and in conventional and strategic forces. Farm and forest reform were most critical in East Germany, where land reform and forced collectivization had turned this once prosperous producer of regular surpluses into a land of constant rationing and dearth. Poland, which had not forced collectivization, was free of the shortages and sporadic food riots which plagued East Germany.[14] Agriculture fascinated Khrushchev, and East German farm policy often dominated bilateral talks.

Ulbricht launched the NÖS at the Sixth Party Congress (15–21 January 1963), where Erich Apel and his deputy Günther Mittag were elected to the Politburo.[15] Apel kept key elements from Socialist Reconstruction, the harshly dirigist program of the First Seven-Year Plan, such as the goal of overtaking the West German economy and the emphasis on industrial methods in agriculture and forestry.[16] Ominously, the Plan goals did not change, only the rules.

Apel's design for the NÖS centered on three basic reforms: decentralization, price and interest rate reform, and improved accounting and control systems. All three of these reforms deeply affected the farm and forest landscapes, and their success or failure would be immediately obvious. Decentralization (in line with Soviet reform theory) formally was the principal focus of the NÖS. But first, Apel had to break the local cadres in favor of the scientific and technical elites. So he inserted "new centers of power" grouped by industry, the Associations of the People's Own Enterprises (VVB), between Berlin and businesses, cutting the cadres' power. East Germany's ninety-four State Forest Districts (which previously reported to county-level Party bureaucrats in the fifteen counties) now reported directly to one of five new centralized "VVB Forest Management" bureaus. These new VVB Forest Management bureaus gave district foresters "scientific guidance" and stood above the cadres' control.[17] Forest district boundaries changed yet again to fit ecological boundaries, erasing the political boundaries imposed in 1952 with the creation of the State Forest Districts and further diminishing the cadres' power. Finally, Apel stripped the county councils of oversight of cooperative and private forests. Apel's NÖS decentralization, far from devolving control, gathered forest management more tightly under the State Planning Commission to cut the power of cadres and advance the scientific and technical experts critical to improving economic performance.[18]

Many foresters took heart from Ulbricht's rhetoric of decentralization and cybernetics, itself easily seen as a metaphor for ecosystem ecology with its emphasis on feedback loops, interactions of population and resource flows, and boundaries. Foresters themselves clearly stood to gain greater influence and power from the NÖS. Foresters also welcomed NÖS's reversal of land reform. National Socialist forest management had proved that unified forest management was a waypoint to a close-to-nature forest, and Apel, just as Reichsforstmeister Hermann Göring did, restored ecological forest boundaries. Most foresters remained committed to traditional forest management and more narrowly to *Dauerwald* despite purge and harassment, and the Party never abandoned ecological rhetoric.[19] So the NÖS bode well at first.

Egon Wagenknecht, a leading silviculturist of the 1950s and 1960s, counseled foresters disillusioned by looming Industrial Production Management after the Berlin Wall: "Many of you will be disappointed that I calmly make concessions to technology and accept the Plan's demand for comprehensive mechanization and thorough rationalization. So it often is with silvicultural theories touched more with religious fervor and fantasy than calm reflection. But what use have we for silvicultural theories, even though they may be quite beautiful, when in practice they are never to be realized? It is only by mechanization and rationalization that *the truly indispensable silvicultural responsibilities may be secured*" (emphasis in original).[20] Wagenknecht urged disillusioned foresters to find hope in cybernetics.

Wagenknecht's "quite beautiful" silvicultural theory was *Dauerwald*, a radical ecological movement whose irrational appeal matched that of cybernetics. It was no accident (as Marxist-Leninists sometimes observed) that Ulbricht embraced both *Dauerwald* and cybernetics: both promised instant success through a magic bullet and release from biological and economic constraints. Wagenknecht urged foresters to recognize *Dauerwald*'s defeat after the catastrophic Menz Resolution and accept clear-cutting as a way station to a close-to-nature forest—one the rational and scientific NÖS would make possible. Johannes Blanckmeister, the founder of *Dauerwald*/Optimal Stocking Forestry and East Germany's preeminent silviculturist after Alfred Dengler's death, was even more optimistic one year later, writing: "The true principles of *Dauerwald*/Optimal Stocking Forestry will be better implemented and the Menz Resolution's plan for a close-to-nature forest will be more successful under Industrial Production Methods. There can be no question of abandoning our beliefs."[21] Yet within half a year the Party crushed forest science and ecology at Eberswalde and again purged its scientists and humiliated its faculty. Blanckmeister could not, perhaps, have expected that the Party would so completely betray its promise, but he was correct that most foresters, particularly those educated before 1955, had not surrendered their belief in close-to-nature forestry and *Dauerwald.* The debate within the forestry profession between close-to-nature and industrial forest management, between custom and innovation, went underground.

As the NÖS unfolded, foresters soon saw that central control, and thus improved "Plan discipline," was the real goal. Rudolf Rüthnick, the future chief of the East German Forest Service, announced in 1964 that the five new VVB Forest Management bureaus would provide "economic leadership" for the State Forest Districts. Agriculture ministry bureaucrats at the same time praised the VVB Forest Management bureaus as "organs for centralized management of forestry, hunting, and environmental protection." A senior Party bureaucrat defined them as "economic leadership organs constructed on unified growth districts," linking the forest districts' new ecological boundaries with maximum production.[22] "The main concern of forestry," the chief of the State Forest Service wrote in 1964 (at the height of NÖS reform and cybernetics), "is to provide maximum raw wood production. The current level of production [temporarily lowered to rebuild stocking] neither satisfies nor corresponds to the needs and potential of the People's Economy."[23] As the director of East German forest management admitted in 1990, "It was impossible to create a sustainable forest structure under these conditions."[24]

But a new, harsher form of industrial forestry loomed at the same time the Party released the admirable 1962 Silvicultural Guidelines (which called for a close-to-nature forest): Industrial Production Methods, introduced at the Party's 1962 Central Committee meeting.[25] The Party wanted to be scientific, but not if it conflicted with maximum production and control. Similarly, the 1966 Silvicultural Guidelines granted district foresters greater freedom to manage their forests according to local site conditions. But the Plan quotas still mandated one management strategy: clear-cutting and planting.[26] The guidelines' formal freedom meant only that "where and how a district forester came up with the wood was up to him." Foresters called this *Narrenfreiheit,* fools' freedom.[27] Foresters had seen comparable conflict between the formal message of conservation and maximum production in 1951–54 when the Party leadership formally embraced radical ecological (*Dauerwald*) management while accelerating clear-cutting.

"Economic" factors, not "political" ones, governed forest management according to Mittag, Apel's deputy and the future economic czar. Yet the five forestry VVB bureaus were political tools of the Council of Ministers—"fulcrums of the proletariat in the countryside." The Party also inserted Central Committee Party organizations and Party unions in each VVB Forest Management bureau to enforce the "Principle of Production," strict compliance with Plan production quotas, and to police orthodoxy.[28] Control was in fact far more concentrated by 1965 than before; all that changed in practice was that the local cadres lost power to the center. The State Forest Districts won no new independence or authority, although the burden of paperwork soared with the NÖS's complicated reporting rules.[29]

Price and interest rate reform, the second of the three NÖS reform programs, meant higher prices and interest rates to correct the inefficiencies of the gross (*Brutto*) accounting system and stop the prodigious waste of raw materials.[30] Marx's sacrosanct "Labor Theory of Value" forbade valuing natural resources before labor had been applied. Apel finessed this by reaching into cybernetics to simulate, rather than fix, prices and allocate resources. Increased prices, interest rates, and taxes would work as "economic levers," a term drawn directly from cybernetics, and enforce Plan quotas. The new prices still reflected government priorities rather than real costs; price reform masqueraded as a "different system of indirect controls."[31] "East German prices," as a senior West German economist wryly noted, "are admittedly nonsensical."[32]

Price increases also targeted ruinous state subsidies; State Forest District directors now had to rely on increased cash flows to cover costs, investment, and losses.[33] As direct subsidies ended with higher prices, foresters focused on speeding up cash flow and meeting short-term Plan goals.[34] Central control and price fixing remained undiluted, reflecting what one scholar called the NÖS's purpose: "not to junk central planning but to make it more effective."[35]

Modern financial and managerial accounting were the third NÖS reform—and should have been the most important. Ulbricht ridiculed accounting as a "trivial departmental task of businessmen's staffs" at the final NÖS Economic Conference in June 1963. He huffed that accounting lacked analytic power, was "insufficiently connected with collective and material interests and not grounded in science," the Law of the Economy of Time which commanded: "always meeting production quotas better and achieving higher yields with lower expense."[36] No reporting system could contradict the Law of the Economy of

Time; accounting was its "servant and subordinate." In the forest economy, accounting could not be used for evaluating overall productivity, only for quality control of raw wood production. Plan bureaucrats evaluated foresters' performance from their bureaus in Potsdam and Berlin without the benefit of objective data from the field. Political reliability became ever more the principal criterion by which managers were rewarded or punished.[37]

Still, few could question that the existing reporting and information systems were corrupt and illegible. So Apel determined, despite Ulbricht's scorn, to modernize accounting. Apel hoped that reformed accounting, together with new raw material prices in line with world markets, would partly resolve the management nightmare and waste of socialist bookkeeping. He staged an economy-wide revaluation of capital assets so managers could charge depreciation against revenues for the first time.[38] Getting indirect, noncash charges onto profit and loss statements was essential to monitoring performance. But Marxist "economic science" foiled his efforts.

Accounting reform threatened the Party, as shown by Ulbricht's intuitive fear of information. Transparency would have drained power from the center and limited the Party's control. Most of all, modern managerial and cost accounting directly challenged a shibboleth of Marxist-Leninist economics, Marx's Labor Theory of Value, which held that "value is determined through the creation of useful value by socially necessary labor." According to Marx, "the earth is not the product of labor and therefore has no value." Marxist-Leninist economic theory permitted no mechanism to price natural resources and release costs to the income statement. Charging the cost of natural resources was considered a form of added value and thus exploitation.[39] It was impossible to quantify the direct cost of raw materials, much less such indirect costs as pollution damage.[40] Natural resources were treated as free goods to the nationalized industries, with all the waste that implies, and had value only after sale and payment.[41]

This made it almost impossible for foresters to invest in capital assets, such as stocking, or to incur costs to build structure and stability. Even had there been no Industrial Production Methods, East German accounting systems would still have made a close-to-nature forest unattainable. The ideological blindfold of Marx's Labor Theory of Value left NÖS reformers in the position of early Christian clerics trying to justify money lending despite scripture's prohibition of usury.[42] Gerhard Schröder, a respected senior forest economist who defended close-to-nature forestry in the 1950s, acknowledged the failure of the accounting reforms in 1966, observing that even for raw wood production "the problem of profit had been neither theoretically nor practically solved."[43] Without accurate prices or reliable reporting no one ever knew what they were doing.

The Party leadership did collect data obsessively—only the loss of control over data interpretation bothered them—to "serve the demands and goals of the Plan economy."[44] The central Forest Management Department in Potsdam kept forest inventories at levels of detail far surpassing Western inventories, and forest ecologists mapped every square meter of East Germany's territory.[45] Raw inventory data were value-free and thus posed little threat to primitive Marxist-Leninist control systems. Foresters suspected the inventories were "a secret weapon of overseers in Berlin to control and maintain output," a waste of an asset of great potential.[46] Forest ecologists could use inventory data only to plan harvests, never to understand the causes of forest decline. Similarly, East German statisticians collected the world's best cancer data but used them to forecast future demand for hospital beds, not for oncologists' research. Searching for the causes of cancer, or of forest decline, was forbidden.[47]

Forest managers came to resist the NÖS despite their initial hopes, perhaps as much as the Marxist-Leninist cadres whom it disempowered. The reforms were too complicated and riven with conflicting demands. Twenty years of the Party's erratic and arbitrary economic policy had raised members' self-interest to the highest level, and managers throughout the Soviet bloc played the NÖS reforms to their own advantage. As Goronwy Rees noted, "The East Germans manage to combine a Teutonic passion for bureaucracy with a Russian capacity for infinite delay."[48] Apel, perhaps naively, wondered at managers' passivity and cynicism: "many managers actually believe, obviously, that science exists solely to justify weak plans."[49] Battered by the Party's erratic economic policies, irrational Marxist-Leninist "economic science," and years of coercion and threat, managers naturally were skeptical about the grand scope of Ulbricht's NÖS.

Gross production targets so drove managers that they christened the Party leadership's fetish for maximum production the *Tonnenideologie* (tonnage ideology). Gross output was, after all, the only meaningful measure the Party's late-Ptolemaic system of price fixing and resource balancing could take. Party cadres drove district foresters to meet their Plan quotas, stressing managers' personal responsibility. Naturally, foresters found strategies and manipulated the control systems to meet quotas. Conservationist foresters were accused of sandbagging, of holding onto scarce timber resources to play later in the Plan cycle. Given the central planners' paranoia and refusal to accept ecological and biological constraints, conservation smacked of bureaucratic fraud or, more ominously, of "sabotage." Yet sandbagging is a normal bureaucratic response to control familiar to Western executives. As a leading scholar of Soviet bloc economies observed: "The prime criterion was always the physical target. Managers tend to hide their capabilities today, to amass reserves of inputs and capacity which will help them in the future."[50]

The cadres in turn recognized the NÖS as a program which both devalued Marxist-Leninist currency and threatened their careers. Frequent characterizations of the NÖS as a "genuine revolution" unnerved the cadres. Revolution in the third world or in the "capitalist camp" was one thing, but hadn't the "socialist camp" already had their revolution? Revolution in East Germany had serious, "ambiguous implications for the structure of political rule."[51]

Still, the cadres' opposition was not based entirely on ideological grounds or on their loss of power; the NÖS was also an administrative nightmare. As Apel warned Ulbricht just as the NÖS got under way, "planning methods from top to bottom have become extremely complicated."[52] The NÖS's new, highly detailed rules turned routine tasks into convoluted, unfamiliar chores. This challenged the talents of the poorly educated Party bureaucrats. The NÖS's complexity raised the relative importance of the scientific and technical elites who alone could interpret its rules and regulations. Soviet bureaucrats had similar complaints about Khrushchev's reforms: Leonid Brezhnev complained in 1963 of the "numerous, now and then far-fetched, reorganizations [which] gave rise to an atmosphere of nervousness and vanity, deprived leaders of the [necessary] perspective, and undermined their faith in their own abilities."[53]

Party members resented the new power and precedence given to the rising scientific and technical elites—privileged elites who criticized ideology and the cadres' performance.[54] These elites saw themselves as new Prometheans bearing advanced science endowed with an authority comparable to Marx's.[55] Orthodox Marxist-Leninists argued that the demotion of the worker contravened Marxist thought. Ulbricht countered that the new professional class was "at least the ideological coequal of the working class."[56] But the scientific-technical elites threatened the establishment; they formed an "institutionalized counterelite" within the state that challenged not only the cadres but Marxist-Leninist ideology itself, the Party's sole source of legitimacy.[57] This in turn defeated a core purpose of the NÖS: to harness cybernetics and the IT Revolution and weld modern science onto Marxism-Leninism's nineteenth-century ironwork. But there was no elasticity in Marxism-Leninism and Ulbricht's innovation, with its "unintentional consequences for the pattern of authority which it is expected to legitimate," ended up challenging rather than legitimizing the Party's sole role.[58]

The ideological struggle did little to stop the continued destruction in the countryside. Whether Plan quotas were handed down from Ulbricht's scientistic wing or from Honecker's religious wing of the Party was irrelevant. The primacy of ideology, whether of class or of information, and central planning drove forest decline. True reform did not mean better science; it meant releasing control and freeing individuals to measure and accept risk.

The Party leadership could not accept that the origin of forest decline was either in industrial forest management or a function of biological constraints. Forest decline had to be political in origin, the result of the "regular flouting and destruction of sustained yield under capitalist relations."[59] Forest decline was impossible under socialism because of "the objective conditions of constant agreement between economic and natural growth and the healthy qualities of the workers' and peasants' power. Socialist ownership of the means of production ensures that the forest serves only the interests and needs of workers, no longer the large estate owners and the capitalist state. The planned development of our forests follows in a great leap forward to an inevitable flowering such as has never before been seen."[60] Marx and Engels explained how socialist administration and the People's ownership guaranteed conservation and productivity equally: "*Capitalist* society is only interested in nature for profit. *Socialist* society places human well-being at the center of all policy and therefore nature protection is unconditionally guaranteed" (emphasis in original).[61]

Industrial Production Methods, the highest expression of Marxist-Leninist farm and forest policy, by itself also guaranteed perfect satisfaction of economic and ecological goals.[62] Socialist administrative structures and state ownership created forest health, not the work of experienced forest ecologists and silviculturists. This so-called *Kielwassertheorie* ("ship's wake theory") held that maximum production automatically optimized ecological and social benefits since it "served both the People's economy and the People's society because *all* sectors of socialist society depend decisively upon a growth and results oriented economy" (emphasis in original).[63] A leading Party theorist declared in 1987, "Social and production relations and the economic interests growing out of them determine a society's relation to nature." Constantly increasing production was the sole measure of ecological health; not only were forest science and ecology useless, they were heretical.[64]

An internal Party report went so far as to blame close-to-nature forestry for forest decline, attacking "forest management which focused on site quality and the biological and silvicultural considerations that dominated planning. Even now the consequences are not fully overcome. 'Forestry-only' thinking and many foresters' actions still reflect this today."[65] Foresters who wanted to restore a close-to-nature, diverse landscape now faced anathema as enemies of the people.

Figure 6. Political and administrative map of East Germany, 1970s–1980s.

Forest scientists could not criticize either industrial forestry or pollution, both defining features of Marxist-Leninist economies. The Party leadership directed research to reactive strategies, particularly ineffective fertilization to counteract the massive pollution loadings and development of new pollution-tolerant strains of trees. They ignored the prime mover in forest decline: the artificial nature of the industrial forest itself, not external pollution.

Adjustment of the natural landscape to industrial pollution had been a constant feature of the Party's response to decline, as recommended in 1956: "The future forest of the central German industrial region will only be able to maintain itself when it is predominantly composed of smoke resistant species [hardwoods]."[66] The East German statistical journal noted in 1958 that "forestry research concentrates on the effects of industrial pollution."[67] But as pollution began more and more to darken people's lives and poison the air, popular attitudes toward pollution changed. The editors of Halle's *Liberal-Democratic Newspaper* singled out the industrial sources of pollution in the early 1960s in an article titled "What's the Matter with Our Air?" Resistance to the Party's industrial policies and pollution centered in the heavily industrialized regions of the Southern Uplands where public health damage was most severe.[68]

As pollution damage and popular protest increased in the 1960s, the Party leadership again cracked down on the Forest Service. In the mid-1960s political officers reprimanded Erich Kohlsdorf, a Party member and respected district forester in Saxony, for "damaging friendship with Czechoslovakia" by recording acid rain damage in the Erzgebirge Mountains near the Czech border. They instructed him: "You may not say what is not allowed!" When Kohlsdorf applied for permission to warn hunters and foresters not to eat red deer (*Cervus elaphus*) liver, a prized delicacy that carried dangerous levels of heavy metals, Kohlsdorf was again slapped down: "Are you for peace?" the Party's political officer demanded, "Then be silent!"[69]

The forest also became a cold war battlefield. Rüthnick, the future chief of the Forest Service, set forestry's principal task as "above all the achievement of the highest national defense capability and support of the West German people's resistance against Bonn's policy of forward deployment and revanchism, military dictatorship and nuclear armaments."[70] Cold war politics increasingly set foresters' tasks: to support Soviet bloc competition with the West and to serve as the Party's "fulcrum of the proletariat in the countryside."[71]

The Party moved rapidly to complete the subjugation of the "bourgeois" Forest Service begun in previous purges and to impose Marxist-Leninist norms through Industrial Production Methods. Industrial Production Methods, the perfection of socialist forestry, was launched in 1964 at the Eighth

Peasants' Congress at the height of the NÖS.[72] Everyone could see that the forest would collapse as ecological and economic entities within several generations if harvest volumes continued to increase 5 percent each year. Yet a senior forest policy scholar was firmly silenced when he warned in 1960 that Industrial Production Methods would "lead to the destruction of the forest. One cannot take more from the forest than the annual increment." A Central Committee representative ended the brief debate, writing: "No one will dispute this. But it is quite inconceivable that forestry's fundamental task—the resolution of political and economic problems—will ever take second place to sustained yield. [The principal determinant of forest policy is] the strategic and tactical policies of the Communist and Workers' Party."[73] Ulbricht moderated Industrial Production Methods only later in the decade when peasants demonstrated against collective farm policy, food shortages returned, and popular anger over forest death surged. By then Ulbricht's political power was exhausted and he could save neither the farm and forest economy nor himself.[74]

The climate that tolerated limited debate disappeared by the mid-1960s. Rüthnick's opening address at the 1966 Potsdam forest conference marked a significant change in tone from his combative stance at the 1964 conference. Rüthnick did not need to stress increasing production after the 1962–63 purge and show trials, only to demand that foresters now conform ideologically.[75] Rüthnick asserted that Industrial Production Methods had resolved forestry's problems despite continuing shortages and missed quotas. The Party leadership, personally, would define good forest management. Ulbricht, "our most highly esteemed chairman," glowed the chief of the Forest Service H. Heidrich in 1967, "has given forestry its fundamental plan of management. The main lesson to be taken is for all active in forestry to develop to the maximum the universal strengthening of our socialist fatherland."[76] Politics, not cybernetics, economics, or ecology, set forest policy.

Marxism-Leninism increasingly formed the core of forestry education after 1961. By 1966 political officers administered comprehensive political examinations at each level of a forester's career.[77] Heidrich located "the heart of all management competence in constant political-ideological work with men. The leader is first and foremost an engineer of men's hearts. In his behavior we must always see that he hates monopolists who, for example, most horribly murder innocent women and children in Vietnam. When our workers recognize this attitude as honorable it builds the leader's authority and respect." In a later article he set forestry's task as "to educate every forestry employee to the love and public acknowledgment of our socialist fatherland. Every woodsworker, every forester, must realize that East Germany alone has drawn the correct lessons from history. Everyone must be convinced of the certainty of socialism's victory throughout all of Germany and deeply hate the West German government and its revanchist politics."[78]

The Party did reduce harvests temporarily in a failed attempt to stabilize stocking by 1985. The 1965 harvest, the lowest in East German history, approached growth for the first time since the late 1930s. But it was a one-year phenomenon and harvests well in excess of growth resumed after 1965 despite increased wood and pulp imports from the Soviet Union. As Apel had forewarned, integration in the Soviet economy was far from an unalloyed blessing. In return for what Schröder termed the "exploitation of the virgin forest of the taiga," East Germany had to pay in barter goods, in hard currency for freight and transfer costs, and invest in Siberian pulp factories.[79] East Germany also doubled its investments in domestic factories to process Soviet pulp, institutionalizing the forest's poor age class distribution through the new plants' dependence on smaller-dimension wood.[80] Once the factories were built it was more difficult to argue that the harvest of low-value wood and pulp, "free" to the socialist economy, should be reduced for returns twenty-five to thirty years out. From a business point of view, the relationship was one-sided, as Apel and Heidrich feared. From the point of view of forest ecology, imports from the Soviet Union made no difference in the Party's exploitative harvest strategy.

Nor was the NÖS successful in completing the afforestation of Soviet occupation era clear-cuts, despite the official completion date of 1959/60. Heidrich, chief of the Forest Service, was still scolding foresters in 1964 for their "dishonest and inflated afforestation reports. Does not such a way of working show a lack of responsibility and dishonesty toward our Workers' and Peasants' State? Before you argue with me, I must put a question to you: 'Who among you does not feel they can meet the Council of Ministers' orders?' Anyone who doesn't ought to stand up now and explain why not. Now is the time we can still discuss this. The matter is of such great importance to me that we will take extra Plan measures." Five State Forest District foresters rose immediately to pledge their "restless commitment to eliminate unproductive area by 1965." [81] Yet despite Heidrich's threats the forest only reached normal levels of unplanted forestland at the end of the decade.

Forest management in the 1960s revealed that the NÖS was neither an attempted "reconciliation of planning and market" nor "the last serious attempt to introduce the market," which some scholars claim.[82] Despite its theory and design, the NÖS served the absolute imperative of control. By the time the Party bureaucracy implemented Apel's system of levers, balancing, and feedback, the NÖS emerged as a more complicated iteration of the command economy, a reprise of Socialist Reconstruction through hidden price, interest rate, and tax hikes. Apel's NÖS did release some decision making authority to managers. But the Party bureaucracy held on firmly to its control over raw materials and components and fortified the tightly centralized supply system at the heart of the Stalinist economy. The Party leadership never intended to surrender this control but sought only to simulate an open supply system by the complicated balancing of resource flows made possible, in theory, by computers.[83]

The NÖS's increased emphasis on central control also reflected the Party leadership's extreme risk aversion. In an unintentional coincidence of contradictions, they wanted total control and predictability as well as the efficiency and political benefits of a liberal policy. Liberalism and decentralization were window dressing, as was apparent in the emergence early in the 1960s of the Party's emphasis on Lenin's Democratic Socialism, "at the summit of all methods" and defined in *Socialist Forestry* as a management system where "all activity flows out from the center to a hierarchical ranking of the people below. The people vote responsibly and accountably to each power center while the masses show their widespread acceptance of state leadership and the independence of local organs."[84] The end result of Democratic Socialism, as a senior district forester and Party member wryly noted, was that "everything was foreordained."[85]

The NÖS was doomed from the start: it disenfranchised the very cadres who were to implement it—so not unnaturally they thoroughly subverted it—and never released central control or opened information and resource flows, steps which Nemchinov and other Soviet economists knew were essential since they saw the West pull ahead in the 1950s.

The Party bureaucracy and regional bosses systematically sabotaged reform in the Soviet Union and East Germany to protect their power and privileges and to safeguard the sole source of their legitimacy: the primacy of Marxist-Leninist faith. A leading scholar on the Soviet economy noted: "To the degree that it was administratively successful, the 1965 reform harmed the economy. But for the most part, the reform was not administratively successful, i.e., many of the reforms measures have scarcely been tried."[86] Innovation and change stalled from the start in the Soviet Union as well as in East Germany.

The Party leadership formally abandoned the NÖS in favor of the "Economic System of Socialism" (ÖSS) in December 1965 after Apel's dramatic suicide on the eve of the Eleventh Plenum of the Central Committee. Apel was depressed, probably for many complex reasons, but many months of defending cybernetics against orthodox Politburo members had worn him down. The collapse of the NÖS represented the end of his dreams for a more flexible and innovative East German economy. The new ÖSS crushed his hopes for the "special path for East German socialism" he shared with Anton Ackermann, and he dreaded the ÖSS's tighter coordination with the Soviet economy.

Günter Mittag, Apel's old deputy, took over Apel's post at the head of the East German economy following his suicide, adopting the orthodox line and rejecting cybernetics. The Party launched an even more openly dirigist economic policy in 1967, a "Developed System of Socialism" that stressed the Party's sole role, imposed a stricter cultural policy, and increased censorship.[87] The new penal code of 1 July 1968 raised the punishment for treason and espionage, even for sharing unclassified information such as forest inventories, to "death by shooting." Party judges sentenced in absentia West German judges and citizens who defended West Germany's claim to be the sole legal authority for all Germany to five to ten years at forced labor. Apel may have looked forward to the future and despaired.

The economy's persistent problems worsened as ideology increasingly conflicted with economic reality. After the Soviets invaded Czechoslovakia in 1968, the Party leadership grew more repressive and withdrawn, finally surrendering its longstanding goal of reunification with West Germany. The second "socialist" constitution of 6 April 1968 called for the "comprehensive building of socialism," completing the transformation of East Germany as a polity discrete from West Germany and reasserting the Party's sole role in a socialist, Marxist-Leninist state.[88] Environmental protection was proclaimed as a core national goal, but this was no better observed than other articles which conflicted with the Party's political and ideological agenda.[89] Later, the "Law on the Plan for Socialism's Improvement of the Environment in East Germany" required that environmental solutions be integrated into all sectors of the economy.[90] But these solutions were solely reactive and the law itself legalized pollution. Pollution damage to forests began to take a more serious and noticeable toll on higher elevation forests, on people's health, and on their sense of well-being as the natural world around them seemed to be dying.[91] Even if pollution was not the principal cause of forest decline (the artificial nature of the industrial forest was), pollution was an outward and visible sign of the Party's extreme materialism and the leadership's fetish for maximum industrial production coupled with minimum investment.

Ulbricht's power had eroded steadily throughout the 1960s. He had survived many challenges to his power, the most serious one coming after the Workers' Uprising in June 1953. Ulbricht's survival for two decades seems almost miraculous given his failures and incompetence. He certainly wasn't a leader anyone (except for Stalin) would have chosen. He was supremely self-confident yet not particularly intelligent, fascinated by technology but without personal qualifications. He nurtured his personality cult but was rude to his colleagues and without charm—a stubborn autodidact and a boor.[92] When the Communist Party of the Soviet Union condemned personality cults at its 1961 Congress, the East German Central Committee declared that they had nothing to learn from the call for "a general discussion of mistakes" and affirmed the Politburo's unqualified allegiance to Ulbricht. Socialism's enemies had resorted to "an old trick" by saying that the people's "love and respect for our leading comrade was a personality cult."[93]

Shortly before Ulbricht's ouster in June 1971 Brezhnev complained of Ulbricht's boorishness six years earlier when Brezhnev visited his villa to report on Khrushchev's ouster. Brezhnev, still indignant six years later, fumed to Honecker: "You know, back then in 1964 at Ulbricht's dacha—he sent my delegation off to the side, pushed me into a small room, and started telling me how everything is wrong in the Soviet Union and everything exemplary in the GDR [East Germany]. It was hot. I was sweating. He didn't care. I noticed only that he wanted to tell me how we Soviets have to work, to rule, and wouldn't even let me get a word in."[94] Ulbricht did not let up, scolding Brezhnev in mid-1970 that East Germany "deserved the respect of a genuine German state. We are no Belorussia, we are no Soviet state."[95] The next month in Moscow, Ulbricht stunned assembled Soviet and East German leaders by repeating that East Germany "was not Belorussia" while berating the Soviets for their "huge apparat" and for "failing to recognize the computer age."[96] Ulbricht then offered his cybermarxism to his Soviet and German comrades as a more advanced ideological model than Soviet Marxism-Leninism.

Less than six months later, assembled Party leaders at the lavishly produced Twenty-fourth Soviet Party Congress heard Ulbricht speak at length again on the virtues of the East German model.[97] Ulbricht did not condemn China (the only East bloc head of state apart from Romania's Nicolae Ceaușescu who failed to do so), he also reminded the massed delegates that he had met Lenin in 1922. Lenin, as Ulbricht recounted to the Soviet leadership's discomfort, had instructed the Russians to learn from their German comrades.[98] One can almost hear the Soviet leadership grumbling, "So this is why we fought the Great Patriotic War? To be lectured by a German?" After the Warsaw Pact invaded Czechoslovakia, Ulbricht returned to Ackermann's theme from the late 1940s of a "special German road to socialism." Ulbricht called for a "socialist community of men," minimizing the role of class, after Honecker attacked him at the May 1971 Central Committee meeting.[99] Ulbricht's fetish for innovation and cybernetics must have reminded the Soviet leadership of Khrushchev's impulsive behavior, and Ulbricht's conversion to humanist socialism after the invasion of Czechoslovakia (led by East German troops and urged on by Ulbricht) must have seemed bizarre.

Ulbricht most dangerously pushed a "counter-détente strategy," stalling détente, threatening stability in Europe, and blocking a key Soviet strategy.[100] The Soviets had long tried to engage European countries, East and West, in détente through a security conference. Ulbricht's independent *Deutschlandpolitik*, his aggressive Berlin policy, and challenging of Four Power rights in Berlin undermined this primary Soviet strategic goal. Brezhnev wanted stability on the Warsaw Pact's long frontier with West Germany so he could attend to domestic problems and meet China's military and ideological challenge. The Wall laid to rest Berlin's usefulness to the Soviets in 1961.

Ulbricht's clumsy negotiation with Willy Brandt, the shrewd Social Democrat West German chancellor, over Berlin's status in 1970 not only stalled détente but also threatened to usurp Soviet rights as a victor of the Second World War. Berlin's future, Ulbricht asserted to Brandt without advance notice to Soviet diplomats, was for West and East Germany to decide, not the Four Powers, a right which the Soviets carefully preserved. Brandt refused to accept Ulbricht's demands, and as long as Ulbricht remained inflexible on Berlin, Brandt declined politely to join the Soviets in a security conference. Without West Germany, as the editors of the *Times* of London observed, a security conference would be "like Hamlet without the Prince" and détente would fail.[101] Brandt urged Khrushchev to control Ulbricht, and his *Ostpolitik* drove a wedge between the Soviets and East Germany—something Ulbricht could have prevented but instead handed to Brandt through his maladroit schemes.[102]

Ulbricht did come to see that forced collectivization and Industrial Production Methods had failed and regretted his overhasty Socialist Spring in the Countryside in April 1960. Food shortages plagued East Germany throughout the 1950s and 1960s, and if he did not revive farm production, the weakest sector in the economy, he could lose everything. So Ulbricht finally took on key planks of *The Communist Manifesto* and Soviet practice, retreating from the *Bündnispolitik*, Lenin's Union of Workers and Peasants, even condemning Industrial Production Methods as "this dangerous development."[103] He then overruled Politburo decisions to promise farmers that further collectivization would not be carried out "hastily or against the will of the peasants."[104] Ulbricht's retreat from the Party's war on the countryside and orthodox Marxism-Leninism alarmed Honecker and other members of the orthodox "Brezhnev-Fraktion" in the Politburo. Ulbricht had already alienated many Party members by elevating the new scientific and technical elites over the workers and raising cybernetics over Marxism-Leninism. If technology was paramount then why was the Party needed and what was the role of Marxism-Leninism?[105]

One could not believe both in the central role of the worker and class struggle and in the primacy of science and technology. Cybernetics was indifferent to class, and technology was an elitist enterprise. A liberal Party member complained, "There can be no science without ideology." A senior Politburo member centered in on the core problem of the NÖS and cybernetics: "pseudoscientific formulae must never undermine the meaning of socialism. We must stay on guard to prevent the takeover of the political language of the Party by the language of specialized science. Then the Party would cease to be a Marxist-Leninist party."[106] Honecker reported that cybernetics was "useful, but limited and not a replacement for socialist political economy as a theoretical foundation."[107] A Party economist offered a postmortem: "Systems analysis' supporters labeled cybernetics 'Marxist-Leninist' but it is far from clear what is Marxist-Leninist about this science, quite apart from the fact that it raises the organizational structure above central control [*Leitung*]. They adapted cybernetics partly from capitalist problem-solving techniques. But it can never override what is most essential in socialism: the leading role of the workers and their Party, or the role of the unions and youth organizations in the factories!"[108] Orthodoxy and religious Marxists emerged, almost from the beginning of the NÖS, as the victors.

Despite the link between legitimacy and ideology which the NÖS threatened, and despite the administrative revolt of the cadres which derailed NÖS reforms, it was Ulbricht's intrusion into Soviet foreign policy that triggered his ouster. To remove Ulbricht's informal veto on détente, Ulbricht himself had to go.[109] Brezhnev met with Honecker on 28 July 1970, telling him that the Soviet Politburo wanted Ulbricht removed from office and that Brezhnev wanted daily progress reports. To hide the Soviet role, thirteen of the twenty-one East German Politburo members of the Brezhnev-Fraktion wrote Brezhnev on 21 January 1971, asking him to remove Ulbricht before the Eighth Party Congress later that year. They charged Ulbricht with ignoring Politburo decisions on farm and forest collectivization, following a personal political agenda, and presenting himself as the equal of Marx, Engels, and Lenin. Brezhnev-Fraktion members, including the future minister of agriculture and champion of Industrial Production Methods, Gerhard Grüneberg, further singled out Ulbricht's "technocratic" leanings and rudeness toward his colleagues.[110] Brezhnev met with Ulbricht and Honecker in the Kremlin on 12 April 1971, after Ulbricht's arrogant speech at the Soviet Twenty-fourth Party Congress, and ordered Ulbricht to resign as first secretary. Fifteen days later the East German Politburo voted unanimously to accept Ulbricht's resignation.[111]

The Party leadership formally marked Ulbricht's removal from power and raised the Soviet candidate, Erich Honecker, to the Party chair at the Eighth Party Congress on 16 June 1971, with the pledge "decisively to confront the unnecessary complications and overdone application of cybernetics in planning."[112] In his main speech to the Party Congress the next day, Brezhnev called for speedy settlement of outstanding Berlin issues.[113] Honecker's closing speech on 19 June 1971 showcased his loyalty to Moscow's line on Berlin and détente. He loyally praised the Moscow-Bonn Treaty of 12 August 1970, an agreement which paved "the way for comprehensive, mutually useful cooperation." Honecker closed with his hopes for an "all-European conference" on security, the central goal of Soviet policy.[114] With Ulbricht went the main roadblock to Soviet foreign policy. Détente was finally realized through the Conference on Security and Co-operation in Europe (CSCE, 1973–75), ending with the Helsinki Final Act (1 August 1975).

Honecker's Soviet-sponsored coup against Ulbricht, rather than being a sharp break with the past, a *Zäsur* (as important scholars have suggested) when the Central Committee abandoned Ulbricht's "reform concept" to reimpose central planning, made little difference to the system qualities of the East German polity, to the peoples' lives, or to the health and productivity of the East German landscape.[115] The scientistic vs. orthodox Marxist power struggle had meaning only for the political elites: for most people, the issues dividing the two factions were distinctions without a difference. The true turning point in East German history was the final closing of boundaries with the Berlin Wall in August 1961—Ulbricht's ouster in 1971 was incidental to the continued dominance of the Party's absolute truth, control, and extreme materialism.

Delegates to the Eighth Party Congress also set forest management policy for the next eighteen years.[116] Honecker's "central pillars" of the Union of Workers and Peasants (the *Bündnispolitik*) and Industrial Production Methods remained the "absolute priority and principal idea" for farm and forest policies for the regime's final two decades.[117] When the Tenth Party Congress (11–16 April 1981) met a decade after Honecker's coup, Honecker proclaimed it as the "Party Congress of Continuity," continuity not just with the Eighth and Ninth Party Congresses, but with all the Stalinist economic plans since the founding of the East German republic in 1949.[118] It could still be reported in 1984, "East German forest management conforms fully to the demand of the Tenth Party Congress of the Central Committee: Increase domestic production!"[119]

A dominant question of the 1960s, whether capitalism or socialism would prevail, remained unresolved as the 1970s began, but the future probably looked favorable to the East German leadership in 1971 after Ulbricht's ouster even as the Great Golden Age of Economic Growth ended as commodity prices rose with the Vietnam War in the late 1960s and then exploded with the two oil crises of the 1970s, triggering global stagflation and recession.

With the passing of the Great Golden Age, a unique period of opportunity also passed for the Soviet bloc leaders. The economic environments of the 1970s would be far harsher and unforgiving than any could have anticipated. So, the East German elites, unconscious of the future maelstrom ahead of them, probably looked forward to the seventies with confidence. Their house was in order now that the ideological infighting was over and Soviet support secure. They were on the way to Helsinki and an ultimate goal: international recognition of the postwar borders and affirmation of the legitimacy of the Party's rule. Then they could go to the East German people and assert international recognition not only of East Germany's borders, but of the Party's legitimacy.

Westerners entered the seventies battered by the backwash of the Vietnam War and social unrest. A "national malaise," in President Jimmy Carter's words, gripped America in the 1970s, and he fretted that symptoms of the "crisis of the American spirit are all around us." It seemed possible that the American economy might sink into chaos, leaving final victory either to the more confident communist elites who ruled through command and absolute control, or to Japanese and European finance ministry bureaucrats who planned and managed their national economies.[120]

Nevertheless, the natural landscapes of East Germany that emerged in the previous decade held the telltales of the West's success in the turbulent seventies. As the Honecker-Fraktion reestablished ideological orthodoxy, they buried what little resilience remained in the economy and in the rural landscape. The Party leadership firmly reasserted its commitment to eradicate all diversity—economic, social, and ecological—in the farm and forest landscapes. There would be one class in the countryside, the worker, and one class of ownership, the state, and even one species of tree farmed in uniform blocks of even age. Honecker insisted on a simple, heterodox society where authority ran linearly from the top to the bottom: Lenin's "democratic socialism." The Party leadership demanded not just maximum, annually increasing production from farm and forest; they craved extreme levels of control to suit their extreme levels of risk aversion. Without diversity in all its forms—not just biological, but economic, social, and cultural—East Germany would suffer in the global chaos of the seventies, a period of social and economic revolution when having a Plan and absolute vision of the future was not just foolish, but the first step on the path to dissolution and oblivion.

"What We Plant Today We Will Harvest under Communism! Socialism and Communism Will Be Victorious!" This first cover of *Die Sozialistische Forstwirtschaft* (*Socialist Forestry*) in 1962 featured a forester in Soviet-style habit directing the gaze of two girls over a uniform young pine plantation toward the future, a pose familiar from Lenin's many statues. (AFZ-DerWald)

Nikita Khrushchev caressed a corn stalk while he lectured the crowd around him, including East German Party chief Walter Ulbricht to his left, on the great future for maize in Soviet bloc farming during his tour of the socialist farm collective "Hakeborn" on 11 August 1957. (Keystone Pressedienst)

Erich Honecker, builder of the Berlin Wall and future East German Party boss, thanks border troops on 15 September 1961 for their services in building the Berlin Wall in August. His deputy in planning and executing the Berlin Wall, Gerhard Grüneberg, was rewarded with a seat on the Politburo and the agriculture and forestry portfolio. (Deutsches Historisches Museum)

Walter Ulbricht and his wife visit Neubrandenburg on 1 September 1965 to celebrate the twentieth anniversary of the Soviet land reform. (Deutsches Historisches Museum)

Forest death (*Waldsterben*) in the Bavarian forest. Nitrogen pollution from the high temperature exhaust of western car engines and stack and chimney effluent from power plants and industry in East Germany's Southern Uplands accelerated forest decline throughout Central Europe, particularly in higher elevation forest stands. (Keystone Pressedienst)

Helicopter spraying insecticide over a pine nursery in 1975 as the Ära Grüneberg and Industrial Production Methods intensified. The chemical is probably DDT. (Keystone Pressedienst)

Soviet officers' wives transplanting pine seedlings in Brandenburg, Möncheberg State Forest District in 1991. (A. Nelson photograph)

East German forest management drew a full range of products from the forest as a part of the campaign for "total tree utilization." Dr. Albert Milnick points out damage from tapping pine sawlogs to fill turpentine quotas (*Hartzen*) on the Eberswalde forest preserve in Brandenburg. (A. Nelson photograph)

An immature pine stand in Brandenburg typical of the young age classes dominating East Germany's forest. (A. Nelson photograph)

A plant manager stands in the loading yard of his "latest generation" forest products factory in 1989. Carts laden with small-dimension roundwood stand ready for processing. (Corbis)

Erich Honecker welcomes his guests to the hunt, 1983, Magdeburg. (Hans-Georg Schumann photograph)

Senior forestry degree candidates dressed in the uniform of the National People's Army celebrate the trophies of Honecker's 1983 Magdeburg hunt by blowing traditional hunting choruses over the bodies of more than one thousand dead hares. (Hans-Georg Schumann photograph)

Tanks and rocket launchers paraded before the guest of honor, Soviet Party Chairman Mikhail Gorbachev, and Erich Honecker to celebrate the East German republic's fortieth anniversary on 8 October 1989, one month before the Berlin Wall fell. The rogues' gallery on the reviewing stand included Romanian dictator Nicolae Ceauşescu; Yao Yilin, fresh from planning the Tiananmen Square massacre four months earlier; PLO Chairman Yasir Arafat; Nicaraguan strongman Daniel Ortega; the hard-line Czech leader Milos Jakes; Bulgarian Party boss Todor Zhivkov; Polish junta chief Wojciech Jaruzelski; Vietnamese Vice-Premier Nguyen Van Linh; and the senior hierarchy of the Soviet army in Germany. (Time Life Pictures/Getty Images)

A girl carries lighted candles to her parents during an evening vigil in East Berlin's Erloser Church organized by the leaders of New Forum, part of the "church-environment" popular revolt against the Honecker regime, on 6 October 1989. As the family prepared for the next day's protest march, Honecker and Gorbachev sat at a state banquet nearby. (Corbis)

Erich Honecker's final trophy stag, a specimen red deer culled from the oppressive game herds of the East German forest, late 1980s. (Editorial staff, "unsete Jagd")

Berliners celebrate German unification on 3 October 1990, less than a year after Honecker and Gorbachev marked the fortieth anniversary of the founding of the East German state. (AP/ Wide World Photos)

# 8

# *The Grüneberg Era and the Triumph of Industrial Production Methods (1971–1989)*

*He that steals a Bell-weather, shall be discover'd by the Bell.*
*—Miguel de Saavedra Cervantes,*
The Adventures of Don Quixote de la Mancha

The 1970s were particularly hard on the resource-poor East Germany. Raw material prices started rising slowly in 1965 and then surged 14.5 percent following the Yom Kippur War and the first oil shock in 1973, then a further 30 percent in the wake of the Iranian Revolution in 1978–79. Not an economy in the world was untouched by global inflation and recession by the end of 1979. As energy prices rose, the Soviets cut their subsidized oil shipments to East Germany and reoriented their exports to hard currency markets in the West. The East Germans were forced to tap their large domestic deposits of lignite to replace lost Soviet oil. Lignite, a subbituminous coal with properties between peat and hard coal, was easily taken from open-cast mines in the Southern Uplands. But lignite has a high water content, stores poorly and degrades in the atmosphere, is unstable, and ignites unpredictably. Worst of all, lignite throws off far more toxins and sulfur than higher-grade coal when burned.

The Party nonetheless converted most of East Germany's power generation to lignite fuel. By 1989, lignite-fired power generation plants met 83 percent of

East German electric power demand. East Germany ate up 40 percent of global production of lignite in 1990, a huge volume for such a thin economy and fragile landscape.[1] As lignite-burning power plants came on line, their unfiltered exhaust created a new, more lethal stage in the ongoing crisis in public health and forest decline.

West Germany faced a similarly chaotic economic environment in the 1970s, but took a radically different path, cutting its sulfurous oxide pollution (acid rain) in half at a cost of $5 billion, more than $17 billion in 2002 value. The Party leadership spent even more than West Germany did on cutting acid rain emissions—$20.4 billion in 2002 value—to convert its power generation to lignite, sinking more than five times its net external debt into burning a more primitive and dirtier fuel.[2] The financial burden of this oil substitution strategy and foreign borrowings fell upon the landscape physically, in the form of millions of tons of toxic ash, particulate, and dust. It literally ripped the forest apart as destructive harvesting for *Devisen*, hard currency, increased to service East Germany's foreign debt. Pollution damage, which rose with oil prices, hastened the pace of forest death and gave the people a powerful outward and visible sign of the leadership's corruption and incompetence. This combination of economic and ecological decline and increased popular awareness would doom the East German republic to oblivion.

As orthodoxy triumphed in 1971, so did Industrial Production Methods. Foresters dubbed the years after Honecker's coup "the Grüneberg era" (*Ära Grüneberg*) after Gerhard Grüneberg's single-minded assault on the farm and forest landscapes.[3] Grüneberg's career tracked the history of the orthodox, hard-line wing of the Party. A tough, abrasive careerist, he rose through the ranks steadily, allying himself early on with Günther Mittag to become first secretary of the Party's Frankfurt am Oder county headquarters, building connections with the cadres whom Ulbricht's reforms would threaten. Grüneberg served in a high-profile Party job in Potsdam in the late 1950s, moving closer to the center of power, and rose to become secretary to the Party's Central Committee in 1958. He then served as Honecker's deputy in building the Wall in August 1961.

Grüneberg's fortunes blossomed with Honecker's. He joined the People's Chamber (*Volkskammer*) and in 1963 entered the Council of Ministers with responsibility for agriculture—the highest profile portfolio in charge of the economy's most difficult sector. His signature policy, Industrial Production Methods, was the perfect mechanism to crush peasant resistance to collectivization (and to Marxist-Leninist ideology) and raise gross farm and forest harvests to "industrial norms." Grüneberg did well and he rose to the Politburo in 1966 to conspire with Honecker against Ulbricht in 1971 as a member of the "Brezhnev-Fraktion." After Ulbricht's overthrow, Grüneberg won even greater power over the rural economy and landscape.

Grüneberg, who nurtured close ties to the cadres, first reversed Ulbricht's NÖS reforms, bringing forestry under the Ministry of Agriculture. He asserted that forestry and agriculture were "naturally close together," despite vast differences in economic and ecological cycles and the longstanding friction—almost antipathy—between foresters and farmers.[4] This sent a clear message to foresters that Grüneberg, not ecologists or forest managers, was in charge. Before Grüneberg's decision, forestry was managed out of its own department and its experts favored by Ulbricht's reforms. Oversight and political control returned to the cadres in the counties as Grüneberg dissolved Erich Apel's five central VVB Forest Management bureaus along with ecological forest boundaries.[5] Industrial Production Methods, imbued with Marx's "revolutionary power," ruled forest management by the Ninth Party Congress in May 1976.[6] "Foresters must," commanded Grüneberg, "follow our line!"[7] Food production crashed, just as it always had, in the wake of the Party's "revolutionary" interventions in the farm and forest economies, recalling the chaos following land reform, *Dauerwald* radical ecologism, and the Socialist Spring in the Countryside. A West German observer commented later: "the failure of these stupidities [Grüneberg's Industrial Production Methods] was so clear that soon after Grüneberg died in 1981 the entire top echelon of the Party agricultural bureaucracy was changed."[8] But these personnel changes were cosmetic. Grüneberg, a detested and violent man, died on the eve of the Tenth Party Congress of Continuity in 1981, but his policies continued with even greater force into the 1980s.[9]

Once brought under the Ministry of Agriculture's control, forest harvests started to mimic the dynamics of farm harvests. A senior silviculturist lamented: "One was not squeamish about the size of clear-cuts during the height of Industrial Production Methods." The link with agriculture forebode the shift to full industrial practice, with the factory rather than the ecosystem as the management model. Farm management made heavy use of energy, inorganic chemical fertilizers, and herbicides to pump up annual production over biological constraints. This industrial model soon suffused all aspects of forest management as forests were torqued into "factories on the land" and foresters stripped of their power. The report of the Ninth Party Congress in 1976 commanded: "deeper cooperation with agriculture and industry. Forestry must harvest in technical complexes and increase the size of management units."[10] A secret directive revealed the precedence of production in Grüneberg's forest: "forestry's task is to supply the People's economy with constantly increasing raw wood deliveries while preserving and expanding the forest at the same time," with the conflict resolved through the "advantages of socialism" — industrial methods.[11] The harvest rose 20 percent, to 8.5 million cubic meters in 1975 and to 9.8 million cubic meters by 1980, 1 million cubic meters more than 1975's ruinous harvest and 50 percent higher than the 1965 harvest.[12] The depredations of direct Soviet occupation returned.

Soviet experience was asserted once again as the model for farm and forest policy — a return to the subordination to Soviet "experts." Erich Apel's caution about closer economic ties to the Soviet Union also went unheeded as the Ninth Party Congress not only charged foresters to "develop domestic raw wood resources thoroughly and move toward complete utilization of trees" but to move toward "ever closer cooperation with the Soviet Union."[13] Lenin returned as the ultimate source of farm and forest policy, as Grüneberg declared, "Lenin gave us, above all with his Plan for the Collectives, the decisive theoretical framework for the transition to socialist agriculture."[14] Honecker declared, "We armed ourselves with classic Marxist-Leninist teachings on agricultural policy and took maximum benefit from the enormous practical experience of the Communist Party of the Soviet Union in their revolutionary transformation of agriculture," an explicit rejection not only of forest ecology, but of Ulbricht's cybermarxism.[15] Even with its great wealth of soils and climate, Soviet agriculture could not meet domestic demand in the 1970s. How could a policy which failed for a resource-rich country work for poor East Germany?

Despite window dressing of "Marxist economic science," shortages were the constant fellow traveler of the command economy. This would end, as Gräfin von Dönhoff commented in 1975, only when the Soviet bloc leaders chose grain over dogma (*Dogma oder Weizen*).[16] Marxism-Leninism drove out rational planning in a political analogue to Gresham's Law. Thus Grüneberg renewed the Party's attack on the forestry schools, particularly Eberswalde. This campaign began with the 1971 Party directive (*Beschluß*) ordering Eberswalde scholars to "pursue more practical research."[17] Rudolf Rüffler, the Party bureaucrat installed as dean "to end the destruction of the Eberswalde faculty," slashed ecological funding severely and attacked "bourgeois" foresters for their loyalty to "old ways." Research must focus on increasing annual production.[18] The Party's campaign against woodsworkers and farmers, to "eliminate class differences between worker and peasant," also revived. Thirty different forestry trades were eliminated and woodsworkers were routinely "loaned" for farm work to weaken their bonds with the forest.[19] The traditional pace and tenor of rural work were liquidated to change woodsworkers and farmers into doppelgängers of the iconic industrial worker.

A new level of subservience was demanded from foresters in the academy and in the field, as a Marxist theorist declared impatiently: "The time has come for a comprehensive conformation of forest science and policy to Marxist principles. It is wrong to limit the relevance of Marxism for forestry, even though Marx rarely discussed forestry directly. Marx's complete work has much more significance for forestry and forest science than foresters realize, because bourgeois forestry scholars buried Marx's teaching. And despite all the Party's efforts and extensive application of Marxist forestry in previous years, East German forestry still does not reflect full application of Marx's teachings. Now we must implement a comprehensive application of Marx's teaching!"[20] That meant Industrial Production Methods, Soviet practice, and the rebirth of farm and forest into "factories on the land."

Despite official pressure to cite Marx—even in daily correspondence and in technical literature—few recalled Marx's first published essay, "On the Law on Thefts of Wood" written in 1842, the first year the words "socialism" and "communism" appeared in German. Marx attacked the closing of the forest commons and abolition of peasant rights, early stages of the great ecological revolution of the nineteenth century heralding the emergence of the modern industrial forest.[21] Marx's bitter complaint that "trees have been given rights to which the rights of people were sacrificed" was even more true in the East German republic than it was in the preindustrial world of the early 1840s Rhineland. The enclosure of the forest commons angered Marx, and despite his distaste for the countryside he savaged the liberal Rhenish reformers. Marx in turn failed to see that the clearing away of the medieval coppices and peasant rights sounded the tocsin for the Industrial Revolution and the triumph of capitalism, whose energies and rationality he admired. Marx didn't understand the enormity of the ecological change under way, yet it is clear from the power of his argument that he intuited the larger importance of enclosure and refabrication of the natural landscape. Had the scale of change he witnessed in the Rhineland forest taken shape in an industrial or urban environment, he may have been more sanguine.

Exegesis of Marx's writings, however, yields contradictory readings that support scientism as easily as the dark orthodoxies of Marxism-Leninism. This early essay reveals Marx's profound humanism and talent for polemic—but he had scant sympathy for the distinctive qualities of the rural economy or of rural life. Despite Marx's empathy for the dispossessed peasants, the central role of materialism in his thought led directly to the Party leaders' war on the countryside. Following Marx, they believed that abstractions and administrative forms, whether class or ownership, and material production determined not only farm and forest production, but the development of culture and history. Cybermarxists and the members of Honecker's junta alike shared an abiding belief in the absolute truth of Marx's historicism despite their party-political differences.

Many foresters, under the cadres' constant watch and cowed by purges, came to deny the primary importance of biology in farm and forest production. A collective of Marxist theorists wrote in *Socialist Forestry* in 1968: "As Marx taught, and as history confirms, nature may in fact be necessary but it is not the most important condition for the existence and development of human society. Far more decisive is material production."[22] A forestry cadre asserted that industrial methods would "lessen dependence on nature in the forest production process," and a senior forest economist emphasized that chemical and technological factors were "fast becoming more important than biological factors. Previous forest managers were obsessed with the biological aspects of raw wood generation. We must now concentrate forest management far more on the economic and technological aspects of forestry." Foresters must "banish old, traditional practices that retard progress and reach into the industrial production sphere of our economy to eliminate the essential differences between urban and rural societies," to wage "uncompromising battle with outdated views and practices of the past."[23] Even Schröder, who defended close-to-nature forestry in the 1950s and 1960s, weighed in to attack *Dauerwald* foresters—still a thorn in the Party's side: "The forest is not self-regulating and not an organism. There is no sense in a 'return to nature.' The forest is meant to be directed and ruled by man. In the forest we are not obliged to follow nature—our task is much more: to direct nature and to learn to control nature."[24] As the Forest Service chief, H. Heidrich, explained, "If we want higher growth we can talk all day about the trees themselves, but they will still grow only as well as men know how to apply the laws of nature and create favorable conditions."[25] Ideology and a rigid scientism even more powerful than that under Ulbricht's NÖS continued to guide socialist forest and farm management.

Marxist-Leninists saw the relation between humans and nature as a dialectic. One economist cited Marx to assert that humans were at "war with nature," which they struggle to "control and rule."[26] Humans had a responsibility to control and order nature and stood apart from nature. As Marx wrote in *Labor and Capital,* "only where industrial production takes place, within social relationships and connections, can man find his relationship to nature."[27] The Party leadership drew their farm and forest policies, as Honecker boasted, "directly from the realization of the classics of Marxism-Leninism," which formed the ideological basis for their war on the countryside.[28] The 1945 land reform and forced collectivization in April 1960 satisfied point one of *The Communist Manifesto,* the "abolition of property in land." Industrial

Production Methods fulfilled point nine, the "combination of agriculture with manufacturing industries; gradual abolition of all the distinction between town and country by a more equable distribution of the populace over the country."[29] Formulated as the *Bündnispolitik,* this demanded, according to a senior bureaucrat, "the elimination of differences between urban and rural life."[30] Marx's and Engels' call for the abolition of the countryside and for "a more equable distribution of the populace over the country" are artifacts of late nineteenth-century thought and reflect *The Communist Manifesto*'s sense of "the idiocy of rural life."[31] Communist forestry bureaucrats, therefore, took as one of their main tasks "the dissolution of the social differences between agriculture and industry and between city and country," leaving one class standing, the worker, and one landscape, the urban.[32]

Ideology enabled the leadership to deny pollution damage that, despite official denials, was seen as a growing threat to public health and industrial production. "Socialist" administrative forms and Marxist-Leninist ideology guaranteed protection of the environment. Western environmental concerns, in contrast, were "an expression of late-bourgeois ideology" and proof of the "general crisis of capitalism."[33] To the Party leadership, Western interest in environmental protection reflected a pessimism which foreshadowed capitalism's imminent collapse.[34] Liberal, free market economies and private property not only caused pollution in the West but were responsible for Soviet bloc pollution: "historical in nature, the cumulative product of the capitalist production process' one-sided exploitation of nature."[35] A Soviet environmental policy expert insisted as late as 1986 that "state ownership of the means of production and of natural resources and the planned character of the socialist economy are the most decisive factors in environmental protection."[36] A senior Party agronomist even defined clear-cutting, socialist forestry's hallmark, as "capitalist forest management"—an eerie Marxist reprise of antiliberal green Nazi environmentalism. The "legacy of capitalist relations," cold war hostility, and Western jealousy caused forest decline, not industrial pollution.[37]

The Party leadership even claimed a place in the vanguard of environmental policy by virtue of its political alignment in "the socialist camp." Propaganda cited East Germany's "leading role" at the U.N. Conference on Environmental Protection in Stockholm in 1972 and participation in the environment conference in Helsinki in 1973 as sources of international legitimacy.[38] Leading Marxist-Leninist theorists explained: "Environmental protection today means above all: our battle against the monopolist bourgeoisie and their conscienceless plundering of nature's riches and their equally conscienceless defilement of man's environment with capitalist waste. We socialists fight for international agreements to halt further degradation of the environment on the basis of coequal peaceful cooperation, even between states with different social orders, for revolutionary change in social relationships in the capitalist lands, for the transition to socialism."[39] "Coequal peaceful cooperation" meant subsidies and support for East German repression, such as turning away refugees from the East.

Ideology not only justified the *Tonnenidologie,* it explained why the natural world and public health were collapsing: "Socialist environmental conditions improve only slowly despite environmental policies superior to those in the West. The Pentagon's arms race forces peaceful socialist lands to waste huge sums on armaments, money that would be better spent on environmental protection. [The Pentagon's arms race is] the greatest environmental threat."[40] A Forest District political officer declared in 1986: "Our declaration of belief, 'I fight for peace where I work,' binds us with concrete deeds and creative initiative toward filling—and even overfilling—Plan quotas. With all these initiatives, dear Comrade Erich Honecker, we wish to make our contributions to the strengthening of our socialist state on all sides, to make it an even more powerful instrument of peace on the battle line between socialism and imperialism."[41] The Party bosses wielded peace as a stick with which they beat down rational argument. A senior district forester and Party member recalled: "The bosses refuted every factual argument with ideology. They would demand, 'Are you for peace or not? If you *are* for peace then deliver the wood! When East Germany is strong then peace is strong.' "[42]

Pollution was linked to the central political event of the 1970s, détente and the signing of the Helsinki Final Act (1 August 1975). These were primary Soviet foreign policy goals over which Ulbricht had tripped fatally. Honecker, who owed his power in large part to Soviet irritation with Ulbricht over détente, was even more interested than Brezhnev in Western affirmation of postwar frontiers. Honecker saw a signal victory in the basket one articles of the Helsinki Final Act legitimizing Stalin's postwar "border adjustments." In return for the West's retreat from its policy of freedom for eastern Europe, Soviet bloc leaders accepted basket three commitments to protect human rights and the environment, what must have seemed a trivial concession. Honecker's reading of the Final Act's meaning (expressed in a September 1975 front-page declaration in *Neues Deutschland*) was clearly political: "the acknowledged inviolability of borders created the conditions for peace"—regime stability. Honecker, with the basket one victory safe in hand, easily signed onto basket three's affirmation that "experience has shown that economic development and technological progress must be compatible with the protection of the environment and the preservation of historical and cultural values; that damage to the environment is best avoided by preventive measures; and that the ecological balance must be preserved in the exploitation and management of natural resources."[43] Basket three called for youth particularly to be educated in their "responsibilities" to protect the environment, an interesting mandate for Honecker (and Egon Krenz, his short-term successor as Party boss in 1989), who began his political career in the communist Free German Youth.[44]

Few countries in the world were as much in violation of this commitment as East Germany. Although Honecker won international recognition of his borders, he lost legitimacy at home when he agreed to respect human rights, allow open debate and religious freedom, and protect the natural environment—commitments he had no intention of honoring. The decline of the natural landscape was ubiquitous, physical, and manifest; forest death and deteriorating public health could not be explained away or hidden, and the border became more rather than less porous to competing ideas and ethics.

Because Helsinki's basket three clauses were "commitments" without the force of treaty or international law, it made sense for the Party leaders to believe they could ignore them while they used détente to convince their people that they were not only legitimate, but there to stay. But as Western diplomats and press pressured the East bloc governments to honor their human rights and environment commitments, the Party bosses bristled, denouncing this criticism as invidious attempts to "delegitimize and discredit socialism."[45] As late as 1989 the Forest Service chief declared: "Certain powers in West Germany for whom East Germany's growth and prosperity has always been a thorn in their eyes, paint again and again a black picture of our forests. Yet the truth is, and one can from all aspects assert, that our forests rest in good hands!"[46] Opaque socialist accounting hid the collapsing industrial sector, but the forest was dying before the people's eyes. Pollution attacked everyone's health no matter what the propagandists asserted. Helsinki stripped the final veil of secrecy and the Party leadership's ability to suppress discussion with impunity.

Secrecy, despite its Helsinki commitments to openness, remained the Party leadership's sole tactic to combat growing popular anger. The Party classified environmental data under the second highest secrecy in 1982, up from the secrecy level imposed in 1977.[47] Workers in environmental laboratories now needed security clearances, effectively cutting the number of environmental research labs by half, and foresters were forbidden to conduct research into forest decline. Starting with the 1971 inventory, foresters were cut out to prevent them from gaining a clearer picture of the damage.[48] District foresters were ordered to keep their records and inventories in locked safes, and copying machines were shackled. Elaborate logs were kept of copying machine use to secure data that might escape in the wastebasket. Remote sensing data were "killed" before forest scientists could analyze them; even Party members complained that aerial photography should have been labeled "Burn before reading!"[49]

After Helsinki, the Party leaders could no longer punish foresters with impunity. Frithjof Paul, a senior forest policy scientist, allowed himself to be quoted in the West Berlin press in 1983, saying, "Industrial pollution is of great concern and can only be fought by attacking the cause."[50] In a way Paul was right, because the same materialist orientation which bred massive pollution also led to Industrial Production Methods and to the artificial forest. Paul was censured for this direct attack on the Party's economic policies yet saved despite criticisms that only a few years earlier would have led to either exile or forced labor for him. The Party leadership had to be cautious after Helsinki about wielding the repression which defined the regime, particularly as West German aid was keeping the economy afloat. Western opinion now mattered. Still, it took considerable courage for Professor Paul, a committed Marxist, to speak out.

Public recognition of environmental damage started to crystallize in the late 1960s.[51] As pollution increased, people began migrating internally away from pollution hot spots in the "iron triangle" in the industrialized Southern Uplands.[52] East Germans, at the epicenter of the Soviet bloc's environmental catastrophe, formed the earliest environmental movements in the early 1970s under the shelter of the Protestant Church, giving environmentalism the name *kirchliche Umweltpolitik* (church environmentalism).[53] The Protestant ministry served as an educated and independent body of support within the state, independent men and women who patiently challenged the Party and its materialist philosophy. Churches were the only public meeting places outside the Party's control and offered printing and copying facilities. Protest was benign at first, focusing on individual activism, such as planting trees, discussion groups, and nature excursions. After Helsinki, priests gathered citizens in small groups to discuss environmental issues, just as the Helsinki Final Act allowed and despite constant harassment.[54] Environmentalism combined with the spiritual values of the Protestant Church fomented powerful opposition to the regime's extreme materialism.

A secret 1980 Party survey of popular opinion documented anger over pollution and the Party's industrial policies. This protest was more dangerous than earlier peasant unrest over collectivization in the late 1950s, a protest that was isolated in the countryside and poorly coordinated.[55] Now the East German people felt existential dread in the face of forest death and deep fear of pollution's health consequences, as the study's authors acknowledged: "Pollution increasingly affects forest production and cultural values. Foresters' morale is poor and the people see in forest death a threat to their own existence."[56] Such fears spilled over into criticism of the Party's economic policies and therefore directly attacked Marxism-Leninism itself as the people saw the Party leaders' malignant incompetence manifest in the dying forest.

By the mid-1980s, the Party leadership could no longer ignore the citizens' anger over forest death and the harm the command economy wrought upon their health.[57] At a Protestant Church Dresden conference in July 1983 more than one thousand East Germans heard a chilling report from the "Ecological Working Group."[58] Their fears were confirmed: sulfurous oxide emissions went far beyond official reports. Children in Bezirk Halle, where I. G. Farben's (suppliers of Zyklon-B gas to the Nazi death camps) Buna-Werke Schkopau synthetic rubber and fuel works lurked, suffered twice the national incidence of lung disease. The Party bosses might preach environmentalism, but Werner Felfe, minister of agriculture and Grüneberg's successor, showed the Party's true self in a 1987 speech welcoming Fidel Castro and a large Cuban delegation to the toxic Leuna chemical works: "Leuna is a symbol of our victory!" he declared.[59] But, as a West German journalist reported in 1986: "Ever more East bloc newspapers report on secrets which once were among the most closely guarded. The Party can no longer blame river pollution, forest death, or the poisoning of the air on the profit-oriented production process of capitalism. Now the causes are written in the smoke stacks."[60] "Real, existing socialism" was poisoning the people and killing the forest, not capitalism or the Americans, and Marxism-Leninism's direct connection to pollution and forest death could not be kept secret. The news flowed in ever-greater streams from the chimneys of the decrepit and aging factories despite secrecy.

Yet many Westerners accepted Honecker's surreal denial of pollution, just as they accepted official data of the republic's economic prowess. Honecker harangued West German environmentalists in 1984, just as East German industrial and agricultural pollution was peaking, with denunciations of the United States: "Everything we do is directed toward the betterment of man and is not determined by the law of profits." He slyly suggested that if the environmentalists really wanted to protect the environment, they should pressure West Germany to accept the "Warsaw Pact's constructive disarmament proposals."[61] His narcissistic appeal to ideology must have mystified even the credulous environmentalists: "It is precisely our policy of the Union of Workers and Peasants [*Bündnispolitik*] which spurs on our workers to ever higher productivity in science, technology, and in all areas of social life. We see in all seriousness the dangers which threaten our world. But the philosophy of world decline is foreign to us. Socialism is directed toward peace, life, and a good future for the people in a world free of threats."[62] No matter how dirty the air, or how sick the forest, the East German environment was nevertheless protected—because "socialism served mankind." The nobility of the Party's goals rendered their consequences unassailable.

The West Germans nodded their heads, seduced by Honecker's stress on the threat nuclear weapons posed to the environment and by his attacks on the United States—a surrogate for liberalism and modernity. Honecker found unexpected allies among the ranks of Western environmentalists and pacifists, ultimately his staunchest supporters in staving off reunification. More surprisingly, West German journalists praised him as a "German realist."[63] Honecker bragged to the editors of *Die Zeit,* the highly respected paper allied to the Social Democratic Party: "We have no need for talk of forest death. And we don't have any 'acid rain' in East Germany. Recently I made a long excursion along the Baltic coast with the Swedish prime minister, Olof Palme, who asked; 'How is acid rain impacting your forests?' To this I replied, 'I regret to inform you, Herr Minister President, that we have no experience with acid rain.' " Palme's reaction was not recorded.

*Die Zeit*'s editor-in-chief, Theo Sommer, passed by this smug lie to glow: "In the main Honecker showed off his realism. Reason and realism—this pair of concepts surfaced again and again in his conversation," an encomium disturbingly close to Werner Felfe's, Grüneberg's successor, praise for Honecker's "reason and realism." Other Westerners lauded Honecker as "a flexible and skilled leader" who ruled with "solid achievements which outweigh setbacks and failures." East German propagandists parroted back such praise to their captive people as proof of their regime's legitimacy.[64]

*Die Zeit*'s board of editors challenged Honecker only on transboundary pollution, ignoring the more toxic pollution internal to East Germany—and the steady traffic of Western hazardous waste flowing into East German dumps in exchange for deutsche marks (DM). West German environmental aid was a political gift to the Party bosses that could have cooled the population's rising anger over forest death as well as reducing transboundary pollution. Defusing popular unrest in East Germany to ensure stability and maintain the status quo may well have been as important a motive for the West Germans as curbing transboundary pollution. Honecker and Mittag certainly jumped at the offer of West German aid, much of which they diverted, declaring their "fundamental readiness to cooperate with West Germany" in 1984.[65] And there is no evidence that the Party used West German environmental aid effectively—toxic emissions continued unabated. Foresters knew better and observed that Honecker had invented a "divided heaven" separating East Germany from Western Europe, another wall that many in the West defended.

Western environmentalists, in addition to their very real concern for the environment, reflected the general malaise Honecker saw in the West, a pessimism on which the Party's propagandists focused. Westerners of all political views felt anxiety and doubt: about the deteriorating natural environment, about the threat of global thermonuclear war and regional conflict, about the health and meaning of their institutions. Compared to this anxiety, Honecker's certainties may have seemed comforting. But Western apologists forgot Thomas Szasz's aphorism about the distinction between certainty and doubt: "Doubt is to certainty as neurosis is to psychosis. The neurotic is in doubt and has fears about persons and things; the psychotic has convictions and makes claims about them. In short, the neurotic has problems, the psychotic has solutions."[66] Honecker wasn't a "realist"; he was a psychotic.

The Party leadership slowly admitted to forest decline after decades of denial in the late 1980s. The leadership had no choice, as decline—what the Germans called "forest death," or *Waldsterben*—was linked nearly universally to pollution and the Party's industrial policies even though the artificial, industrial nature of the forest was the prime cause of forest decline, not pollution. Yet the industrial forest and pollution were both by-products of the leadership's political philosophy. Felfe reported to the Politburo in 1987 that forest decline was "the result of a suddenly developed dynamic of pollution which could be controlled"; forest death was a recent phenomenon.[67] Dr. Hans Barciok, director of the State Forest Service's Statistical Office, claimed to have learned of regional forest decline only in 1988. Barciok later defended his silence: "Damage estimates were state secrets," an awkward statement for the senior forest statistician. He insisted in 1991 that two-thirds of East German forest decline came from Czechoslovak pollution, still denying the Party's responsibility for the public health and ecological damage that their policies wreaked.[68]

In 1985, forest growth fell by 20 percent as air pollution damage loaded on top of secular decline from the artificial structure increased.[69] West German government pressure (coupled with DM payments), media coverage, and popular pressure finally forced the Party leadership to make a proper forest damage survey in 1987 at West German insistence.[70] Felfe, Grüneberg's successor, finally admitted publicly in December 1987 that 33 percent of East German forests had "long-term severe damage" from industrial pollution.[71] Yet the chief of the Forest Service still rejected "traditional protection," noting, "Our policies will emphasize environment-adjusted production techniques." "Traditional protection" meant the pollution-reduction strategies of the hated Social Democrats or regulations like Richard Nixon's Clean Air Act of 1970 and Clean Water Act of 1972, which vastly improved air and water quality in West Germany and the United States.

The East German landscape itself bore the highest burden of sulfurous oxide in the world, four times West Germany's per capita deposition. East Germany also led Central Europe in pollution exports, depositing 120,000 net tons of sulfurous oxides ("acid rain") each year in Czechoslovakia.[72] Other Soviet bloc countries suffered even more: East Germany exported 450,000 tons net of sulfurous oxides to Poland and even more acid rain to West Germany — 762,000 metric tons vented westward annually.[73] Acid rain, and ransomed political prisoners, were Honecker's only export successes.

East German foresters had understood the link between declining growth and a syndrome of pollution and artificial management since at least the early 1970s.[74] East Germany's senior economic geographer conceded in 1978: "Smoke damage is currently the most important source of damage in forestry. Sustainable recovery will come only when damaging emissions from the source itself are reduced, despite introduction of smoke-resistant species and fertilization."[75] The secret 1979 "Forest Management Guidelines" revealed the internal consciousness of the link between forest decline and pollution: "The increasing amount of pollution damage affects not only wood production but also increasingly social and cultural values of the rural landscape. Therefore the chairman of the State Forestry Committee has reached an agreement with the other ministries for wide-reaching coordination of the forest management cadres on pollution accommodation."[76] Horst Paucke admitted in 1981 that forest decline "demands, much more, that we regard this part of the human environment *ecologically,*" although the Party forced him to retract this statement in 1987 (emphasis in original).[77]

Günther Mittag, the Party's economic czar, decreed cuts in pollution and controls on industrial emissions at the smokestack. Yet he also ordered that the forest accommodate pollution through fertilization and planting of smoke-tolerant species.[78] Mittag never funded pollution controls, and forestry got neither the cash to plant hardwoods nor any relaxation of Plan quotas. As late as 1987 the chief of the Forest Service declared that it was the "responsibility of forests to adapt to the demands of industry."[79] As a senior forest policy scientist commented: "Ecology was always well theorized. But when concrete actions were required the Primacy of Economics was always invoked."[80] The first step to reversing forest decline was to restore the close-to-nature forest and to cut the game wildlife herd that smothered regeneration, conclusions Mittag's decree implicitly acknowledged.

Although the Party managed the forest resource centrally under Mittag's oversight it never enacted a national forest law, an important goal of National Socialist forest reformers.[81] Mittag personally killed the project; the Party would not give formal status to any branch of the rural economy which implied parity with industry.[82] In any event, a national forest law was superfluous; "the Plan was law" (*Plan war Gesetz*), as foresters said. The State Planning Commission set total harvests through a ten-year budget which mandated clear-cutting and harvesting in excess of growth.[83]

Official estimates of forest damage reached 54 percent by 1989 when the editors of *Neues Deutschland* finally admitted that "East Germany is not only one of the most heavily polluted lands in Europe, but for a long time now has been a paradise for environmental criminals."[84] The estimates were wildly inflated; not only did investigators find the catastrophic levels of destruction they expected to find, but they were unfamiliar with sampling forest damage. Industrial forest management lay at the root of forest decline, and the destruction of the forest's economic value outstripped the ecological decline.

Despite rhetoric that smoke-tolerant hardwoods, such as poplar and birch (*Populus spp.* and *Betula spp.*), would stabilize forest structure, this meant an unthinkable retreat from cash flow and pure conifers.[85] Policy instead relied on aerial fertilization to counter acid rain, a strategy that, incidentally, supported the chemical industry at staggering cost for a prescription which at best did no harm (see figure 7).[86] Production of agricultural chemicals increased in an ecological Ponzi scheme in the faith that heavy applications of inorganic fertilizer, pesticides, and herbicides would lift farm and forest production beyond biological constraints.[87] Rüthnick dismissed the cost of 1,500 east marks (OM) per hectare with a simple "it's worth it," even though fertilization of forest stands brought neither benefit nor relief to the battered forest.[88] Such a sanguine response was possible only under socialist accounting where "when costs had to be covered the price followed."[89]

The belief that industrial policy complemented farm and forest policy captivated even sophisticated Western observers who accepted Marxist claims for a special synergy between socialist industry and agriculture. Kurt Sontheimer, a senior West German political scientist, lavished praise on East German farm policy in 1976 "as a model for solving the problems of agriculture in comparable industrial societies."[90] Western journalists in 1997 still cited East German farm policy, and therefore Industrial Production Methods, as superior to Western practice.[91] Party officials inevitably claimed pollution was a valuable by-product of socialist industry. Nitrous oxide emissions did fertilize entire districts, prompting the Party to welcome pollution as "a net contributor" to the economy. "Environmental Damage Serves the Welfare of the People" declared the perplexing title of a 1972 article.[92] Nitrous oxides pollution did heat up forest growth, but it also delayed the hardening off of trees in the fall, worsening winter stress and accelerating decline. Party theorists argued that

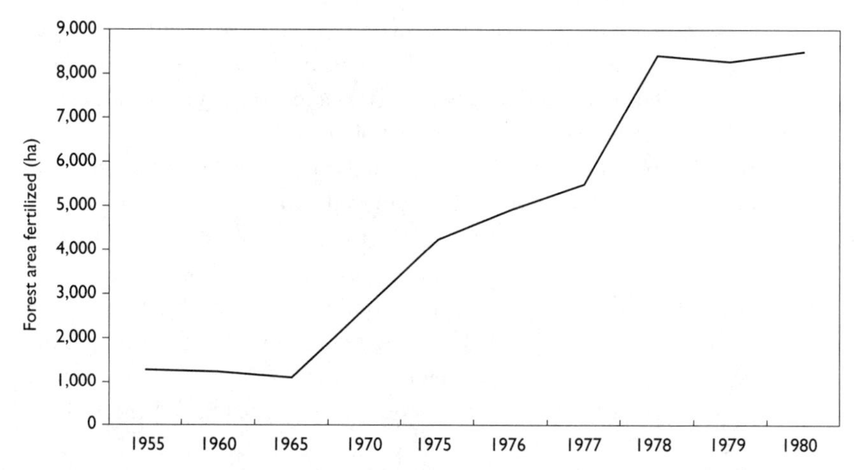


Figure 7. Aerial fertilization of the East German forest, 1955–1980. Data gathered from Staatlichen Zentralverwaltung für Statistik (SZWS), ed., *Statistisches Jahrbuch der Deutsche Demokratische Republik,* 26th ed., 195; Cornelson, "Umweltprobleme und Umweltbewußtsein in der DDR," 48.

pollution controls, or any constraint on industrial production, short-circuited the industrial-environmental synergy and were thus heretical. Finally, growth in East German forest and farm production stalled by the mid-1980s and then crashed as the exhausted soil could no longer incorporate the heavy fertilizer loads, which increasingly poisoned rather than nourished the forests and fields of East Germany.

Despite the fixation on decline in the industrial forest of pure pine and spruce, pollution took its heaviest toll on hardwoods in the rural Northern Lowlands where sandy, poor Pleistocene soils magnified stress.[93] Massive chalk and phosphate ash deposition blown in from power plants in the Southern Uplands lowered the soil acidity critical to oak vigor, killing more than 2 million cubic meters of oak on the North German Plain between 1988 and 1990: "the worst case of oak damage in forest history."[94] Native heath plants, which thrive in acidic soils, disappeared over wide areas, replaced by an infestation of thick, invasive sand grass (*Calamagrostis epigejos*) that erupted on forest soils sweetened with calcium carbonate pollution. This tough grass smothered natural regeneration of oak and beech, limiting the future of the East German forest as surely as Industrial Production Methods.[95] Despite high levels of sulfurous oxide and basic ash deposition, spruce in the Southern Uplands suffered lower damage due to better climate and soil conditions.[96] The abrupt increase in spruce harvests in the Southern Uplands in the late

1980s ascribed to salvage of acid-rain-killed trees probably had more to do with *Devisen* than disease.

The East German popular reaction to forest death and existential dread mirrored the reaction in West Germany where forest decline eclipsed all other issues, even the stationing of Pershing II intermediate-range missiles on West German soil.[97] West German Greens reprised the timbre of traditional German romanticism in Luddite rages against technology, venting *völkisch* fear of modernity and angst about the future, what Fritz Stern called the "politically exploitable discontent which for so long has been embedded in German culture."[98] Arnulf Baring noted this return of romantic environmentalism: "Germans' deep-seated and deeply felt anxiety over forest death is only one of many signs of the return of other peculiarly German worldviews. Since the end of World War Two a hidden, presumably lost German special consciousness grows anew."[99] East and West Germans shared this worldview.

Günther Haaf's extraordinary front-page article in *Die Zeit* in 1983, "Will There Be a German Forest in Twenty Years?" indulged in an anti-intellectualism and irrationality worthy of Weimar radical conservatives: "The threat of death to our most important ecosystem—the forest—comes not from a pollutant but from technological civilization."[100] Haaf's article sparked a broad response echoing Marxist-Leninist scorn for the natural sciences as "hairsplitting" and "a waste of time." A forest director from Hamburg solemnly agreed: "Air pollution has quietly become a threat to humanity of the highest order. We are threatened by nuclear war only by probability, while pollution threatens us with quantifiable certainty." Another West German trembled: "We don't need to worry about atomic war. Before that mankind will destroy itself through environmental catastrophe."[101] Radical environmentalists and orthodox Marxist-Leninists found that objective science interfered in the irrational solutions to be found only in their ideologies—whether deep ecological or Marxist-Leninist.

Such West German intellectuals as Hans Magnus Enzensberger echoed this renascent cultural despair and fear of modernity coupled with the chiliastic fervor of radical conservatism. He dwelt on capitalism's faults and asserted moral equivalence between Marxism-Leninism and liberal market economics. He wound up a long piece lamenting the fall of the East German republic with an attack on Western business: "by the end of the century the question (will) no longer be one of improving the world but of saving it. The most difficult retreat of all will be in the war against the biosphere which we have been waging since the Industrial Revolution. It is time for our own diminutive statesmen to measure up to the demolition experts. Certain large industries—ultimately no less threatening than one-party rule—will have to be broken up."[102] Somehow the scale of the relative achievements of the liberal democracies were forgotten and the fall of the poisonous East German dictatorship measured with regret.

Rita Ökten also developed the underlying theme of moral equivalence common to many European intellectuals, lecturing, "To avoid the looming ecological catastrophe East and West both must take a radical path, change their thinking to go beyond the foundations of the industrial rationale."[103] Antimodernist, anti-Western rants blended naturally with such Marxist-Leninist pieties as: "Environmental protection means above all; battle against bourgeois-monopolists and their driven, heedless plundering of nature and regardless destruction of man's world by waste from capitalist industrial production. We battle for fundamental change in social relationships in capitalist countries and for the victory of the socialist social order."[104] Little wonder, then, that West German Greens and many leftist intellectuals fought reunification in 1989.

Anti-Americanism infused propaganda. The invidious timbre of forest policy inflated to include "peace," blaming the Soviet bloc's pollution on the imperialists. Peace, as well as environmental protection, yet again became synonymous with increased raw wood production and the Party's industrial policies. Grüneberg had defined the main goals of forestry and agriculture as "preserving and ensuring peace," and an economist lectured farmworkers, "We fulfill our international class duty by increasing agricultural production."[105]

Honecker made his last attempt to resuscitate the moribund farm and forest economy with the Agricultural Price Reform of 1984. He faced a dilemma: how to revive the rural economy without abandoning Marxism-Leninism. He turned once again to a mixture of coercion and incentives, "economic levers," but rejected the decentralization reforms characteristic of the NÖS. Despite enthusiasm in the West, the price reform merely rehashed orthodox policies such as the *Bündnispolitik*.[106]

Farming and forest management costs protected under the NÖS increased as subsidies shifted to consumer prices.[107] Stable food prices were one of the few bright points of the economy. As Honecker boasted in 1975: "In weighing our agricultural policy's success it falls particularly on the positive end of the scale that the East German people have seen essentially no increases in the price of basic foodstuffs. One kilo of rye bread costs as much today, 52 pfennig [about $0.15], as it did twenty years ago."[108] A constant food price policy supported by subsidies was economic idiocy in an environment of global inflation, and East German standards of living continued to plummet despite the constant price of bread.

And the people might well disdain their plentiful loaves of cheap, dark rye bread when faced daily with television signals from the West that wafted over the frontier. The airwaves bore shimmering images of white bread and exotic croissants, of fresh fruit and ices, of tomatoes and lettuces, all casually shown in advertisements and television soap operas. Shortages in the *Mangelwirtschaft* (the shortage economy) of the "thousand little things" blighted peoples' lives perhaps as much as pollution or the Wall. Honecker pledged to meet "legitimate consumer demand" but refused to satisfy "petty-bourgeois [*spießerhaften*] desires or anything alien to socialist ideology."[109] He alone determined what was good for the people. The East German people had scant supplies of fresh vegetables in the winter; cabbage (*Blaukohl*) gave them the nourishment they needed. Anything more—oranges from Israel, tomatoes from Spain, or bananas—were luxuries. Similarly, Honecker refused to import cleaner and more powerful Volkswagen engines for Trabant and Wartburg automobiles despite favorable terms. He advised a West German journalist that his people preferred the dirty Trabi to the neat VW Polo: "I must tell you that, naturally, two-stroke engines have many friends. They are very much loved"—particularly for their cold weather starting performance.[110] If one had no garage, then a two-stroke engine might even be a luxury.

At the same time the Party leadership introduced the Agricultural Price Reform in 1984 (following the "principles of Lenin's Plan of Collectives"), it transferred even more power to the county organizations and the cadres. Industrial Production Methods was expanded to forest and nature preserves, throwing all reserves into the "international struggle for peace" as the economy entered its terminal stage of decline.[111] Every farm and forest hectare was now marshaled for exports to earn *Devisen* and stave off economic collapse.

After Grüneberg's death on the eve of the Tenth Party Congress in 1981, the ferocity of Industrial Production Methods shifted from gross production to integrating forest management directly with forest products factories, ostensibly to follow Swedish practice. The Swedes, however, enjoyed vast acreages of native pine and spruce forest, and their industrial model of forestry was no more suitable for the thin East German forest than was the Soviet Union's. District foresters were ordered to integrate operations vertically and to combine forests in ever-fewer, ever-larger units.[112] Soon saw mills, oriented strandboard and plywood plants, gravel works, and preparation yards generated more than half of the average Forest District's income.[113] Forest management and the quotidian tasks of timber stand improvement (TSI), thinning, and repair suffered as investments in machinery and capital goods increased. The conflict of interest between foresters' stewardship and their duties to the Plan and industry was never resolved as they faced "too much pressure to overexploit both to feed their own plants as well as to export," as a senior forest ecologist recalled.[114] Foresters were no better than others at running industrial plants under the Plan's Byzantine structure of levers and penalties, and it was well known that State Forest District plants were inefficient even by East German standards.[115]

"Intensification," the theme of East Germany's last Party Congress in April 1986, continued the emphasis on vertical integration. Yet forest policy still followed the Eighth Party Congress line from 1971. Now for the first time the exhausted forest stands could not deliver the volumes demanded.[116] The final Five-Year Plan called for a harvest of 55 million cubic meters (more than 4 cubic meters per hectare), twice growth.[117] There was not enough volume in mature trees, so the Party lowered harvest standards to take trees four centimeters in diameter, about the breadth of a man's upper arm. The 1989 sawlog harvest was only 30 percent of the total harvest in 1989, half the 60 percent share (the normal harvest percentage of sawlogs) last reached in the mid-1950s.[118] The Party ordered foresters to meet their Plan quotas through "scientific-technical advances," absorbing the immature wood through development of "completely new products."[119] This approach reflected the downward trend of forest products' quality as forest structure steadily deteriorated between 1945 and 1989. Even with increased predation on immature trees the State Forest Districts failed to reach their quotas of the final Five-Year Plan. Harvests in the 1980s peaked at just over 10 million cubic meters in 1986, well below the 11 million cubic meters ordered (see figures 8 and 12).[120]

Forest management in the 1980s displayed the same chaos as the general economy.[121] "Rational, orderly planning collapsed" after 1985, a senior East German forest policy scholar reflected.[122] Siberian pulp shipments became unreliable as transport and production problems in the Soviet Union, itself on the road to dissolution, mounted. Forest Districts were ordered to make up short deliveries of Siberian pulp without regard to cost.[123] Rüthnick, the *Generalforstmeister,* telephoned junior foresters over the heads of their district foresters to order immediate harvests of sawlogs for pulp, a return to the disorder and destruction of Soviet reparations harvests.[124] By the mid-1980s the East German forest had fallen back to medieval conditions: abused, brutally simple fields of small-dimension wood growing weakly on degrading forest soils. The forest was on the verge of disappearing into savanna or a fiber farm for pulp, poles, and saplings as the age class distribution skewed radically towards the very youngest classes, an economic as much as an ecological catastrophe (see figure 9).

Heavy storm and insect damage reminiscent of the late 1940s returned at the end of the decade as political collapse loomed.[125] The head of the Statistics Bureau noted the anomalous frequency of storm events: "In forty years we had six to eight 'hundred-year events.'"[126] Salvage harvests surged, but they may

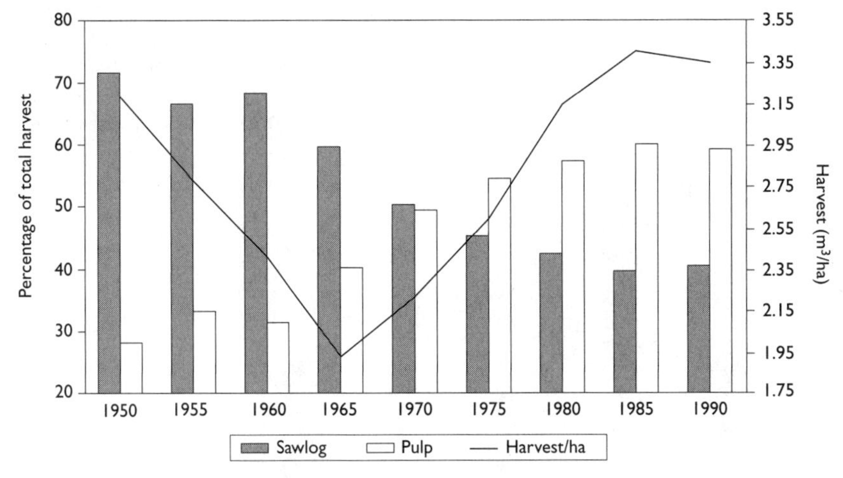


Figure 8. Pulp and sawlog harvest plotted against total yield per hectare (1950–1990). Data gathered from Wünsche and Schikora, "Der Waldfonds der DDR," 75; Schikora and Wünsche, *Der Waldfonds in der Deutschen Demokratischen Republik,* secs. 5–5.1.2, pp. 129, 131.

also have masked increased exports to hard currency markets.[127] Dollar earnings from spruce exports so pleased the export czar, Alexander Schalck-Golodkowski, that he set a premium on wood export earnings and split them 40:60 with the State Forest District directors, who used export earnings to import Stil chainsaws and Volvo skidders, tools essential to "socialist competition."[128] Exports for hard currency had been a feature of the forest economy since 1945; in the late 1980s, however, Schalck-Golodkowski sold ever-scarcer and thinner wood into a world timber market in free fall.

Three benchmarks guide evaluation of the Party's stewardship of the forest resource: reversal of endemic forest decline through conversion to a close-to-nature forest, protection of the forest from industrial and agricultural pollution, and cutting the oppressive game wildlife herd—necessary conditions for a close-to-nature forest as well as for a more equitable forest policy. Long-term stability and the health of the ecosystem, rather than cash flow and control, were the goals of rational management. The Party leadership failed in these tests, and the forest structure remained resolutely an even-aged monoculture alien to the soils and climate of Central Europe. Simplification of forest structure and the rural economy went hand in hand with reduction of the rural population to conform to the leadership's class fetish. The ecological system qualities of the East German forest—its diversity, hierarchies, resource and population flows, and resilience—declined steadily in the Party's forty-year rule, matching a comparable impoverishment in the broader East German

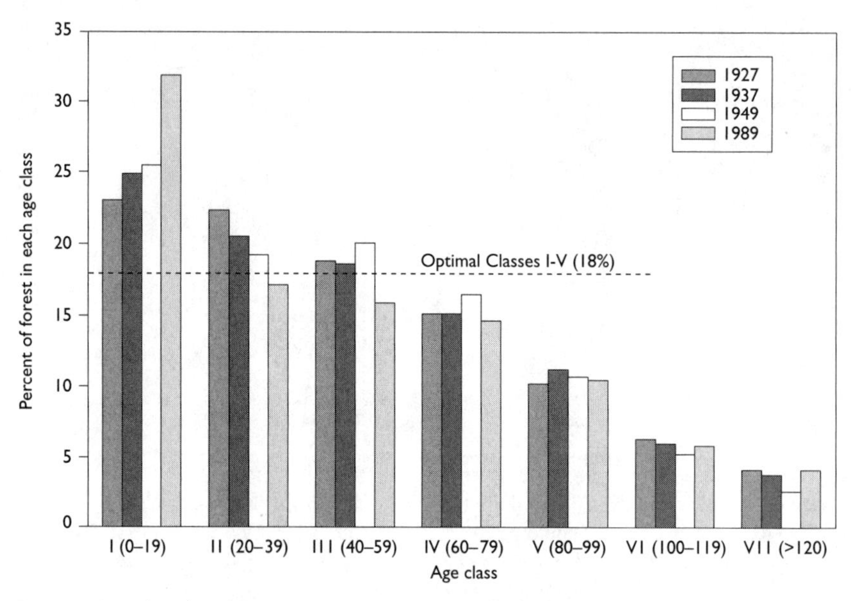


Figure 9. Age class distribution, 1927–1989. Data gathered from Hildebrandt, "Die Forsteinrichtung in der DDR 1950 bis 1965," 123; Statistisches Reichsamt, *Statistik des deutschen Reiches,* vol. 592; Statistisches Reichsamt, *Statistik des deutschen Reiches,* vol. 386; Raab, *Die deutsche Forstwirtschaft im Spiegel der Reichsstatistik;* Reichenstein, "Entwicklung von Vorrat und Zuwachs in den vier Besatzungszonen Deutschland seit 1945"; Reichenstein, "Die forstwirtschaftliche Lage Deutschland vor und nach dem 2. Weltkrieg," 30; Office of the Military Government for Germany (U.S.), *Special report of the military governor: The German forest resource survey,* 6, 7; Heitmann, "International Sammlung von Forststatistiken," 473; Wiebecke, "Zum Stand der deutschen Forststatistik," 1.

economy and society that anticipated political and economic decline. Even by the metrics of industrial forest management that the Party leadership chose, however, the Party failed miserably.

The Party claimed an increase in stocking of almost two and a half times between 1956 and 1989 as a signal achievement of socialist forestry. A forestry official bragged in May 1989, "The strategy of the Party and the government to protect and preserve the forest receives the accolades of our people and also high international recognition." The actual increase in East German forest stocking, however, was only just over one and a half times (not two and a half as claimed officially), even if one accepted 1956 as the base year. West German spruce stocking, although probably below eastern Germany's in 1945, increased almost threefold between 1949 and 1987 from a far higher basis.[129] Ten years of Soviet reparations harvests before 1956, a chronic failure to replant, and land reform destruction artificially depressed 1956 stocking

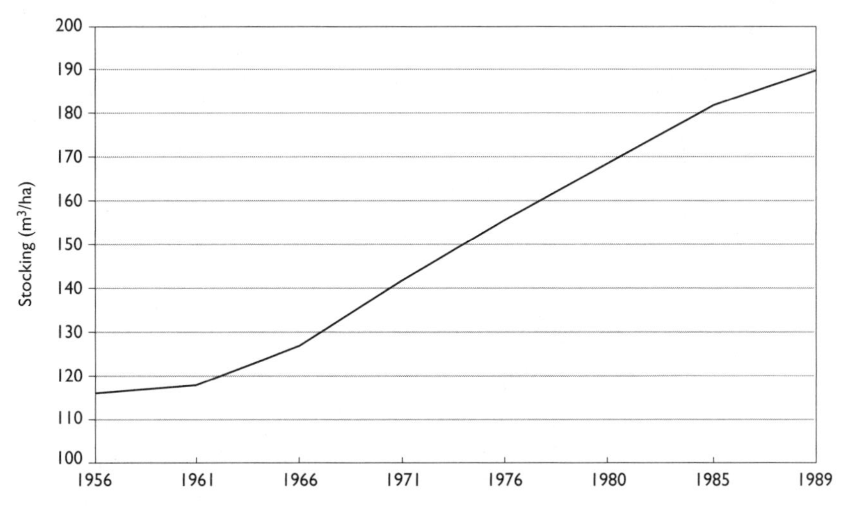


Figure 10. Stocking growth, 1956–1989. Data gathered from Schikora and Wünsche, *Der Waldfonds in der Deutschen Demokratischen Republik,* secs. 5–5.1.2, p. 74.

and created a low basis that was an ideal point from which to calculate future success. Choosing a 1956 basis granted the Party a "big bath" in stocking. Had the Party bureaucrats taken 1945 as the base year, stocking levels may even have declined.[130] Certainly, had the Party simply done nothing, growth would have been many times greater.

East German forest growth also was not of the same quality as West German growth, but rather was concentrated in precommercial stands under twenty years old—37 percent of the 1989 forest.[131] Figure 8 shows the steady deterioration of forest structure over the history of the republic as the Party stripped the forest of capital and vigor to churn out ever shorter-term returns. The forest they inherited in 1945, still predominantly artificial and in decline, nevertheless was more complex and diverse culturally, economically, and ecologically. The development of the forest's age class distribution mirrored the state's economic and social decline: a radical simplification of structure which drastically reduced diversity, resilience, and productivity.

East German forest growth plateaued in the mid-1980s (see figure 10). The modest 2–3 percent annual gains of the early 1970s, partly fuled by nitrous oxide pollution, were not sustainable.[132] In the republic's final Five-Year Plan, just as the slow growth in stocking since 1956 began its reverse into losses, the Party raised harvests to the highest levels since 1949.[133] Stocking growth dropped below 2 percent in the early 1980s, and in 1986 it started falling by 0.2 percent per year, dipping below 190 cubic meters per hectare—

Figure 11. Emergence of the German nation, 1937–1990.

BALTIC SEA
Tilsit
TO SOVIETS
Köningsberg (Kaliningrad)
EAST PRUSSIA
TO POLAND
Stolp (Słupsk)
Danzig (Gdańsk)
Elbing (Elbląg)
Köslin (Koszalin)
Marienburg (Malbork)
Allenstein (Olsztyn)
Vistula
Narev
EASTERN POMERANIA
Bromberg (Bydgoszcz)
Bug
NEW MARK
Warthe
Vistula
Warsaw
Landsberg (Gorzów Wlkp)
POLAND
Posen (Poznań)
Piliza
Grünberg (Ostrów Wlkp)
TO POLAND
Neiße
Warthe
Breslau (Wroçlaw)
SILESIA
Oder
Oppeln (Opole)
Vistula
Krakow
Königgrätz
GALICIA
Elbe
Prague
CZECHOSLOVAKIA
Moldau
Map Area

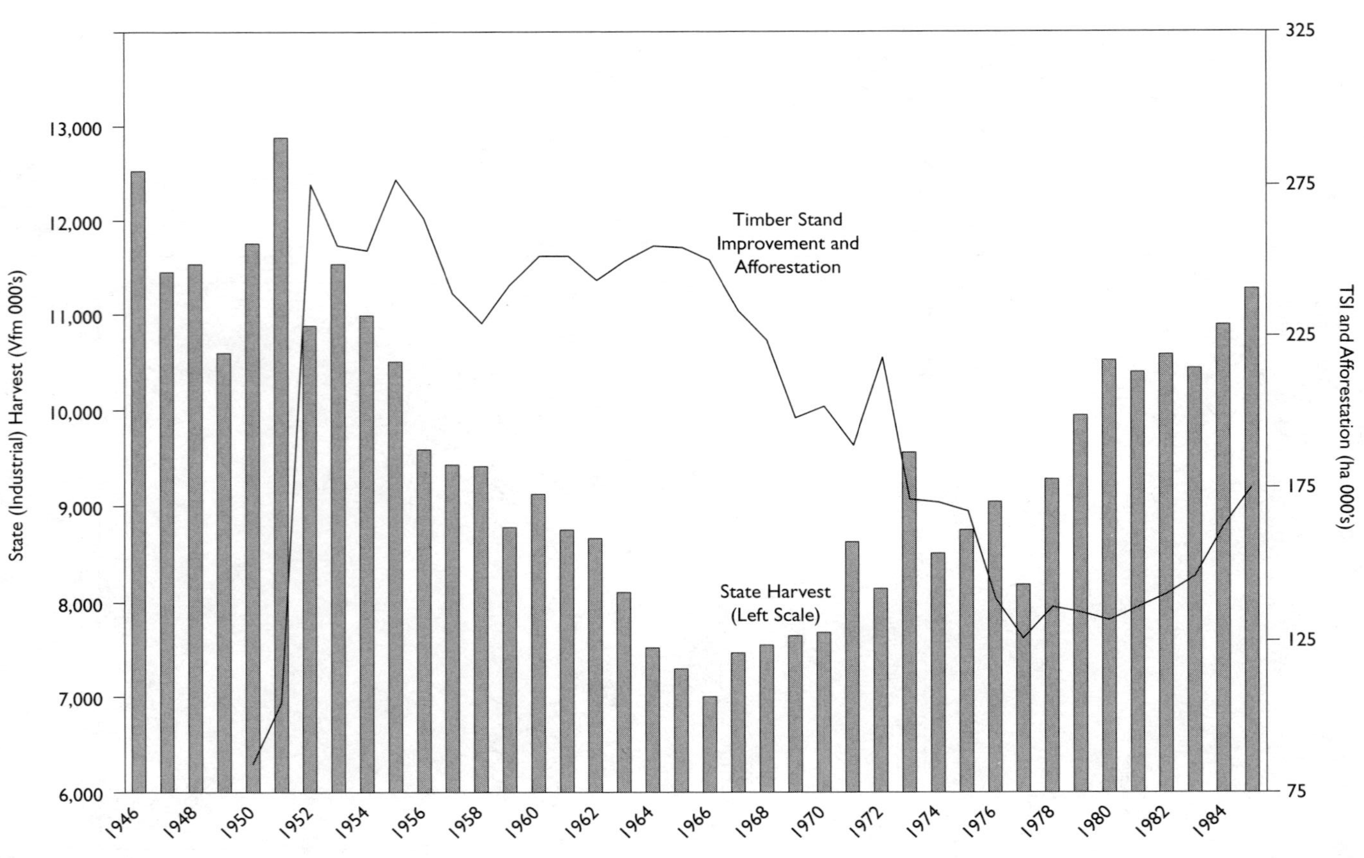


Figure 12. State harvest plotted against afforestation and TSI investment, 1946–1985. Data gathered from Schikora and Wünsche, *Der Waldfonds in der Deutschen Demokratischen Republik*, secs. 5–5.1.2, pp. 129, 131.

well below the anemic Plan goal of 220 cubic meters per hectare.[134] Even when faced with the imminent destruction of this vital natural resource, the Party leadership would not retreat from the Marxist-Leninist *Diktat* of "maximum, annually increasing production."

The story of East German forest management is also told through harvest levels and cash outlays for replanting and timber stand improvement (TSI). Harvest levels reveal the relative precedence of orthodox Marxism-Leninism and Grüneberg's Industrial Production Methods; expenses illuminate the relative importance of the forest resource to the Party leadership and indirectly measure the leadership's competence. Harvests fell into a concave U shape. Steep shoulders of heavy extraction in the late 1940s and the 1980s (see figure 12) bookended the history of the republic's forested landscape, marking the peaks of Soviet reparations and Honecker's final retreat to ideology in the 1980s. The saddle in harvest levels in the 1960s reflects Ulbricht's decision to let harvests drop for one season when forest management was roughly consistent with classic, even aged forestry.[135] Then, with Honecker's ascendance at the Eighth Party Congress in 1971, the Grüneberg era began and harvests steadily rose back to the levels of Soviet direct occupation.

Afforestation and timber stand improvement outlays tell the rest of the story. Investment ran down from its peak in the early 1950s to a low point in 1977 as the Ära Grüneberg peaked. Expenses fell in the 1950s as foresters finished clearing the huge backlog of Soviet clear-cuts, then rose again in the early 1960s as Ulbricht's NÖS reforms took hold. Once the orthodox Party members reasserted control in the mid-1960s, investment again plummeted as woodsworkers and foresters were exhorted to take up the slack. As the 1980s and the Party's power waned harvests and expenses climbed together as pollution damage control costs escalated on top of already excessive fixed costs in personnel and overhead. A more rational economic policy in the Grüneberg era would have been to slash all expenses, just as the Soviets did when they mined the forest for reparations. This would be the policy of postreunification forest management, although harvests also came to a near total halt.

Wildlife and privileged preserves for the Party and military elites (and Western business leaders and politicians) claimed more than 10 percent of the East German forest area and devoured many millions in hard currency.[136] Where hunting was concerned, neither the Primacy of Economics nor the "socialist camp's war against imperialism" counted. Honecker and Mittag, both avid hunters, demanded a high deer herd and an unnatural proportion of Frankensteinian trophy animals. Hunting specialists supervised the selection and import of specimen red deer from farther east to stiffen the herd, and district

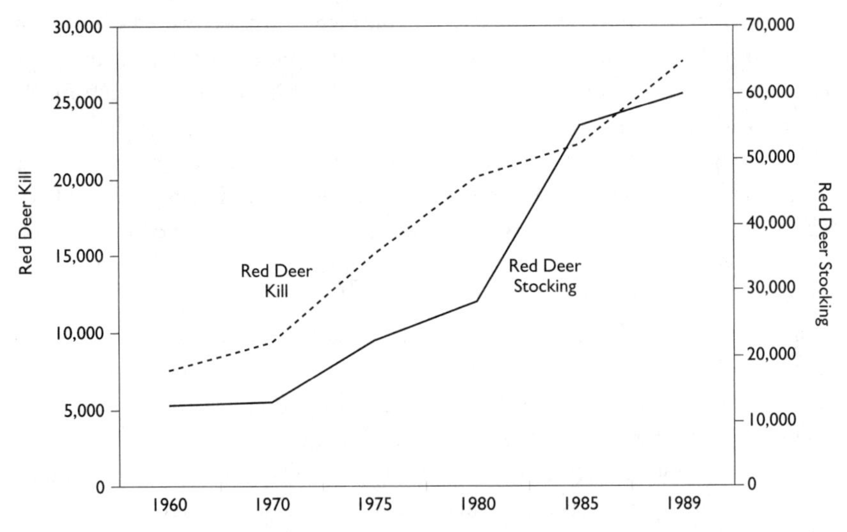


Figure 13. Red deer kill and stocking, 1960–1989. Data gathered from Staatlichen Zentralverwaltung für Statistik (SZWS), ed., *Statistisches Jahrbuch der Deutsche Demokratische Republik,* "Wildabschuß und Wildbestand," sec. X, Land- und Forstwirtschaft, 34, 245.

foresters with the best hunting fields won valuable import licenses for exotic West German animal feed to build red deer body mass and antler span: "only milk was needed to make the feed a complete baby food" snorted a disgusted feed mill manager.[137] The Party sank scarce foreign exchange into the unsportsmanlike and ghoulish project that was socialist wildlife management while denying the people fresh vegetables or fruit out of season. Foresters eager to curry favor with the Party bosses in Berlin bred trophy animals many times the habitat's carrying capacity and managed the forest for red deer: "Who has good hunting has good guests!" was a common saying among district foresters. Marx would have had bitter words for the elites who pampered animals for their sport while humans suffered deprivation.

East German hunting districts were reserved for the Politburo, favored guests from the Soviet bloc leadership, the loyal cadres, or—and this particularly infuriated East Germans—for wealthy Western businessmen. The new hunting law of 1984 restricted hunting licenses to "those who display a high sense of duty to society and contribute to the protection and strengthening of our state and actively represent the state's interests," a demand for purity reminiscent of Göring's expulsion of Jews from the *Jagdverein.*[138] Hunters had to display "increased class consciousness," and licenses were reserved for the faithful in much the same way the "children of workers," often the offspring of

Party bureaucrats, enjoyed preference in admission to university and professional schools.

Hunting values and the fetish for semidomesticated specimen red deer conflicted directly with a close-to-nature forest. Browse pressure from the heavy deer herd made natural regeneration impossible, as deer destroyed young hardwood growth and regeneration. The diverse, close-to-nature hardwood forest of mixed ages could not emerge from the simple industrial forest as long as there was such a burden of game wildlife. Foresters comfortable with industrial forest management accepted wildlife management's limitations; it won them privileges from the Party bosses, but East German game wildlife management also fit perfectly the control and production imperatives of industrial forestry. Conservationist foresters argued in vain for a balanced wildlife policy, if only to reduce the damage and expense of wildlife management. The State Commission for Forestry coyly noted the conflict: "In determining and maintaining the optimal wildlife densities certain subjective differences of interest arise between foresters and hunters."[139] The high game herd underscored the absolute failure of the Party leadership to restore a close-to-nature forest ecology.

In the end, politics saved the East German forest, as orthodox Marxist-Leninists had always said it would, but outside their control and through a political catastrophe: reunification with West Germany and oblivion for the "First Workers' and Peasants' State on German Soil." One often heard, in the years around the collapse of the East German republic, thoughtful Party members almost come to despair over the battered farm and forest landscape, sometimes saying, "But I must be optimistic. I am a communist." At 6:53 p.m. on Thursday, 9 November 1989, Günter Schabowski (secretary to the Party's Central Committee and boss of the Berlin Party organization) gave the exhausted response "Ab sofort" (immediately) to a reporter's question of when the Berlin border would open. Schabowski's comment released thousands of citizens who pushed against barriers, almost overwhelming the guards who finally let go at half an hour before midnight. The Wall evaporated overnight. And then the East Berliners came home, taunting the Party elites with contemptuous chants of "we're back!" The next day, Schabowski pleaded with the people to stay loyal to Marxism-Leninism, begging them, "As a communist . . ." This last call to a faded idealism failed, of course. The Party stayed in power only with Soviet force behind them, and with the Wall to keep their people down.

East German industry and farming burnt up almost overnight in the oxygen-rich atmosphere of liberal markets and freedom. Once the Party's control and power blew off, ecological structures began slowly to re-emerge: qualities of economic and ecological diversity, and of stable hierarchies in depth supplanting the Party's monoculture. The destruction of economic vigor and structure turned out to be far worse than the ecological decline. In the wake of the Party's stunning incompetence and destructiveness, the lives and careers of thousands of woodsworkers and foresters were shattered. The landscape did not blossom economically as Helmut Kohl, the West German chancellor, promised in 1990, but a greater change passed ecologically: the slow, sometimes halting, return of complex economic and cultural environments as the forest reknit itself seamlessly into the fabric of German national culture and memory. The era of socialist forestry vanished almost without a trace, apart from the grievous human cost, as foresters returned to oversee Central Europe's century-long process of fulfillment of Cotta's design for introducing diversity, stable structures, vigor, and high productivity to the forested landscape: a close-to-nature forest grown for the wealth, pleasure, and spiritual nourishment of all people.

# 9

## *Reunification*

*Here pause: the poet claims at least this praise,*
*That virtuous Liberty hath been the scope*
*Of his pure song, which did not shrink from hope*
*In the worst moment of these evil days;*
*From hope, the paramount "duty" that Heaven lays,*
*For its own honour, on man's suffering heart.*
*Never may from our souls one truth depart—*
*That an accursed thing it is to gaze*
*On prosperous tyrants with a dazzled eye;*
*Nor—touched with due abhorrence of "their" guilt*
*For whose dire ends tears flow, and blood is spilt,*
*And justice labours in extremity—*
*Forget thy weakness, upon which is built,*
*O wretched man, the throne of tyranny!*
*—William Wordsworth, "Here Pause: The Poet Claims at Least This Praise"*

*Life is more complicated, more mysterious and more convoluted than the party, the army, the police. Let us detach ourselves a little from this truly horrible reality and try to write about doubt, anxiety, and despair.*
*—Zbigniew Herbert, "Advice to Young Writers"*

The sudden collapse of East Germany in the weeks following the regime's fortieth anniversary on 9 October 1989 halted the Party's war on the countryside. After reunification one year later western German lawyers and foresters spread out into the five new eastern states to remake forest policy, furloughing almost the entire upper strata of the forestry bureaucracy. Most senior officials were close to retirement and had grown up professionally within the Party to reach positions of influence during the Ära Grüneberg. In the last months of his life, Professor Richard Plochmann came from Munich to bring forest policy and history to his colleagues and students at Eberswalde and Tharandt, whose traditions he so admired. A few East German foresters helped manage the transition to democratic institutions, men whose careers had been spent outside the Party. Robert Hinz, a sixth-generation chief forester of the Protestant Church forest, took over Brandenburg's forest, and Alexander Riedel, formerly in charge of the Catholic Church's forest, took over Saxon forestry.

Dr. Hans-Friedrich Joachim, now codirector of the Eberswalde Forest Research Institute, set generous pensions for retiring Party officials despite their past animosity to him and his family. I would come early in the still-dark morning to his office in the institute's imposing main building, a massive block of concrete and brick put up during Hermann Göring's tenure as Reich chief forester (*Reichsforstmeister*), a monument to Göring's twin passions for *Dauerwald* and hunting.[1] I would knock and stand in the doorway as Dr. Joachim came forward cheerfully to greet me, "Guten Morgen, Herr Fleißig!" ("Good morning, Hard Worker!"), the young assistants around him shining with energy. Natural light alone brightened the office once the sun came up, and the overhead lights were turned off to save electricity. We sometimes shared midday dinner in the institute's cafeteria, a "one pot" stew of peas and potatoes thinly larded with fatty lumps of ham, a remnant of shortages since the Second World War. When I left Eberswalde for good and we looked across the wildflower meadows in Berlin's Botanical Garden we sensed the complex landscape emerging, one balanced between *Dauerwald*'s ecological utopia and the modern industrial forest of Marxist science. The ever-changing landscape and memory of the native woodland meadow before Alfred Möller's villa evoked models for the European landscape and metaphors for political and economic policy.

The West German foresters fanning out through the eastern pine and spruce forest gradually found that the reports of forest death (*Waldsterben*) were exaggerated. Forest health was not much worse than in West Germany: declining but fitting for an industrial forest planted off-site. The first damage surveyors reported what they expected to find: massive acid rain trauma and a syndrome of pollution-induced disease and decline. Damage estimates steadily dropped as forest workers learned how to estimate damage and as the initial panic subsided. The forest was in decline, but the primary cause remained its artificial structure.

Alexander Riedel, Saxony's new minister of agriculture and forests, came to the Tharandt forestry school on a clear day in early March 1991 to talk with faculty and students about forest reforms.[2] Teachers, foresters, and chatting students in crisp green uniforms modeled on the National People's Army's filled the hall. Most in the audience expected there would be few changes to their lives. They looked for Riedel to affirm their status and the survival of socialist forest administration.

Riedel arrived a few minutes late in an old Trabant sedan, sitting in the front passenger seat next to the driver. Tired from the long morning's work, Riedel got out of the car stiffly, the plain, loden green jacket of the old Saxon Forest Service tightening slightly around his waist as he straightened. He entered the crowded hall without the retinue of aides who customarily attended senior Party officials. Sun streaming in through the hall's large windows reflected patches of shine on the worn woolen fabric of his jacket and highlighted mending along the sleeve cuffs. Riedel seemed alone against the serried ranks of professors and bright students in the formal lecture hall.

Horst Kohl, dean of Tharandt and East Germany's leading forest management scholar, rose from his chair to greet Minister Riedel. Riedel calmly looked over the audience and began, clearly and slowly, to speak of socialist forestry's economic waste and destruction. Industrial Production Methods were finished; the task of Saxony's foresters now was to restore a close-to-nature forest, no longer to produce "maximum, annually increasing tonnages of wood." The forest economy was bankrupt and harvesting had to end, as much for economic as for ecological reasons. Stunned western German overseers faced "an enormous rebuilding enterprise." Socialist forestry had "decimated the older age classes," leaving some State Forest Districts with few trees older than eighty years. Forest Districts lost between 40 and 80 DM per hectare annually and posted an annual deficit of more than 1 billion DM.[3] The Forest Districts needed to harvest 14.5 million cubic meters annually just to break even, a biologically impossible level of exploitation. After the final massive harvest in 1989 of more than 10.5 million cubic meters, foresters took 6.6 million cubic meters off the forest in 1990, close to the 1965 harvest. Then, massive salvage harvests throughout Central Europe in 1990 flooded the market. Europe's timber market chased its own tail in a downward spiral of oversupply and a market that seemed to have no bottom. Students argued from the floor that market prices of 40 DM per cubic meter were "wrong" when wood cost five times as much to deliver. Prices could not be less than cost; they angrily demanded that Riedel reinstate socialist accounting and the *Ausgleich*, fixing prices to cover costs.

Riedel shifted topics, gently advising the students that most would never work as foresters. Disbelief percussed through the suddenly nervous audience. Several students angrily demanded that the state fulfill the career promises made to them. Staffing had risen more than 50 percent in the Honecker era, to levels many times higher than in comparable West German forests, also overburdened with staff, overhead, and expenses.[4] Yet the return on East Germany's forest workers had been poor. The forest economy's annual turnover was only 420 million DM and the return per worker was 7 percent below even East Germany's low norms.[5] Clearly the Party leaders desired more from forest workers than returns: control of the countryside through foresters' service as "fulcrums of the proletariat in the countryside" and satisfaction of Honecker's fetish for autarky, economic self-sufficiency within the East German state.

Riedel then turned his suddenly sharp focus to Tharandt's faculty and the older foresters in the audience, speaking quietly of their complicity in the destruction of the countryside and their failure to protect the people: "You knew of the destruction of our homeland (*Heimat*), and the damage to our peoples' health from pollution. Our people trusted you, yet you said nothing. One word from you would have alerted our people. Because you were silent their suspicions of the Party's lies never caught fire." Riedel stopped, waiting for questions that suddenly did not come. Horst Kurth rose after a moment and brightly said to the audience, "You have your tasks. To work!"—the ritualistic closing heard at countless Party meetings. Riedel gathered his papers, fumbling just a bit, to depart through the lecture hall's sudden silence, a world vanishing behind him.

The People's Chamber adopted the West German Federal Forest Law in summer 1990, and by mid-1991 each of the five new states had passed individual forest legislation. The Forest Service was thoroughly remade through an *Umgestaltung* (reorganization). By 1991 only 3,500 forest workers remained in Brandenburg out of the 12,000 employed in 1989, and Saxony's forest workers' count dropped by two-thirds, from 10,000 to less than 3,000.[6] The farm economy suffered even greater loss, its workforce collapsing from 850,000 to under 180,000 in 1994.[7] National forest management, a signal reform of the Nazi era, ended abruptly along with East Germany's national inventory system and ecological mapping. The massive State Forest Districts, already doomed by the currency and economic union, dissolved. Even most socialist foresters recognized that the State Forest Districts were narrow "economic units" (*Wirtschaftsbetriebe*) oriented almost solely toward timber production and motivated by *Autarkie*.[8]

With reunification, close-to-nature forestry came home to eastern Germany. Lothar de Maizière, East Germany's last president, declared: "We need a totally new concept of the forest. Ecological and cultural values must have precedence over wood production."[9] Large clear-cuts were forbidden and close-to-nature forestry (*naturnäher Waldwirtschaft*) replaced Industrial Production Methods. A program modeled on West Germany's was introduced to increase the share of hardwoods to 40 percent, up from the existing anemic 20 percent, a huge task when spruce and pine rotations were lengthened beyond one hundred years and only 1 percent of the total forest was replanted each year.[10] The expensive and laborious task of underplanting hardwoods in conifer plantations will continue at a glacial pace in the postreunification environment of severe budget constraints.

Restoring economic and social diversity to the farm and forest landscape may be even slower. West German Chancellor Helmut Kohl claimed that undoing land reform and collectivization were the "most urgent" tasks of reunification, and the Federal Forest Law required that the five new states restore a balanced and diverse ownership. The Treuhand, the trustee for former East German state property, took over about 75 percent of the forest to restore private and communal forest ownership. But, as a senior Treuhand official admitted, it "has proven to be one of the most difficult problems the Treuhand has had to face."[11] Restoring the diversity and structures flattened by the Soviet land reform and collectivization will be far more difficult than smashing them apart.

Preserving land reform was one of the many unaccounted for costs of reunification. Richard Schröder, SPD whip in the People's Chamber, declared: "Had we announced after the 1990 elections that land reform would be repealed, massive popular resistance to reunification would have arisen. We wanted to avoid that."[12] The Kohl government claimed that the Soviets had demanded the inviolability of land reform as a condition of their consent to reunification. Lothar de Maizière declared, "the results of the land reform will stand," citing the "decisions of the occupier."[13] In fact, this issue was of no consequence to the Soviet leadership. Eduard Shevardnadze, the Soviet foreign minister, dismissed any suggestion of Soviet interference, and Gorbachev repeatedly denied either interest or interference in the disposition of land reform property.[14]

The Unification Treaty of 3 October 1990 had left final disposition of land expropriated under Soviet land reform to the parliament, which legalized land reform in Article 142 of the new constitution. Owners of land expropriated between 1945 and 1949 lost their rights to the farms and forests taken from them, unlike those who lost land under National Socialism or after 1949.[15] Norman Stone called it "a nasty piece of sharp practice. Something has gone very wrong in the state of Germany."[16] Despite conservatives' hatred of collectivization, Stalin's plan for a permanent shift in the social and political makeup of the central German landscape succeeded, as did his "border adjustments," both changing the nature of the German polity forever.

Land reform's distortion of the central German landscape and rural society endures, along with another twist of political geography: the loss of the German provinces across the Oder-Neiße line, lands now irretrievably Polish and Russian. Revanchist anger had defined a major strain in European politics ever since Germany took Alsace and Lorraine west of the Rhine in the 1870 Franco-Prussian War. It endured in France well after the Second World War as reflected in France's reparations withdrawals, its attempt to annex the Saar, and its demand for control over the Ruhr Valley. Recovery of the Oder-Neiße territories was a goal widely shared across the German political spectrum and by no means confined to nationalists or conservatives. Most of the Ulbricht Group sent into the Soviet zone in June 1945 supported recovery of the Oder-Neiße territories, and a return of these lands was also a cornerstone of early cold war American policy. Yet a *revanchard* movement failed to emerge in postwar Germany, an early war conundrum.

German reconciliation with the past may supply the answer. German war guilt and responsibility for the ultimate horror of the twentieth century, the Holocaust, probably conditioned most Germans to accept the geographic status quo in 1945 almost without complaint—as well as defeat, the division of their country, reparations, and harsh occupation policies in all four zones. More practically, the loss of the Oder-Neiße territories granted Germany several advantages that were not immediately apparent. The addition of the Oder-Neiße territories to the geographic and economic heft of Germany in 1989 would have raised Soviet and European opposition to reunification to unendurable levels. It is hard to think of Germany reunifying within its 1936 borders, at least not in 1989. Although the Oder-Neiße territories would have weakened a unified Germany—they had always been a drain on the Reich's treasury—the symbolic force of a German nation weighted so far to the east would have been unacceptable.

Once Poland entered NATO in 1999, Germany's security frontier shifted 650 kilometers to the east. Germany then reaped the geopolitical benefits of its lost eastern provinces without having to support the Polish population or economy. Financing reunification of the poor rump of Prussia that had been the East German republic has been massively expensive and painful. The additional burden of carrying the Oder-Neiße territories, an area one-third larger than the former East Germany, would have bankrupted Germany. By 1999 the geographic and military buffer of the Oder-Neiße territories strengthened German security at minimal cost.

The Oder-Neiße territories are also no longer German. German politicians would never have brought the Slavic populations living there into Germany, and no democratic government could have expelled them. In any event, it was hard enough to convince western German managers and civil servants to move to the former East Germany; few bureaucrats would have decamped from their comfortable villas in the leafy Düsseldorf suburbs for the dry, poor, and destitute lands of the New Mark and Silesia to live among a foreign people.

The loss of the Oder-Neiße territories, however, distorted Germany's geographic and ecological balance. Pine dominated the German forest before 1945, when Prussia alone held more than 58 percent of its forest.[17] West German characteristics dominated after reunification, particularly spruce managed uniformly under contract with the State Forest Services.[18] Germany lost the historic influence of Prussia's strong, independent private ownership, a counterbalance to the state and a historically productive base for ecological experiment and innovation. The wholesome influences of pine, Prussia, and large independent private forests—key elements in the evolution of ecological forestry—are muted in modern Germany. Other Prussian influences were also lost. Tharandt in Saxony was maintained at a university level whereas Eberswalde in Prussia was demoted to an applied science *Fachhochschule*. The Christian Democrat Kurt Biedenkopf ruled Saxony, so "red" Tharandt, even though in a smaller state growing spruce like most of West Germany, was favored. Eberswalde specialized in pine ecology and was better poised both to serve Germany as well as to carry on its historic research in the ecology of eastern Europe. Political considerations, and perhaps an unconscious bias against Prussia, won out.

The Soviet bloc collapsed in only a few weeks in late 1989 although the Soviet superstructure tottered on for some months. Long-term patterns of reduced structures and entropy foretold the Soviet empire's eventual collapse, but the timing of the sudden and catastrophic moment of change itself emerged from an ecological process of bottom-up organization of individuals into communities leading to unanticipated change. Collectivization, forest death, and pollution showed people their existential peril in the Party leaders' extreme materialism, and the Helsinki Final Act's basket three clauses diminished the Party's ability to suppress resistance, particularly protest against forest decline and pollution. But as long as the elites supported the leadership, resistance had limited scope. For a popular revolution to be successful, the elites would have to abandon their own fear and privileges and join the people in opposition.

This moment came as Soviet prestige reached its postwar apogee in the glow of Gorbachev's peace offensives, burnished by his relative youth and promises of reform and openness. Authoritarian regimes are at their most vulnerable, as de Tocqueville observed of France's ancien régime, when they start to reform: "It is almost never when a state of things is the most detestable that it is smashed, but when, beginning to improve, it permits men to breathe, to reflect, to communicate their thoughts with each other."[19] Gorbachev was not alone in his belief that he could save Soviet communism, but as de Tocqueville noted: "The very moment when our governments have appeared the strongest is when they are stricken by the malady that made it perish. The Restoration began to die the day no one any longer spoke of killing it."[20] But Western politicians, scholars, and journalists read continuity in what they saw as Gorbachev's energy, courage, and vision.

Gorbachev mixed his fumbling attempts to fine-tune Soviet communism through perestroika (restructuring) with cycles of repression. Gorbachev, as David Remnick observed, was "floundering, zig-zagging between liberalizing impulses one day and authoritarian impulses the next, steeped in the habits and history of the Communist Party apparatus."[21] He sought, in the manner of Khrushchev and Ulbricht in the 1960s, to rescue the bankrupt institutions of Soviet communism, never to replace or truly to reform them. Gorbachev's failure to complete the process of change he had begun and his holding onto control while tinkering with the command economy's clockwork signaled enlightened men and women in the Soviet bloc hierarchies to revolt. Maybe the old Brezhnev-Fraktion hard-liners were wiser than Gorbachev, for without absolute fidelity to ideology and an unyielding commitment to use power to crush opposition the Soviet bloc regimes were doomed.

Honecker's extravagant celebration of the republic's fortieth anniversary in early October 1989, capped by a triumphalist, barbaric panoply of goose-steeping soldiers, tanks and rocket launchers, and warplanes, revealed his understanding that his power rested on fear.[22] Honecker welcomed Gorbachev, his guest of honor, and the cohorts of Soviet bloc satraps, clients from nonaligned states, and third world dictators with speeches at the Palace of the Republic on 6 October 1989, declaring his fidelity to Marxism-Leninism and determination to uphold the Wall. Honecker mocked the demonstrators in the streets of Leipzig, Dresden, and East Berlin, threatening to repeat the Chinese communists' massacre of hundreds of students and civilians in Tiananmen Square four months earlier on 3–4 June 1989. "Tiananmen," Honecker reminded East Germans in a front-page article in *Neues Deutschland*, "taught the fundamental lesson of adhering to the basic values of socialism and at the same time further perfecting socialist society."[23] At the height of the fortieth anniversary festivities on 7 October 1989, Honecker ordered his successor-to-be, Egon Krenz, to be ready to use the "Chinese solution" against demonstrators in Leipzig. Krenz, even though he admired the Chinese for their firmness at Tiananmen, hesitated and held back the police as he saw Honecker's power fading.

Honecker turned to other threats, declaring refugees to be counterrevolutionary "victims of large-scale provocation by West German revanchists and neo-fascists." He advised the thousands of demonstrators flooding the streets to "ask themselves if they are prepared to put at risk the safety of their families — and not least of all their children — for the sake of West German revanchists and neo-fascists." Honecker asserted a self-evident justification: "It is a matter of advancing further the unity of economic and social policy": the hoary old *Bündnispolitik*. Demonstrators, Honecker chided, should not attack the "accomplishments of socialism" but return to their "battle stations for socialism and peace" to work even more diligently and enthusiastically.[24] The Party had always used threats, from withholding education to the forced labor camps, to command parents' loyalty. Now Honecker revealed that he was prepared to murder children to hold onto power.

Gorbachev urged reform on the East German Politburo in a 7 October 1989 meeting, prophesizing, "Life will punish the man who comes too late" (Wer zu spät kommt, den bestraft das Leben).[25] Honecker, his behavior recalling Ulbricht's sparring with Brezhnev, retorted with denunciations of West German "revanchism" and a catalogue of industrial successes, such as the Robotron four-megabyte chip. In a speech at a state banquet later that evening Honecker continued, "Socialism will be halted in its course neither by ox, nor ass." The parade the next day, choreographed to the music of crashing military bands, was Honecker's final answer: if the Soviets left the fate of Germany to "Berlin," as Gorbachev proposed, he would use force to hold onto power. Ominously, almost as soon as Gorbachev left for his flailing government in Moscow, East German intellectuals burst out in unprecedented criticism, as senior members of the Academy of Arts joined with ranking Party members to demand freedom of speech.[26] Honecker's Politburo colleagues deserted him, and he was out of power only ten days later, on 18 October.

Gorbachev's commitment to reform was no stronger than his influence on his client states' leaders. During his stay in East Berlin, Gorbachev also met in private to urge his perestroika on the gathered Soviet bloc and nonaligned leaders. Daniel Ortega left East Berlin to stun a Central American presidential summit meeting in Costa Rica with his decision to renew fighting with the U.S.-supported contra rebels.[27] Two weeks after the Czech Party bosses returned home, secret police arrested leading dissidents and Charter 77 human rights advocates, seizing Vaclav Havel from his sickbed on 28 October 1989. Western diplomats' requests to visit Havel in prison were denied until the Czechoslovak regime's own anniversary celebrations were safely over.[28] Czech Party leaders had seen how demonstrations had nearly wrecked Honecker's celebration and led to his fall.

Gorbachev preached reform to his satraps and clients, but he remained at heart an authoritarian fixated on control at home and revolution abroad. Despite the peace offensive Gorbachev mounted in the West, Soviet foreign policy and support for terrorists and international liberation movements continued on the same aggressive path that Brezhnev took.[29] So the Soviet bloc leaders and nonaligned clients who left East Berlin to put down rebellion and forestall change were doing as Gorbachev did, not as he preached.

Events in Moscow spun as quickly out of control for Gorbachev as they had for Honecker in East Berlin. In asserting "continuity," particularly the permanence of Communist Party power, Gorbachev alienated the Soviet bloc's middle class without winning the support of right-wing hard-liners. With his reform rhetoric, however, Gorbachev surrendered full use of the only tool remaining to hold onto power since Marxist-Leninist ideology had been discredited: the threat of force. Gorbachev's natural constituency, the white-collar and the scientific and technical elites, began publicly to vent their deep frustration with perestroika, openly criticizing Gorbachev himself. Once intellectuals, business managers, and government officials lost their fear, they joined with citizens exhausted from decades of mismanagement, threats, and deprivation to overthrow the Marxist-Leninist regimes of the Soviet bloc. Now the elites and the people had lost their fear of the Party and belief was cold, the revolution could finally proceed.

The imminent fall of the Soviet empire thus burst into the open at the Soviet Diplomatic Academy in Moscow on 27 October 1989, ten days before the Berlin Wall fell. The Soviet Foreign Ministry had invited Zbigniew Brzezinski, the bête noir of the Soviet empire and former national security adviser under President Jimmy Carter, to speak to five hundred senior diplomats and academicians. Brzezinski illuminated a scene familiar to most in the audience: the instabilities of the Soviet system, its intractable problems with nationalities, and the waning power of Marxist-Leninist ideology and fading memories of the Stalinist Terror.[30] Soviet economists, such as Nemchinov, had warned of the inherent weakness of the command economy and central planning since the 1950s; now an even more serious threat joined with economic decline: the demographic crisis in the Soviet Union, a crisis heavily influenced by Soviet destruction of the natural environment.[31] The life expectancy for Soviet men had fallen from sixty-seven years in 1964 to under sixty-two in 1980 at the end of the Brezhnev era. By 1989 it fell below that of Bangladesh, India, and some countries of sub-Saharan Africa—a population crash for an industrialized nation without precedent in human history in the absence of famine or war.[32] The existential threat was inexorably linked to the regime's extreme materialism and Marxist-Leninist political philosophy.

The atmosphere in the hall was electric; Marin Strmecki, Brzezinski's assistant, described the energy as echoing the high drama of the 1917 March revolution when Tsar Nicholas II abdicated and the imperial government was overthrown.[33] Brzezinski, speaking under a large statue of Lenin, marked the tides of history running against Marxism-Leninism and urged the Soviets to join the Western community. He recalled that this was the fourth time he had stood publicly while unrest raged in the "socialist camp"; he had spoken in János Kádár's Budapest after the Hungarian Revolution of 1956, in Prague as Alexander Dubçek pursued "socialism with a human face" in spring 1968, and in Poland in August 1989 just before Solidarity was recognized and invited to join a coalition government, all events that led to regime change.[34] Cheers punctuated Brzezinski's comment: "This visit to Moscow is my fourth!"

An American diplomat present described the reception to the speech as "extraordinary." A senior Soviet diplomat said that Brzezinski's analysis was "actually mild in comparison to the current tenor of debate in Party and academic circles and that most of the ideas he mentioned had adherents in the Party and government structure."[35] The massed elites gave Brzezinski a standing ovation twenty-three months before an abortive Soviet army coup against Gorbachev on 19–21 August 1991 finally opened the door for Boris Yeltsin. Enlightened individuals were building communities and joining the groundswell of popular revulsion and anger at Marxism-Leninism's destructiveness, de Tocqueville's preconditions for revolutionary change.

The Foreign Ministry then granted Brzezinski permission to visit the site of the Katyn massacre, an atrocity with special meaning for an American born in Warsaw in 1928. In 1940, a KGB execution squad had shot as many as fifteen thousand senior Polish officers captured when the Soviets invaded Poland in partnership with Hitler under the terms of the Molotov-Ribbentrop (Nazi-Soviet Nonaggression) pact.[36] A television crew was dispatched to Katyn to film Brzezinski's visit for broadcast on Soviet television, challenging Gorbachev's control of the media and anticipating official acknowledgment of Soviet responsibility for the mass murders in 1992. Brzezinski arrived at the forest clearing on the margins of the mass graves just as four hundred relatives of the

Katyn martyrs unexpectedly arrived for a solemn ceremony of remembrance, the first allowed them by the authorities. Three priests celebrated a requiem Mass on a rough altar in the forest, Solidarity banners and a rough timber crucifix rising behind them over the flower-covered mass graves. This was not the glasnost (openness) that Gorbachev had envisaged; it was spontaneous, deeply revealing, and it forged strong bonds not just between opponents of the Soviet empire, but between conscious individuals within the Soviet apparat and the people of eastern Europe.

For months after the fall of the Berlin Wall many expected that the socialist economy would persist within a new confederated German nation. But just as the economic decline in the forested landscape was a more immediate problem than the ecological crisis, so too were economic problems most severe in East Germany, a state whose economy in 1990 suddenly seemed to be below the productivity and living standards of many its Warsaw Pact allies. No one had seen the depth of economic decay before. The finance minister's report that the deficit ran over 73 billion DM, dwarfing production, sparked "gasps of amazement in the hall." External debt stood at 29 billion DM, almost twice previously reported amounts. Estimates early in 1990 cut per capita production from 50 percent of West Germany's to 30 percent, and that was probably still too generous. A Party economist confessed that inflation was far above 12 percent, shattering the myth of stable prices.[37] These reports did not touch the bottom; there was no way to draw up a balance sheet for such total economic failure.

The burden of absorbing the failed East German state and economy (virtually derelict after forty years of neglect and exploitation) still dominates the finances of Germany. East German recovery ate up 180 billion DM by 1992, 25 percent of total federal spending and 6.5 percent of the West German gross national product. This enormous transfer did not include postal and rail expenses, or the Treuhand's, the receiver for the East German state's property, costs of over 270 billion DM.[38] Net transfers to the east continue to increase, more than 1 trillion West German marks by 2000, not counting private and corporate investment, and still rising. A Bundesbank report pessimistically forecast in 1994 that these costs would be "redeemable only in the course of a generation," yet another optimistic estimate.[39] Reunification by the end of 2004 cost more than 1.5 trillion euros (over $1.9 trillion). Despite this huge expense, eastern Germany remains dependent on annual injections of 90 billion euros ($116 billion), a massive continuing outflow that is beginning to look structural.

The conditions necessary for a successful revolution therefore came into place late in 1989, conditions that had been missing in earlier revolts in Berlin (1953), Budapest (1956), and Prague (1968). The structures and ideas of the Soviet empire, however, had been in decline since at least the end of the Second World War. The energy of the Soviet empire was entropic, not progressive. But why did Western political scientists not discern the clear telltales of the sustainability of Soviet power in the late 1950s, when senior Soviet academicians and economists first warned of the command economy's inherent instabilities?

The irrational appeal of the Party's power overcame most Western analysts, who projected "continuity" with no less conviction than Honecker. They hailed East Germany as the "thirteenth member of the European Community," a Soviet bloc success story. Westerners joked that "only the Germans could make socialism work." Honecker turned to Western ignorance to bolster his prestige at home. In his speech opening East Germany's fortieth anniversary, Honecker hailed the World Bank's ranking of East Germany as "amongst the first ten leading industrialized states, relying only on the purity of socialism." The World Bank had raised East Germany to the same economic class as Great Britain and Italy when its economy probably lagged behind Poland's and Hungary's.[40] The myth of East German economic prowess that he had so carefully created returned with the imprimatur of the world's leading economic institution just as East Germany was dissolving, the only Warsaw Pact country to vanish without a trace.

Social scientists' failure to predict even reunification is particularly damning given East Germany's instability and the "sheer economic incompetence" of its leadership.[41] Diplomats, scholars, and journalists all misread the true, fragile nature of East Germany. Louis Edinger, in his authoritative 1986 study *West German Politics,* declared that reunification was "not even a remote possibility."[42] Another scholar declared in 1987 that "any chance of political reunification is as far away as ever it has been since 1945."[43] Henry Kamm in the *New York Times* in March 1989 praised the skills of the East German central planning apparat, particularly "the leadership of Erich Honecker, strong with a sense of comparative success, [which] affirms the eternal Communist verities." Kamm highlighted East Germany's "Prussian, highly orderly, highly effective system" and the peoples' discipline and loyalty.[44] As late as November 1989 Craig Whitney of the *New York Times* declared, "the two countries are likely to join their industries before their politics," and suggested that German reunification would be as laborious a process as European political union.[45] Even the *Wall Street Journal*'s Terence Roth adopted Marxist-Leninist syntax to downplay reunification due to the Party's "rigid control over the means of production."[46] When reunification came, it came with sudden force and energy, sweeping aside the corrupt and sclerotic institutions of the "First Workers' and Peasants' State on German Soil" that so many Western analysts had found so stable and strong.

Many analysts accepted the Marxist-Leninists' terms of debate, that "capitalist" and "socialist" camps fought one another. But perhaps the struggle was not economic and political but rather intellectual and moral, between people who understood the limitations to human control and those who saw the world as simple and governed by an absolute truth. Goethe asked us to live in the middle landscape between custom and innovation, as if he were defining the landscape between National Socialism, with its belief in race and nature, and Marxism-Leninism, an ideology formed around class and progress. The bipolar model of global security blinded many Westerners to the "complexity of evolving reality," in Oskar Morgenstern's words. It was almost a truism in the years following the collapse of the Soviet empire that security had to be thought of in its political, economic, environmental, and cultural dimensions to account for complexity.[47] Yet this was no less true during the cold war, where perceptions of the world as simple and bipolar led directly to the systemic failure to foresee the Soviet collapse.

The Party leaders' orientation to reality, a key to deciphering the nature of their rule, surfaced first in the natural landscape and is indispensable to analysis of the regime's political history. Landscape change and structure revealed the state's entropic path. The Party leaders' first political initiative, the land reform of 1945, marked the beginning of a forty-year process of simplification and increasing disorganization as they reduced diversity and structure. Forced collectivization in the 1950s, and the parallel tracks in forest policy of extreme ecologism and exploitation, advanced this trend and launched the first major act of popular resistance: mass *Republikflucht* across the inner German frontier to the West, almost causing the collapse of the East German regime. After the Berlin Wall plugged the last open portal to the West in August 1961, Ulbricht schemed to simulate complexity through cybernetics in his NÖS reforms of the 1960s. Rather than a program of reform, the NÖS was a weakly executed, poorly conceived innovation relying on a cybermarxist magic bullet.

After the collapse of Ulbricht's cybermarxism in 1971, entropy and the process of disintegration quickened under the leadership of Gerhard Grüneberg, the principal enforcer of Industrial Production Methods in the farm and forest landscapes. Pollution for the first time in the early 1970s was linked in people's minds to forest decline, a process already under way for one hundred years. These symptoms of the Party leaders' extreme materialism—pollution and continuing forest decline—fused in the popular imagination as forest death, *Waldsterben*. Unable to confront the Party's political power and control directly, the people saw forest death as a metaphor for the anxieties they felt in the face of the increasing disorganization around them and their existential peril. In protesting against forest death, the East German people confronted the inconsistencies and poverties of "real, existing socialism."

Popular anger over forest death ignited the overthrow of the Honecker regime, but perceptions centered on symptoms of the Party's misrule, pollution, and forest death rather than on the problem itself—the Party leaders' extreme materialism and absolute faith in their grasp on historical forces. The prescriptions Party bureaucrats favored to combat forest decline (fertilization and introduction of plantations of fast-growing, smoke-tolerant hardwoods, such as poplar [*Populus spp.*]) might have mitigated the symptoms of decline, but they never addressed the root problem: the forest's artificial system qualities; its lack of diversity; impoverished and flattened hierarchies; weak resilience; and the opacity and paucity of its information, resource, and population flows. West German environmental aid to the East addressed the symptoms of decline in a similar manner, incidentally protecting the Party leaders from the telltales of their misrule. The Party's policies had flattened and impoverished ecological system qualities in the natural landscape; these same dysfunctions existed in civil society and in the economy. Political philosophy ultimately was responsible for forest decline, and it took political change to bring relief to the landscape and its peoples.

The true test of the Party's stewardship was not whether it chose the correct economic theory, whether management was Marxist-Leninist and industrial, or "capitalist," or conservative and ecological; at various times in East German history all these approaches were followed, sometimes simultaneously with near-disastrous effect. Criticism should be based on the degree to which policy reflected complexity. The goals of socialist forest management were less significant than how policy deepened and extended forest diversity and structure, its resilience, and the transparency and strength of its flows.[48] The objective measures of forest structure and stand dynamics discussed in chapter 8, such as harvest levels and inputs over time, growth rates, age class distribution, and changes in species composition and stocking levels, revealed across the board impoverishment of ecosystem quality without the filter of prices or interpretation. The lesson of East German forestry is that we should make policy with an eye toward reducing control to increase individual freedom and autonomy and let communities emerge from the bottom up. Rather than striving to direct these complex systems toward a specific goal or endpoint, even the most enlightened, policymakers should work to create richness and abundance in the systems qualities of economic and political structures.

Neither National Socialism nor Marxism-Leninism had any power that was not at heart destructive and entropic. Their leaders' orientations to complexity and fear of diversity and the individual strongly bind these irrational ideologies. The wars of the twentieth century were not fought between fascism and communism, or even between capitalism and socialism, but between the West's default orientations of complexity and contingency, an aesthetic of messiness and emergence, and the Nazis' and Soviets' materialist orientations of order and control. This struggle was therefore quintessentially ecological and philosophical, neither biological and race driven, as the Nazis proposed, nor economic and class driven, as the Marxist-Leninists asserted. Of course, complexity, diversity, and the individual are also under constant attack in the West. The fall of the Soviet empire did not resolve Emerson's "primal antagonism" of nature versus modernity any more than it brought about an end to history.

Marxist-Leninist ideology was billed as a political philosophy of progress and humanism. More accurately, it reflected an irrational belief in an entropic utopia of disorganization and sameness. An ecological approach, in contrast, does not look toward a steady state or discrete goals, but toward abundance and complexity. Ecosystems do not have purpose but rather evolve from current conditions toward a future stage; if control levels are low, then what Norbert Wiener, the father of cybernetics, called "local and temporary islands of organization and life" will emerge despite a surrounding pattern of increasing disorder.[49] Goethe marked this point of balance critical to life and creativity in *Faust*, in the dynamic "intermediate period when the opposition is still possible," the evanescent and ever-changing region between the certainties of nature and modernity, a point of balance similar to Thomas Jefferson's "middle landscape."[50] Hannah Arendt identified the limits control forces upon productivity. Release of human control to allow complex systems to emerge and adapt will be a better policy than directing them toward a specific goal. Human values still matter deeply, but those values close to Isaiah Berlin's "pluralism," of openness, diversity, and tolerance rather than the absolute truths of ideology or politics.[51] If we nurture such ecological values in our political and economic systems, we will have more productive as well as more democratic institutions.

Social scientists need to confront experience to look at the results of policy in communities and neighborhoods and then temper their observations critically against regimes' formal goals. The natural landscape, a complex mix of biological, abiotic, and human-directed forces and culture, manifests unambiguously leaders' informal goals. The East German leaders' formal goals were "the betterment of mankind" and peace. Their subliminal goals were control, power, and a process of disorganization that masqueraded as ideology. Measuring goals against their manifestation is particularly valuable in the analysis of repressive or authoritarian societies, where secrecy is always a central tool of state power and absolute truth not only gives the elites an imprimatur to crush individuals, diversity, and freedom, but also to assert the validity of their truth as justification for the suffering they cause. Goals may be better used to mark the degree of difference between the Plan and what happens in reality, a direct measure of productivity and resilience.

Resolution of forest decline will not be found in the natural sciences. Foresters have known for almost two centuries that forest decline is a function of reduced, human-made structures, yet they have been unable to complete the great ecological revolution of the nineteenth century and introduce a close-to-nature forest. People do not see natural landscapes as made up of diverse ecosystems of ambiguities and individuals, of temporary, messy structures mediated by sudden change. For most people, an ideal landscape reflects their own preferences for control and risk aversion and their cultural biases. "Natural" more often means structures that are either artificial or governed by abstractions. Many Germans perceive the modern, artificial, park-like forest as something both uniquely natural and expressive of the German spirit at once; the complex, never-before-seen close-to-nature forest is alien to the popular aesthetic. They also read the large game herd as evidence of wildness and nature, even though these animals are semidomesticated and block the emergence of a truly natural and German forest.[52]

Neither the radical ecologism of *Dauerwald* nor the extreme materialism of Marxism-Leninism yielded the flourishing forested landscapes their artificers promised. They instead blocked the emergence of healthy and productive ecosystems through their designers' fetishes for absolute truths and belief in simple historical process. When the ecosystems of Central Europe become more stable and productive it will be because people have come to see them as emerging from complex ecological and geographic constraints and have let go of the assurances of modernity and the comfort of romantic abstractions. People, no less than political and economic leaders, need to release illusions of control and embrace risk and transparency instead—to live with messiness and a contingent world.

Rather than trust a society's well-being to the unique insights of any group, whether green, red, or brown, it is healthier to create political, economic, and cultural structures that foster diversity, to stay out of the way of individuals and let communities and structures emerge from the bottom up rather than imposing abstractions, whether the beautiful constructs of a perfected human relationship with Nature or the realization of the secular heaven of communism. George Kennan, speaking of student radicals, observed in the late 1960s:

> I have seen more harm done in this world by those who tried to storm the bastions of society in the name of utopian beliefs, who were determined to achieve the elimination of all evil and the realization of the millennium within their own time, than by all the humble efforts of those who have tried to create a little order and civility and affection within their own limited entourage, even at the cost of tolerating a great deal of evil in the public domain. Behind this modesty, after all, there has been the recognition of a vitally important truth—a truth that the Marxists, among others, have never brought themselves to recognize—namely, that the decisive seat of evil in this world is not in social and political institutions, and not even, as a rule, in the will or inequities of statesmen, but simply in the weakness and imperfection of the human soul itself, and by that I mean literally every soul, including my own. For this reason, as Tocqueville so clearly perceived when he visited this country a hundred and thirty years ago, the success of a society may be said, like charity, to begin at home.[53]

How will history judge East Germany's forty-year span? Stefan Heym, the East German writer, commented wanly in a television interview at the Palace of the Republic after the stunning conservative victory in the 2 December 1990 elections returned Chancellor Kohl to power, the first free general election in all of Germany since the Reichstag elections of July 1932: "There will be no more GDR. It will be but a footnote in world history."[54] More and more, the postwar division of Germany which once seemed so absolute emerges as temporary, a footnote, as Heym mourned. The destruction of East Germany's diverse and flexible rural society and the damage the forty-year reign of control and fear wreaked on the eastern German people's psyches, not political ideas and structures, are Marxism-Leninism's principal legacies. Polluted rivers, dying forests, and exhausted, nitrate-poisoned agricultural commons are lethal political pentimenti of the communists' authoritarian irrationalism, outward and visible signs of the twisted idealism and arrogance which destroyed the "First Workers' and Peasants' State on German Soil."

# *Notes*

## *Chapter 1. Prologue*

1. Timothy Garton Ash, "Ten years after," *New York Review of Books* (18 November 1999): 16.

2. Möller is best known for his *Dauerwald* (permanent forest) system of ecological forest management. *Dauerwald* evolved into modern Germany's close-to-nature forest management movement.

## *Chapter 2. Landscape and Culture*

1. Cornelius Tacitus (AD 56–117), *Germania,* trans. M. Hutton, rev. by E. H. Warmington (Cambridge: Harvard University Press, 1970), 137. Tacitus speaks of "silvus horrida aut paludibus foeda."

2. Michael Charles Kaser and Edward Albert Radice, eds., "Economic structure and performance between the two world wars." In *The Economic History of Eastern Europe, 1919–1975,* vol. 1 (Oxford: Oxford University Press, 1985); Horst Kurth, ed., *Wissenschaftliche-Technische Darlegung zur Intensivierung der Holzproduktion und zur komplexen und volkswirtschaftlichen effektiven Holzverwertung* (Tharandt: Technische Universität Dresden, 1979), 4.

3. Willy Brandt, "Man darf die Grüne nicht in Schönheit lassen," *Die Welt* (5 February 1987).

4. Richard Wagner (1813–1883), "Ring of the Nibelungen," trans. Andrew Porter (New York: Norton, 1977); Bernard Williams, "Wagner and politics," *New York Review of Books* (2 November 2000).

5. Heinrich Heine (1797–1856), *Harzreise* (English and German), trans. Charles G. Leland (New York: Marsilio, 1995), 85.

6. Johann Wolfgang von Goethe (1749–1832), *Faust,* part 1, trans. Randall Jarrell (New York: Farrar, Straus and Giroux, 2000), 235–39.

7. Arnulf Baring and Volker Zastrow, *Unser neuer Größenwahn. Deutschland zwischen Ost und West,* 3rd ed. (Stuttgart: Deutsche Verlags-Anstalt, 1989); C. Wiebecke, "Zur 150. Wiederkehr von Goethes Todestag (22 March 1832)," *Forstarchiv* 53:2 (1982): 72. Wiebecke examines the influence of the forest on Goethe and German culture and Goethe's relevance for forestry; Franciscus Wilhelmus Maria Vera, *Grazing ecology and forest history* (Wallingford, Eng.: CABI Publications, 2000).

8. Cornelius Tacitus, trans. Charles William Eliot in *Voyages and travels: Ancient and modern,* "The Harvard Classics" (New York: P. F. Collier and Son, 1909–14), vol. 33; Montesquieu, Charles Louis de Secondat, baron de La Brède et de, *Laws,* 11, 6. Cited in Cornelius Tacitus, *The Germania and Agricola,* ed. W. S. Tyler (New York: D. Appleton, 1869), 98, *Making of America books*, University of Michigan Humanities Text Initiative, http://www.hti.umich.edu/t/txtgifcvtdir/aht6987.0001.001/00000110.tifs.gif (accessed 9 March 2002). Montesquieu traced the inspiration for the British constitution to the ancient German reverence for the forest.

9. James George Frazer, Sir, *The golden bough; a study in magic and religion,* 3rd ed. (New York: Macmillan, 1935), 9; *The golden bough,* chap. 9, sec. 1, "Tree-spirits"; Julius Caesar, *De bello Gallico,* 6, 25, cited in Tacitus, *Germania and Agricola,* ed. Tyler. Caesar in *The Gallic Wars* described the Hercynian Forest as "nine days' journey in breadth and more than sixty in length," a claim somewhat more modest than Frazer's. Frazer, the brothers Grimm, Caesar, and Tacitus present the standard take on the historical landscape ecology of Central Europe: that a dense cover of primeval forest, an *Urwald,* blanketed Europe before the late Neolithic forest clearances for agriculture, or the medieval clearances. Modern researchers suggest instead that the combined pressure of peasant woodcutting and grazing created a park-like, more thinly wooded ecosystem than previously thought, one not dissimilar to the North American forest at the time of contact in the early sixteenth century.

10. Goethe climbed the Brocken twice more and made many other excursions in the Harz region, cataloguing its plants and animals, mapping its geology, and making marvelous pen and ink drawings. Nicholas Boyle, *Goethe: The poet and the age,* vol. 1, *The poetry of desire, 1774–1790* (Oxford: Clarendon Press, 2000), 298; Johann Wolfgang von Goethe, Letter to Charlotte von Stein, 10 December 1777, *Briefe und Tagebücher,* vol. 1 (Leipzig: Im Insel-Verlag, 1957), #159; Johann Wolfgang von Goethe, Letter to Clausthal, 11 December 1777, *Briefe und Tagebücher.* vol. 1, #227.

11. Johann Wolfgang von Goethe, *Tagebücher,* ed. Herbert Nettl (Düsseldorf-Cologne: Eugen Diedrichs Verlag, 1957), 34. The last sentence is from the Eighth Psalm.

12. Brockhaus, "Walpurgisnacht," *Die Enzyklopädie*, vol. 23 (Leipzig and Mannheim, 1996), 532. The Walpurgisnacht myth's roots lay in the eighth century, when popular storytellers confused pagan rites celebrating the end of winter and the Celtic fertility goddess Waldborg with the anniversary of the canonization of St. Walburga (710–779), daughter of King Richard of England and missionary to the Germans, conflating contradictory symbols of wildness and ancient custom with Christian order, innovation, and

modernity. In the seventeenth century the Walpurgisnacht myth evolved into its modern form, a fairy tale of the "witches' Sabbath" (May Day Eve), when witches rendezvous over the Brocken summit on broomsticks and churn-crashers, searching the fields and villages below for careless peasants who had not protected their barn thresholds with crossed boughs and twigs of mountain ash and birch. Then, the witches could swoop down unhindered to dry up the milk of the lazy peasants' cows. St. Walburga ruled both monks and nuns at the great Benedictine abbey at Heidenheim and is buried at Eichstätt.

13. Ralph Waldo Emerson (1803–82), "Goethe, or the writer." In *Representative men: Seven lectures* (Cambridge: Belknap Press of Harvard University Press, 1987), 161.

14. Robert D. Richardson, Jr., *Emerson: The mind on fire* (Los Angeles: University of California Press, 1995), 249.

15. John Muir (1838–1914), *The mountains of California, 1838–1914* (New York: Century, 1894).

16. Mendelssohn played Bach and Mozart for Goethe. Mendelssohn's teacher was K. F. Zelter.

17. Johann Wolfgang von Goethe, letter to Felix Mendelssohn, 9 September 1831.

18. See Hensel, *The Mendelssohn family* (1886) for an "exceedingly graphic and delightful description." Mendelssohn described this event in his letters to his family as well.

19. Heine, *Harzreise,* 89–91.

20. Karl Marx (1818–83). G. F. W. Hegel (1770–1831).

21. Isaiah Berlin, *Karl Marx: His life and environment*, 4th ed. (New York: Oxford University Press, 1996), 27.

22. Felix Mendelssohn-Bartholdy, *Selected letters of Mendelssohn,* ed. W. F. Alexander (London: Swan Sonnenschein, 1894), 77. Mendelssohn later again made light of Idealism and Hegelian philosophy in a letter to Fanny, "I have an idea, which is 'more sublime than the whole of Nature'—I mean to go to bed." Felix Mendelssohn-Bartholdy, letter to Fanny Mendelssohn (13 August 1831), in *Letters from Italy and Switzerland,* trans. Lady Wallace, 3rd ed. (New York: F. W. Christern, 1865), 250.

23. Mendelssohn liked Berlioz when they met in Paris in 1843, yet Berlioz seemed coarse to Mendelssohn, who found his music "completely without talent." Friedrich Blume, ed., *Die Musik in Geschichte und Gegenwart,* "Mendelssohn, Felix" (Kassel: Bärenreiter, 1961), vol. 9, p. 69.

24. Felix Mendelssohn-Bartholdy, letter to Fanny Mendelssohn (22 February 1831), in *Letters from Italy and Switzerland,* trans. Lady Wallace.

25. Nicholas Boyle, *Goethe: The poet and the age,* vol. 2, *Revolution and renunciation, 1790–1803* (Oxford: Clarendon Press, 2000), 298. Goethe saw these "last surviving pre-Christian Druids" as primal worshippers of pure nature and its healing powers.

26. Blume, *Die Musik in Geschichte und Gegenwart,* 69; Harold Stein Jantz, *The function of the "Walpurgis Night's Dream" in the Faust drama* (Madison: University of Wisconsin, n.d.), 407. Although Goethe "clearly recognized the possibility of a Christian perversion of something originally dignified and noble," he wished that neither extreme should dominate.

27. Berlin, *Karl Marx,* 44.

28. Karl Hasel, "Die Beziehung zwischen Land- und Forstwirtschaft in der Sicht des

Historikers," *Zeitschrift für Agrargeschichte und Agrarsoziologie* 16:2 (Oktober 1968): 141, 153.

29. G. Hofmann, "Vergleich der potentiell-natürlichen und der aktuellen Baumartenanteile auf der Waldfläche der D.D.R," *Hercynia* N.F. 24 (1987): table 2, 199.

30. Heinrich Cotta (1763–1844), *Anweisung zum Waldbau* (Dresden and Leipzig, 1916).

31. Thomasius, "Waldbauliche Auffassungen," 726; Harald Thomasius, interview by author, Tharandt, 26 March 1991.

32. J. R. Wordie, "The chronology of English enclosure, 1500–1914," *Economic History Review* 36:4 (November 1983): 483.

33. Edmund Wilson, "Karl Marx decided to change the world," in *To the Finland Station* (New York: Farrar, Straus and Giroux, 1972), 146. The agricultural depression of the early 1840s intensified the peasants' anger against enclosure.

34. Ibid.

35. Hubert Hugo Hilf, "Forstwirtschaft zwischen Gestern und Morgen," *Forstarchiv* 31:3 (15 March 1960): 3.

36. Horst Kurth, "Max Robert Pressler—ein Pionier intensiver Bestandwirtschaft," *Sozialistische Forstwirtschaft* 36:12 (1986): 366, 371–74. The influence of liberal economics culminated in Judeich's 1871 book *Die Forsteinrichtung,* where he developed soil rent theory (the *Bodenreinertragslehre*), the direct ancestor of modern industrial forest management. No authoritative scientist stepped forward to advocate restoration of hardwoods after Cotta died in 1844. Instead, influential forest scientists for the next generation were men like Max Robert Preßler and Judeich who emphasized financial analysis and the use of net present value calculations to determine harvests rather than biological maturity. Management science after Cotta was pursued almost to the exclusion of ecology and reinforced the dominance of pure conifer stands.

Forest management's task was to direct biological growth through the control tools of statistical analysis and financial theory to the highest possible financial return. Further, Judeich based forest management on a "normal forest" model of conifer monocultures and clear-cutting, a useful euphemism for an almost wholly artificial human fabrication. Judeich's work completed the justification for modern industrial forest management and the "normal" artificial forest.

A similar split between liberal and conservative worldviews developed within agriculture. Albrecht Thaer, an agrarian reformer and like Judeich a supporter of Hardenberg, saw agriculture and forestry as businesses little different from industry. Thaer challenged the traditional views of farmers' and estate owners' special and intimate connection to the land and questioned the irrational values of nature. He and Judeich both viewed the forest as a collection of individual assets, which had to generate profits for the owners, or a soil rent. Traditional conservatives criticized him as a "Manchester-School liberal" and capitalist.

Adam Müller countered Thaer's drive to rationalize agriculture. Müller, influenced by Ernst Haeckel's ecosystem theory, saw natural systems as organic, not as mechanisms comparable to factories. Forest rent theory (*Waldreinertragslehre*) evolved in Prussia at the Eberswalde Forest Institute to counter soil rent forestry. Close-to-nature ecological forest managers and owners thought of the forest as an organism. They focused on building long-term capital value rather than maximizing cash flow. Soil rent forestry, codified as G. L. Hartig's Eight General Rules and epitomized by clear-cutting, was anathema to close-to-nature foresters.

37. Peter Burschel, "Karl Gayer und der Mischwald," *Allgemeine Forst- und Jagdzeitung* 23 (1987): 587. Karl Geyer (1822–1907) was an early proponent of biodiversity and natural structures through his seminal work on forest ecology, *Mischwald (The Mixed Forest)*. Later Borggreve's passionate defense of natural regeneration in *Die Holzzucht* foreshadowed the increasingly cultural dimensions of the scholarly argument over forest structure, foreshadowing the close-to-nature movement.

38. A. B. Recknagle, "Some aspects of European forestry. Management of pine in Prussia. Management of spruce in Saxony," *Forestry Quarterly* 11:2 (June 1913): 137.

39. Thornton T. Munger, ed., "Observations on the results of artificial forestry in Germany: Excerpt from a trade letter written to an American paper company by their German correspondent," *Journal of Forestry* 21 (1923): 719.

40. Hilf, "Forstwirtschaft zwischen Gestern und Morgen," *Forstarchiv* 31:3 (15 March 1960).

41. Peter Staudenmaier, "Fascist ideology: The "Green Wing" of the Nazi Party and its historical antecedents," in *Ecofascism: Lessons from the German experience,* ed, Janet Biehl and Peter Staudenmaier (Edinburgh: AK Press, 1995), 16. More extreme variants of *Dauerwald* live on in Green Party environmentalism and Rudolph Bahro's "ecosocialism," reflecting "the political volatility of ecology" in German politics, from romantic conservatism to "green" National Socialism through to the radical environmentalism of the left. The leaders of all these movements saw the forest as an organism, which in its natural state, free of human agency, promised an arcadian utopia. This irrational reverence for nature, and almost religious fervor, contrasted starkly with the rationalism of the liberal forest scientists of the mid-nineteenth century, many of whom supported the 1848 revolution.

42. Rudolf Rüffler, interview by author, Eberswalde, 19 March 1991.

43. Edmund Spencer, "An Englishman resident in Germany," *Sketches of Germany and the Germans* (London: Gilbert and Rivington, Printers, 1836).

44. Brockhaus, *Die Enzyklopädie,* vol. 23 (Leipzig and Mannheim, 1996): 532.

45. Derek Victor Ager, *The geology of Europe* (New York: Wiley, 1980): 69. The Brocken straddled the old East-West German border just inside the Saxony-Anhalt border with its western and southern flanks falling into Lower Saxony in West Germany. The elevation of Snowdon, England and Wale's highest mountain, is 1,085 meters.

46. The Main is known in southern Germany as the *Weißwurstequator* ("white sausage equator")—"Everything north of the Main is Prussian" is a common Bavarian jibe.

47. The North German Plain held the states of Brandenburg, Mecklenburg-Vorpommern, two-thirds of Saxony-Anhalt, and part of Saxony. The lowlands dominated East German geography with two-thirds of its area. Only Thuringia lay entirely in the Southern Uplands.

48. Ager, *The geology of Europe,* 12, 121.

49. Thomas Henry Elkins, *Germany, an introductory geography,* Praeger Introductory

Geographies (New York: Praeger, 1968), rev. ed., 15; John B. Garver, Jr., "The military geography of East Europe," in *East Europe: The impact of geographic forces in a strategic region* (Washington D.C.: Directorate of Intelligence, 1991), 61.

50. Statistisches Reichsamt, *Die Besteuerung der Landwirtschaft* (Berlin: R. Hobbing, 1930), 11. Per farm unit the relation is reversed.

51. Richard Plochmann, "Forestry in the Federal Republic of Germany," 451.

52. Gertrude Seidenstecher, "D.D.R.," in *Umweltschutz und ökonomisches System in Oste: Drei Beispiele Sowjetunion, DDR, Ungarn,* ed. Hans-Hermann Höhmann and Gertraud Seidenstecher (Stuttgart: Kohlhammer, 1973), 85. East Germany had the most precarious water supply of any industrial state in the world. Rainfall varies from 15 billion cubic meters to 6.5 billion cubic meters annually. Except for the coastal region most of the lowlands only receive between 500 and 560 mm of precipitation annually, with an average below 500 mm annually. West Germany's average rainfall is 700–800 mm. Rainfall on the Brocken ranged from 1,600 to 1,700 mm per year.

53. Francis Ludwig Carsten, "Origins of the Junkers," in *Essays in German history* (London: The Hambleton Press, 1983), 268.

54. "Loess," *The Columbia Encyclopedia,* 6th ed. (New York: Columbia University Press, 2001); www.bartleby.com/65/ (accessed 4 March 2002); "Europe," Encyclopædia Britannica Online, http://search.eb.com/bol/topic?eu=108573andsctn=3 (accessed 4 March 2002).

55. A. B. Recknagle, "Some aspects of European forestry," *Forestry Quarterly* 11:1 (1913): 41; A. B. Recknagle, "Some aspects of European forestry. Management of pine in Prussia. Management of spruce in Saxony," *Forestry Quarterly* 11:2 (June 1913): 135; VEB Forstprojektierung Potsdam, *Übersicht der Forstwirtschaft in der Deutschen Demokratischen Republik,* Sch. 4.4.6 "Vorrherschendes Bonitätssystem Fichten," Source: NAREWA/PERP bei Anwendung der D.D.R.-Fichten Ertragstafel 1984 (1986); VEB Forstprojektierung Potsdam, *Übersicht der Forstwirtschaft in der Deutschen Demokratischen Republik*, Sch. 4.4.6 "Vorrherschendes Bonitätssystem Kiefer," Source: NAREWA/PERP bei Anwendung der D.D.R.-Kiefern Ertragstafel 1975 (1986); Klaus Schikora and H.-E. Wünsche, *Der Waldfonds in der Deutschen Demokratischen Republik* (Potsdam: VEB Forstprojektierung Potsdam). Spruce in Central Europe demands at least 700 mm of rain annually to grow without water stress. Even the more mesic uplands planted in spruce only get between 550 mm and 720 mm of precipitation annually, barely adequate. Water shortages are particularly limiting to pine and spruce growth in the three central Brandenburg Bezirke of Potsdam, Cottbus, and Frankfurt am Oder (the state of Brandenburg).

56. Ager, *Geology of Europe,* 189. The prosperous western German states of Bavaria, Baden-Württemberg, Rhineland-Pfalz, and the Saar sit between the Bohemian massif to the east and, to the west, the Black Forest and Spessart Mountains, the Ardennes and Eifel massifs and the forested sandstone plateaus and slate uplands of the Rhenish Uplands, also known as the Central Rhine Highlands, the *Rheinisches Schiefergebirge.*

57. J. G. C. Anderson, *The structure of Western Europe* (New York: Pergamon Press, 1978), fig. 9.6, p. 233.

58. Historically it served the interests of invaders from the southwest. The young Brigadier General Napoleon Bonaparte moved up through the Fulda corridor to defeat

the Saxon army outside Dresden in 1793 in the War of the First Coalition (1792–97), winning for France a twenty-year tenure on the east bank of the Rhine. Bonaparte marched his armies through Fulda thirteen years later to rout the Hohenzollern armies at Jena-Auerstädt in October 1806 and occupy Berlin. Napoleon's final passage through the corridor witnessed his retreat to France after defeat at the Battle of Leipzig in October 1813, ending French occupation of the east bank of the Rhine. More recently, Patton drove his Third Army divisions from nearby bridgeheads on the Rhine through Fulda directly into the Leipzig Basin and the Reich's eastern industrial heartland in the closing months of the Second World War.

59. The Thüringerwald (Thuringian Forest) is a narrow, 100-km-long forest running northwest just under 1,000 meters in elevation.

60. The Association of the Third Armored (Spearhead) Division Museum and Archives, Inc., http://www.3ad.net/fulda_gap.htm (accessed 9 March 2002).

61. Francis Ludwig Carsten, *The origins of Prussia* (Westport, Conn.: Greenwood Press, 1981), 75–80, 88, 101, 275. At the end of the seventeenth century the population density in Brandenburg and Prussia was "still less than one-third that of Saxony or Württemberg"; Elkins, *Germany,* 102–3. East Germany's geography reflects a combination of "colonial" and Western traits. The 1945 zones of occupation borders were "in some way an inherent one, reflecting fundamental and continuing contrasts in social geography."

62. H. W. Koch, *History of Prussia* (London: Longman, 1978), 25; Carsten, "Origins of the Junkers," 31, 36; Klaus J. Bade, ed., *Deutsche im Ausland—Fremde in Deutschland* (Munich: C. H. Beck, 1993). Five million Germans, many fleeing the serial agricultural depressions in the east, left Germany between 1840 and 1900, mostly for the United States.

63. Solsten, *Germany: A country study,* 77. Statistisches Reichsamt, 1940, "Besteuerung," 32. Stalin took over one-third of 1939 German Reich's territory; Roy E. H. Mellor, "The German Democratic Republic," in *Planning in eastern Europe,* ed. Andrew H. Dawson (London: Croon Helm, 1987), 140. Stalin's "border adjustments" cut off East Germany from critical sources of raw materials and coal from the Ruhr and Silesia, and from access to markets in western Germany for Saxon textiles and the chemical industry in the Elbe-Saale Basin.

64. Based on the Reich's 1939 territory. Stalin's border adjustments cost Germany more than 12.9 million hectares east of the Oder, perhaps in rough exchange for the 25.9 million hectares of eastern Poland Stalin had already taken under the Molotov-Ribbentrop pact (23 August 1939), through which the Soviet Union took the entire Masurian lake district and Polish lands east of the Narew, Vistula, and San rivers.

### *Chapter 3. Initial Conditions and Reparations*

1. The East German state (the German Democratic Republic [GDR], or Deutsche Demokratische Republik [DDR]) was proclaimed on 11 October 1949.

2. D. Harris Chauncy and Gabriele Wulker, "The Refugee Problem of Germany," *Economic Geography* 29:1 (January 1953): 25.

3. Our Special Correspondent, "Brutality Rife," *The Times* (11 September 1945), 4.

The frustrated Special Correspondent concluded: "There is an urgent need for complete information on these mass expulsions; all the control council could do today was to refer the subject to its coordinating committee for full study." Norman Davies, in "Europe's forgotten war crime," *The Sunday Times* (7 April 2002), concluded, "There can be no doubt that the largest single act of ethnic cleansing in European history was carried out with the full approval of the western powers."

4. Rudolf Binsack, "Zeitläufte: Enteignet, deportiert, geflohen: Ein Augenzeuge schildert das Unrecht der Bodenreform in der Sowjetzone," *Die Zeit* 3 (19 March 1998): 86. The Binsack farm was the Rittergut Kremplin, near Neuruppin.

5. Werner Klatt, "Food and farming in Germany: I. Food and nutrition," *International Affairs* 26:1 (January 1950): 45, 47. Still, it would be several years after the war until the German civilian population again reached the minimum level of nourishment (2,500 calories) it had when the war ended. Part of the reason for Germany's favorable food supply in 1945 was Production Minister Albert Speer's "independent decision of his own that the war was lost and the next year's crop should be protected." Cited in David MacIsaac, ed., *The United States strategic bombing survey* (New York: Garland, 1976). Franklin D'Olier and Henry C. Alexander, editors, *Summary and final reports of the United States strategic bombing survey, 1945–1947 (inclusive),* (Washington, D.C.: U.S. Government Printing Office, 1946) 10, 132.

6. MacIsaac, *United States strategic bombing survey,* 71, and 12, 13, introduction. The British zone in northern Germany was closest to bomber bases in Britain and thus sustained the greatest damage. See also Calvin B. Hoover, "The future of the German economy" (in *Economic Problems of Foreign Areas*) *American Economic Review* 36:2, Papers and Proceedings of the Fifty-eighth Annual Meeting of the American Economic Association (May 1946): 642.

7. MacIsaac, *United States strategic bombing survey,* 71ff.

8. Dorothea Faber, "Entwicklung und Lage der Wohnungswirtschaft in der sowjetischen Besatzungszone, 1945–1953," *Wirtschaftsarchiv* 8:17 (5 September 1953): 5943.

9. Doris Cornelsen, et al., *Handbuch D.D.R.-Wirtschaft.* 4th ed., trans. Lux Furtmüller. Deutsches Institut für Wirtschaftsforschung (DIW) (Farnborough: Saxon House, 1984). Over 1939 levels.

10. Wilhelm Cornides, *Wirtschaftsstatistik der deutschen Besatzungszonen, 1945–1948 in Verbindung mit der deutschen Produktionsstatistik der Vorkriegszeit* (Oberursel [Taunus]: Europa-Archiv, 1948). 38, 39; Karl Christian Thalheim, "Volkswirtschaft," in *Ploetz, Die Deutsche Demokratische Republik: Daten, Fakten, Analysen,* ed. Alexander Fischer (Freiburg: Ploetz, 1988), 86. Saxony, in the central uplands to the south, was more heavily industrialized than most western regions and relatively untouched by air bombardment; David Childs, Thomas A. Baylis, and Marilyn Rueschemeyer, eds., *East Germany in comparative perspective* (London: Routledge, 1989): 140; Karl Willy Hardach, *The political economy of Germany in the twentieth century* (Berkeley: University of California Press, 1980), 112. National Socialist economic planners concentrated investment in Saxony, raising capital investment in eastern Germany 80 percent between 1936 and 1944. The future Soviet zone by 1939 already had a higher net industrial production per capita than western Germany. By the end of 1946, industrial production in the Soviet zone reached 42 percent of 1936 levels while western production lagged at 32 percent.

11. Binsack, "Zeitläufte: Enteignet, deportiert, geflohen." Rudolf Binsack, a thirteen-year-old boy, lived with his parents (fairly typical, non-Junker farmers in the Prussian Old Mark) on their 208–hectare farm. Anton Hilbert, "Denkschrift über die ostdeutschen Bodenreform," Gräflich Douglas'sches Archiv Schloß Langenstein (1946) (Anm. 18), cited in H.-G. Merz, "Bodenreform in der SBZ. Ein Bericht aus dem Jahre 1946" *Deutschland Archiv* 11:24 (1991): 1166, further describes the complex but on balance favorable role of large estate owners in the Southern Uplands state of Thuringia.

12. R. Beyse, "D.D.R.-Bodenreform vor dem Bundesverfassungsgericht," *Holz-Zentralblatt* 117:20 (1991): 305; Norman M. Naimark, *The Russians in Germany: A history of the Soviet zone of occupation, 1945–1949* (Cambridge: Belknap Press of Harvard University Press, 1995), 153, discusses further the often brutal quality of expropriation.

13. Michael L. Hoffman, "The harvest in Europe is hunger once more," *New York Times* (21 September 1947). Hofman quotes the head of the Food and Agricultural Organization, Sir John Boyd Orr, as saying, "Next winter and spring many in Europe will be worse fed than they were during the war." West German civilians' rations were less than half prewar levels while North America's food supply was unchanged. Great Britain's was already 94 percent of prewar levels, still a thin diet of 2,800–2,850 calories per day.

14. Herbert Hoover, "Text of Hoover mission's findings on the food requirements of Germany," *New York Times* (28 February 1947). Hoover, the seventy-three-year-old former president, had unparalleled experience, credibility, and success in international relief work as the former head of the Committee for Relief in Belgium after the First World War and the American Relief Administration (1921–23), where he organized relief for Europe and the Soviet Union.

15. Our Diplomatic Correspondent, "Survey in Germany: The first steps in political re-education, aims of British policy," *The Times* (27 November 1945), 5.

16. Our Special Correspondent, "Brutality Rife," *The Times* (11 September 1945), 4.

17. Naimark, *Russians in Germany,* 10.

18. Merkel and Wahl, *Das geplünderte Deutschland,* 17.

19. SOPADE, Vorstand der Sozialdemokratischen Partei Deutschlands, *SOPADE-Querschnitt durch Politik und Wirtschaft* (Hannover: Vorstand der Sozialdemokratischen Partei, 1947–48, 1949/1950–1952/1954). I rely heavily on the regular reports put out by the Social Democratic Party Executive Committee (SOPADE, the *Sozialdemokratische Partei Deutschlands*) on the Soviet zone and East Germany compiled by economics experts, and by ecology and farm and forest policy scientists.

For my discussion of the forested landscape and of Soviet and East German communist forest policy, I also took great benefit from the Social Democratic Party's monograph on the eastern German forest from 1945 to 1955: SOPADE (Executive Committee, Social Democratic Party, Germany), "Die Forstwirtschaft in der Sowjetzone," Denkschriften, Sopadeinformationsdienst (Bonn: Vorstand der Sozialdemokratischen Partei Deutschlands, 1955).

Sozialdemokratische Pressedienst (Hannover) (21 August 1947), cited in SOPADE, Vorstand der Sozialdemokratischen Partei Deutschlands, "Der Terror in der Sowjetzone geht weiter," *Querschnitt durch Politik und Wirtschaft* #5 (August 1947); *New York*

*Times,* "71 billion penalty seen by Germans: Report on reparations set $28,000,000,000 as value of lost territories," *New York Times* (13 February 1948). U.S. Secretary of State Byrnes valued these lands conservatively at $14 billion, more than $110 billion in current value, dwarfing the Soviets' formal reparations demands at Potsdam; Gustav W. Harmssen, *Reparationen, Sozialprodukt, Lebensstandard; Versuch einer Wirtschaftsbilanz* (Bremen: F. Trüjen, 1948); Gustav W. Harmssen, *Am Abend der Demontage; sechs Jahre Reparationspolitik mit Dokumentenanhang,* ed. the Bremer Ausschuss für Wirtschaftsforschung (Bremen: F. Trüjen, 1951).

20. Stephane Courtois, *Livre noir du communisme, The black book of communism: Crimes, terror, repression,* trans. Jonathan Murphy, consulting editor, Mark Kramer (Cambridge: Harvard University Press, 1999), 205. Stalin seized 466,000 square kilometers of Polish land in 1939.

21. E. Reichenstein, "Die forstwirtschaftliche Lage Deutschland vor und nach dem 2. Weltkrieg," *Forstarchiv* 21 (1950): 32, and Bundesministerium für gesamtdeutsche Fragen (BGF), *SBZ von A-Z.*, 4th ed. (Bonn: Deutscher Bundesverlag, 1958), 49, discuss the postwar forest structure.

For the Party leadership's explanation for the devastated conditions revealed in the early 1950s, see Ekkehard Schwartz, "Die demokratische Bodenreform, der Beginn grundlegender Veränderungen der Waldeigentums und der Forstwirtschaft im Gebiet der Deutschen Demokratischen Republik," *Sozialistische Forstwirtschaft* 20:10 (1970): 289–93; Heinz Kuhrig, "Demokratische Bodenreform legte den Grundstein für stetig steigende Agrarproduktion," *Neues Deutschland* 198 (21 August 1975): 3; Deutschen Institut für Zeitgeschichte, *Jahrbuch der Deutschen Demokratischen Republik* (Berlin: Verlag der Wirtschaft, 1956 and 1957); Alfred Kosing and Richard Heinrich, "Natur-Mensch-Gesellschaft: Das Verhältnis der sozialistischen Gesellschaft zur Natur," *Neues Deutschland* 127 (1 June 1987): 3. As late as 1987 Kosing and Heinrich still cited "wartime destruction" and the "cumulative product of capitalism's one-sided exploitation of nature" for the sustained post–land reform farm and forest economic and ecological crises.

22. VEB Forstprojektierung, *Forsterhebung 1949: Flächen und Vorratsgliederung nach Besitzverhältnisse* (1949). Klaus Schikora, who assisted in this inventory, confirms this.

23. Ryle, "Forestry in western Germany, 1948."

24. Friedrich Raab, *Die deutsche Forstwirtschaft im Spiegel der Reichsstatistik* (Berlin: Verlag von Paul Parey, 1931); Statistisches Reichsamt, *Statistik des deutschen Reiches,* "Die Ergebnisse der Forstwirtschaftlichen Erhebung in Jahre 1927," vol. 386 (Berlin: Puttkammer and Mühlbrecht, 1930); Friedrich Heitmann, "International Sammlung von Forststatistiken," *Zeitschrift für Weltforstwirtschaft* 10:7/10 (1944): 473; Statistisches Reichsamt, *Statistik des deutschen Reiches,* vol. 592 (Berlin: Puttkammer and Mühlbrecht, 1937); C. Wiebecke, "Zum Stand der deutschen Forststatistik," *Forstarchiv* 26:1 (15 January 1955): 1; VEB Forstprojektierung, "Handwritten summary of 1937 Forest Inventory annotated for East German conditions" (1949). Friedrich Raab's 1927 inventory was the only total forest inventory available in 1945 from which to estimate conditions. Raab's 1927 inventory was the first total inventory since 1913 and the most thorough inventory until the Großraum inventories of the 1960s. The 1937 inventory,

although dutifully carried out in the regular ten-year cycle, was not reconciled with the 1927 inventory, did not include the 16 percent of forestland in units under ten hectares in size, and was "too much under the influence of the National Socialist Market Order."

25. These reports are: G. D. Kitchingnam, "The 1945 census of woodlands in the British zone of Germany," *Empire Forestry Review* 26:2 (1947): 224–27; G. B. Ryle, "Germany: Military Government, C.C.G., North German Timber Control (NGTC)," *Empire Forestry Review* 26:2 (1947): 212–23; E. H. B. Boulton, "The forests of Germany: What they can supply on reparations account for the U. K.," *Timber Trade Journal* 173 (1945): 7–8; A. C. Cline, "A brief view of forest conditions in Europe," *Journal of Forestry* 43 (1945): 627–28; Joseph C. Kircher, "The forests of the U.S. zone of Germany," *Journal of Forestry* 45:4 (1947): 249–52; Office of the Military Government for Germany (U.S.) (OMGUS), *Special report of the Military Governor: The German forest resource survey* 17 (1 October 1948) (Office of the Military Government for Germany [U.S.], 1948).

26. Office of the Military Government for Germany (U.S.). *Special report of the Military Governor: The German forest resource survey*, 7.

27. Food and Agriculture Organization (FAO) of the United Nations, *Forestry and forest products: World situation, 1937–1946* (Stockholm: Stockholms Bokindustri Aktiebolag, 1946), 9, 15.

28. Kircher, "Forests of the U.S. zone," 249–52.

29. S. H. Spurr, "Post-war forestry in Western Europe. Part II," *Journal of Forestry* 51:6 (1953): 415–21.

30. G. B. Ryle, "Forestry in western Germany, 1948," *Forestry* 22:2 (1948): 158 (Ryle cites Heske, 1958); E. Reichenstein, "Entwicklung von Vorrat und Zuwachs in den vier Besatzungszonen Deutschland seit 1945," *Weltholzwirtschaft* 1:7/8 (1949); E. Reichenstein, "Die forstwirtschaftliche Lage Deutschland vor und nach dem 2. Weltkrieg," 30. Ryle expected to find that the National Socialists had "looted and thoroughly and scientifically exploited" the forests of central and eastern Europe. National Socialist forestry was if anything anodyne and constructive in its emphasis on smaller dimension lumber, particularly the younger pine age classes for mine timbers and railroad ties and the younger spruce age classes for pulp.

*Der Deutsche Forstwirt*, "Durchführung kriegswirtschaftlicher Maßnahmen in der Forst- und Holzwirtschaft," *Der Deutsche Forstwirt* 21:76/77 (22 September 1939): 909; Früchtenicht, "Leistungssteigerung im Walde," *Der Deutsche Forstwirt* 22:69/70 (1940). Früchtenicht reflected the continued importance of ecological considerations even as the end of the war approached. Michael Charles Kaser, "Interwar policy: The war reconstruction," in *The economic history of Eastern Europe, 1919–1975*, ed. Michael Charles Kaser and Edward Albert Radice (Oxford: Oxford University Press, 1986), 411, 413; Von Dieterich, "Der Ausbau der Forstwirtschaft" (1941). Von Dieterich showed that National Socialist forestry in the conquered lands remained ecological even as the Holocaust gathered momentum and military crises cascaded.

W. Köhler, "Wild game production and harvesting methods in some intensively managed European forests," *Proceedings*, Fifth World Forestry Congress, Seattle, Wash. (1960), 1801. National Socialist wildlife policy remains exemplary in its focus on habitat and long-term values. The Reich Hunting Law of 1934 and the Reich Nature Protection

Law were the most sophisticated and scientific environmental regulations of their time. National Socialist wildlife regulation was unique in working through incentives rather than through prohibitions. Its goal was "to maintain a natural balance of various appropriate and healthy species of wildlife."

See also Heinrich Rubner, *Deutsche Forstgeschichte—1933–1945: Forstwissenschaft, Jagd und Umwelt im NS-Staat* (St. Katharinen: Scriptae Mercaturae Verlag, 1985), 173; Karl Hasel, "Forstbeamte im NS-Staat am Beispiel des ehemaligen Landes Baden," *Schriftenreihe, Landesforstverwaltung und Forstwirtschaft Baden-Württemberg,* #62 (1985): 197; and Michael L. Wolf, "The history of German game management," *Forest History* 14:3 (October 1970): 16.

31. A. Heger, "Aufbau und Leistung von naturnahen Wäldern im Osten und ihre forstwirtschaftliche Behandlung," *Forstwissenschaftliches Centralblatt* 1 (1944): 34–35.

32. Karl Hasel, "Forstbeamte im NS-Staat am Beispiel des ehemaligen Landes Baden," *Schriftenreihe, Landesforstverwaltung und Forstwirtschaft Baden-Württemberg,* #62 (1985): 128; Food and Agriculture Organization (FAO) of the United Nations, *Forestry and forest products: World situation, 1937–1946,* 18. The Nazi forest minister, Alpers, did extend the Second Four-Year Plan's 150 percent of growth harvests from state forests to private and communal forests. Nevertheless, these targets were rarely met.

33. Heinrich Himmler quoted in Peter Staudenmaier, "Fascist ideology: The 'Green Wing' of the Nazi Party and its historical antecedents," in *Ecofascism: Lessons from the German experience,* ed. Janet Biehl and Peter Staudenmaier (Edinburgh: AK Press, 1995), 16; Heinz Haushofer, *Ideengeschichte der Agrarwirtschaft und Agrarpolitik im deutschen Sprachgebiet,* vol. 2. (Munich, 1985), 107.

34. Stephen Haden-Guest, John Wright, and Eileen M. Teclaff, *A world geography of forest resources* (New York: Ronald Press, 1956): 285.

35. Kurt Hueck, "Aktuelle Aufgaben der Forstwirtschaft," speech by dean of Forstfakultät Eberswald at the Agricultural Science Congress, Berlin, 4 February 1947; *Forst- und Holzwirtschaft* 1:1 (1 April 1947): 6. Hueck was the first postwar dean at the Institute for Forest Sciences in Eberswalde.

36. SOPADE, Vorstand der Sozialdemokratischen Partei Deutschlands, "Die Forstwirtschaft in der Sowjetzone," *Denkschriften, Sopadeinformationsdienst* (Bonn: Vorstand der Sozialdemokratischen Partei Deutschlands, 1955), 29.

37. Bieger, "Umfang und Ursachen der thüringer Sturmschäden," *Forst- und Holzwirtschaft* 1:2 (15 April 1947).

38. Roland Barth, interview by author, Eberswalde, 20 March 1991. Barth, East Germany's senior forest statistician, began his career in the late 1940s and had firsthand knowledge of the extent of damage. His information is internally consistent with published data and other interviews on postwar damage and with the forest structure first clearly defined in the mid-1950s. Barth's estimate of damage is authoritative, particularly since the East German government did not acknowledge the vast scale of damage.

39. Rubner, *Deutsche Forstgeschichte, 1933–1945,* 174; Gustav Willenstein, *Die große Borkenkäferkalamität in Süddwestdeutschland, 1944–1951* (Ulm, 1954).

40. SOPADE, "Die Forstwirtschaft in der Sowjetzone," 18.

41. Ibid., 11.

42. Ibid., 15. From 6 million cubic meters at year end 1946 to 11 million cubic meters one year later—half of which was high-quality *Stammholz* (sawlogs).

43. Statistisches Bundesamt, *Statistisches Jahrbuch für die Bundesrepublik Deutschland* (Stuttgart: W. Kohlhammer, 1953).

44. Edward A. Morrow, "Reparations lag in East Germany: Russian officials are warned to let nothing interfere with deliveries to Soviet," *New York Times* (17 October 1948).

45. Senior Soviet zone official, possibly Tulpanov, undated speech between the end of August and the beginning of September 1947, ZPA NL 36/734 Bl 347–62, quoted in Rolf Badstubner and Wilfried Loth, eds., *Wilhelm Pieck—Aufzeichnungen zur Deutschland Politik, 1945–1953* (Berlin: Akademie Verlag, 1994), 162.

46. Wolfgang F. Stolper and Karl W. Roskamp, *The structure of the East German economy* (Cambridge: Harvard University Press, 1960), 5; Stephen F. Frowen, "The economy of the German Democratic Republic," in *Honecker's Germany*, ed. David Childs (London; Boston: Allen and Unwin, 1985), 36; Roesler, "Rise and fall of the planned economy zone," 46–61; Wilma Merkel and Stephanie Wahl, *Das geplünderte Deutschland. Die wirtschaftliche Entwicklung im östlichen Teil Deutschlands von 1949–1989*, 2nd ed. (Bonn: Instituts für Wirtschaft und Gesellschaft, 1991), 16; Bundesminister für innerdeutsche Beziehungen, ed., *D.D.R.-Handbuch,* table 1, "Reparations and other expenses in the Soviet zone and East Germany between 1945 and 1953," vol. 2, M-Z (Cologne, 1985). Estimates of total reparations cost through 1953 run between DM 66 billion (US$119 billion in 2002) and DM 120 billion (US$216 billion in 2002), overwhelming for an economy with a 1950 gross industrial product of only DM 75 billion.

47. Our Diplomatic Correspondent, "Fixing the German reparation," *The Times* (19 June 1945).

48. The Soviet Stock Companies, *Sowjetische Aktiengesellschaften* (SAG), were formed on 30 October 1945 through the Soviet military government's (SMAD) Order 124 (SMAD-Befehl 124) and SMAD-Befehl 126 of 31 October 1945. The Soviets took more than one-quarter of German industrial production for export to the Soviet Union.

49. Barth, interview by author, Eberswalde, 20 March 1991. Dr. Roland Barth, East Germany's senior forest statistician, estimated that the cut for reparations and fuelwood was as high as 25 million cubic meters each year.

50. Arthur Geoffrey Dickens, letter to the editor, *The Times* (9 April 1947) in response to A. Sudachkov's 5 April 1947 letter to the editor, *The Times* (5 April 1947).

51. Kindleberger, *German Economy*.

52. Our Special Correspondent, "Mr. Bevin backs American plan for Germany," *The Times* (16 May 1947), 4.

53. Ibid.; *The Times,* "No surrender in Berlin: Mr. Bevin's review. Ernest Bevin's 30 June 1948 speech to Commons," *The Times* (1 July 1948), 4. Bevin restated his anger at the Soviets' parasitical reparations policy, "We would not and could not agree to the principle of reparations from current production which involved Great Britain and the British people, and other allies, in virtually paying reparations to another ally. (Opposition cheers)."

54. Board of Editors, "Opportunity in Moscow," lead article, *The Times* (7 April 1947).

55. Carl J. Friedrich and Henry Kissinger, eds., *The Soviet zone of Germany* (New Haven: Bechtle, 1956).

56. Günther Bischoff, "History: Introduction" in Kindleberger, *German Economy,* xiv. Twenty-five percent of total Soviet reparations of $10 billion (a figure also under debate) would come from the western zones. In return, the British and American zonal economies would get 15 percent of the value of Soviet withdrawals in food from the Soviet zone.

57. Our Berlin Correspondent, "In the Russian Zone," *The Times* (25 September 1947), 5.

58. Bruno Gleitze, "Zielsetzung und Mittel der sowjetzonalen Wirtschaftspolitik bis zur gegenwärtigen Krisensituation." Speech to the Working Group of German Economists Research Institute. Cited in SOPADE, "Die sowjetzonale Wirtschaftspolitik," #937 (September 1953), 41.

59. "The war against hunger," *New York Times* (17 August 1949); John Boyd Orr, "Program to meet the world's food crisis," *New York Times Magazine* (9 November 1947); D. A. FitzGerald, "Abstract of the FitzGerald Report alleging short-sighted policies in feeding the world," *New York Times* (28 October 1947).

60. *Fortune,* 1945–50, editorials and feature articles.

61. *Economist* 16/2 (12 October 1957).

62. A. J. Nichols, *Freedom with responsibility: The social market economy in Germany, 1918–1963* (New York: Oxford University Press, 1994).

63."Report on reparations," *Telegraf* (Berlin) (7 March 1949), cited in SOPADE, "Raubbau an den Ostzonen-Wäldern" (April 1949), 10; SOPADE, "Die Forstwirtschaft in der Sowjetzone," 15.

64. Julius Speer, "Die Forstwirtschaft im Wirtschaftsgeschehen des Jahres 1948," *Allgemeine Forstzeitschrift* 4:1 (5 January 1949): 1.

65. Fuelwood harvests, which equaled reparations' harvests in intensity, are not included in official harvest figures.

66. Vyacheslav Mikhaylovich Molotov (1890–1986) quoted in the Sozialdemokratische Pressedienst (Hannover) (21 August 1947).

67. Our Special Correspondent, "International timber conference at Marienbad (Marianske Leyne)," *The Times* (12 May 1947). Soviet officials did not attend the International Timber Conference in Marienbad on 11 May 1947; the Soviet military administration was already harvesting and selling for hard currency as much timber as it could from the Soviet zone forest.

68. Speer, "Die Forstwirtschaft im Wirtschaftsgeschehen des Jahres 1948," 1.

69. Pfalzgraf, "Forstwirtschaft contra Holzwirtschaft?" *Forst- und Holzwirtschaft* (1 July 1947): 98. The Potsdam Agreement called for reduction of stocking to eliminate "war potential," sacrilege to any conservative, *Dauerwald* forester. This comment reflects the desire of most German foresters to return management to its roots in forest science and ecology.

70. Reichenstein, "Die forstwirtschaftliche Lage Deutschland," 30; Bundesministerium für gesamtdeutsche Fragen (BGF), *SBZ von A-Z,* 1st ed. (Bonn: Deutscher Bundesverlag, 1953), 48.

71. SOPADE, "Die Forstwirtschaft in der Sowjetzone," 24.

72. SOPADE, "Ostzonenreparationen," #5 (August 1947), 20. These trains were routed through Frankfurt am Oder due to flooding at Küstrin. Marshal Georgy Konstantinovich Zhukov (1896–1974).

73. Ibid.

Manifests of Trains Carrying Reparations through Frankfurt am Oder to the Soviet Union for the Month of April 1946

| Commodity | Locomotives | # Wagons |
|---|---|---|
| Locomotives | 37 | n/a |
| Railroad ties | 38 | 1,152 |
| Industrial goods | 169 | 10,041 |
| Sugar | 11 | 574 |
| Grain | 1 | 52 |
| Autos | 5 | 233 |
| Potash | 21 | 1,017 |
| Vehicles | 1 | 32 |
| Literature | 1 | 52 |
| Cement | 3 | 143 |
| "Special goods" | 1 | 39 |
| Tools of all sorts | 1 | 53 |
| "Special goods" #15 | 2 | 102 |
| Textiles | 1 | 44 |
| Pigs | 1 | 44 |
| Horses | 18 | 10 |
| Political prisoners | 1 | 34 |
| TOTAL | 312 | 13,612 |

74. SOPADE, "Die Forstwirtschaft in der Sowjetzone," 2.

75. Charles P. Kindleberger, Letter "Berlin #7" (August 1946), in *The German economy, 1945–1947: Charles P. Kindleberger's letters from the field* (Westport, Conn.: Meckler, 1989), 36;"Report on reparations," *Telegraf* (Berlin) (7 March 1949), cited in SOPADE, "Raubbau an den Ostzonen-Wäldern" (April 1949), 10; SOPADE, "Die Forstwirtschaft in der Sowjetzone," 25.

76. Working Paper, Forstprojektierung Potsdam, Archives, handwritten, undated, said to be 1949. Normal stocking was 120 cubic meters per hectare. The 1933 level was 113 cubic meters per hectare. The Soviet zone 1946 informal inventory showed a stocking density of 96.81 Vfm/ha and an unplanted area of only 2 percent calculated on an area of 2.6 million hectares.

77. *Neuer Vorwärts,* "Waldraubbau-Holzexport," *Neuer Vorwärts* (Hannover) (8 January 1949), cited in SOPADE, "Raubbau an den Ostzonen-Wäldern" (February 1949), 16.

78. G. Schröder, "Zu einigen Problemen der Forstwissenschaft und -praxis in Prozeß

der wissenschaftlich-technischen Revolution," *Sozialistische Forstwirtschaft* 15:11 (1965): 323.

79. SOPADE, "Die Forstwirtschaft in der Sowjetzone," 26.

80. Bruno Gleitze, "Zielsetzung und Mittel der sowjetzonalen Wirtschaftspolitik bis zur gegenwärtigen Krisensituation," speech to the Working Group of German Economists Research Institute, cited in SOPADE, "Die sowjetzonale Wirtschaftspolitik," #937 (September 1953), 41.

81. C. Wiebecke, "Zum Stand der deutschen Forststatistik," 1, 2; G. Hildebrandt, "Die Forsteinrichtung in der D.D.R. 1950 bis 1965. Ein Beitrag zur jüngeren deutschen Forsteinrichtungsgeschichte aus Anlaß des 80. Geburtstages von Albert Richter in Eberswalde," *Allgemeine Forst- und Jagdzeitung* 160:6 (1989): 123; H. Eberts, "Forstwirtschaft in Ost und West," *Holz-Zentralblatt* (1949): 942. Eberts perplexedly commented on Soviet refusal to release ownership data or share other data necessary for planning. Potsdam forest management was to be carried out by the Quadripartite Forestry Subcommittee of the Quadripartite Directorate of Economics. As a palliative to Western requests for information, the Soviets commissioned the superficial *Waldfondserhebung* forest inventory followed by the unsatisfactory 1949 *Forsterhebung* and 1952 inventories.

82. SOPADE, "Die Forstwirtschaft in der Sowjetzone," 11.

83. Office of the Military Government for Germany (U.S.). *Special report of the Military Governor: The German forest resource survey*, 6, 7. The U.S. military government reported that the "prodigious" afforestation task would take seven to ten years replanting 66,000 hectares annually. The U.S. zone's 1945 unplanted forest area of 5.7 percent suggests that a Soviet zone unplanted forest area in 1945 would have been between 5 and 6 percent. Under normal conditions, 1 to 1.5 percent of a forest managed under a one-hundred-year rotation will be unplanted.

84. Reichenstein, "Die forstwirtschaftliche Lage Deutschland," 30; Statisches Amt für die sowjetischen Zone, cited in SOPADE, "Die Forstwirtschaft in der Sowjetzone," 26; *Darmstädter Echo,* "Report on forestry in eastern Germany," *Darmstädter Echo* (20 January 1949). The total area of bare clear-cuts in 1950 was 400,000 hectares. Unplanted clear-cuts meant an annual loss to the Soviet zone's economy of DM 90 million from foregone growth alone, roughly $21 million, or almost $170 million in annual lost production in 2002 value. Growth = 3.5 cubic meters ("Festmeter")/ha/year. Loss from clear-cuts = roughly 33 million cubic meters at DM 30/cubic meter—an annual loss of DM 90 million.

85. SOPADE, "Die Forstwirtschaft in der Sowjetzone," 10.

86. Army foresters of the 10th Engineers (Forestry) and the 20th Engineer "Lumberjacks" battalions. Kircher, "The forests of the U.S. zone of Germany," 251.

87. Office of the Military Government for Germany (U.S.), *Special report of the Military Governor: The German forest resource survey*, tables 20, 10, 11; *Forestry Abstracts, 1947–1948*, #9 (Oxford: Imperial Forestry Bureau, 1948), 537. By the end of 1946 control was reestablished in the U.S. zone and the total harvest falling to match growth, about 3.3 percent of total volume. The U.S. Zone Forest Survey recommended bringing harvest levels down to this level "within a reasonable time." Growth was not strong and comparable to the Soviet zone's data in 1945, between 2.5 and 2.75 cubic meters per hectare. Harvest reductions were accompanied by additional afforestation to reclaim

cleared forest at the end of the war. True rebuilding of Soviet zone forests began tentatively with the Party's First Five-Year Plan in 1951 and only in earnest with the end of reparations on 1 January 1954.

88. Delbert Clark, "Critic of U.S. leads for German post: Baumgartner of Bavaria, who assailed military regime, may be 2-zone farm chief," *New York Times* (13 July 1947), 13; Jack Raymond, "German demands timber be saved: Bizonal Economic Council hears report Western Allies are destroying all forests," *New York Times* (29 April 1948), 33; Richard Plochmann, interviews by author, Eberswalde, Tharandt, and Munich, May-June 1986, January 1990, and March-April 1991. Dr. Hans Schlange-Schöningen, former *Reichskommisar für die Osthilfe,* chief of land reform under the Brüning government and director of food, agriculture, and forests in the British zone, and Dr. Josef Baumgartner, the Bavarian minister of agriculture in the U.S. zone, complained with great effect.

89. Plochmann, interview by author, May-June 1986.

90. Charles P. Kindleberger to Sara Kindleberger, Letter "Berlin #4," 5 August 1946, in *German Economy,* 20.

91. Erwin Kienitz, *Denkschrift über forstwirtschaftlichsorganisatorischen Reformen, insbesondere des Bauernwäldes der Deutschen Demokratischen Republik: Ein Beitrag zur sozialistischen Umgestaltung der Forstwirtschaft* (Tharandt: Institut für Forstliche Wirtschaftslehre, 1958), 57; Puttkammer, "Forstliche Rechts- und Verwaltungsprobleme der Gegenwart," *Forst- und Holzwirtschaft* 2:3 (1948): 42.

92. O. Dittmar, "Zur Geschichte des Forstlichen Versuchswesens des Institutes für Forstwissenschaften Eberswalde," in *Ehrenkolloquim anläßlich des 150. Geburtstag von Professor Dr. phil. Max Friedrich Kunze am 12. Oktober 1988,* ed. H. Kurth (Tharandt: Technischen Universität Dresden, 1988), 12; "Report on reparations," *Telegraf* (Berlin) (7 March 1949), cited in SOPADE, "Raubbau an den Ostzonen-Wäldern" (April 1949), 10. The Soviet military government's "Forest Management Regulations for the Soviet zone" decreed that "the planning and management of the entire Soviet zone forest economy will come under (the Party's) central authority," Befehl 97 (13 October 1945), the *Ordnung der Forstwirtschaft in der sowjetischen Besatzungszone,* was followed by the 20 October 1945 "Erlaß der deutsche Verwaltung für Land- und Forstwirtschaft," #54/45. Germany's principal forest research institute at Eberswalde was reborn as the Forstliches Versuchs- und Forschungswesen der Deutsche Wirtschaftskommission (DWK). Finally, in 1952, when the Academy of Agricultural Sciences was founded, it became the Institut für Forstwissenschaften Eberswalde (IFE).

93. Office of the Military Government for Germany (U.S.), *Government and its administration in the Soviet zone of Germany,* Civil Administration Division, OMGUS (November 1947), 11; Naimark, *Russians in Germany,* 52–55. National Socialist economic regulations provided the basis for the economic control and planning of postwar eastern Germany and the DWK (German Economic Commission), set up by Befehl 138 (4 June 1947) as an "incipient government bureaucracy" for the future East Germany state.

94. Hildebrandt, "Die Forsteinrichtung in der D.D.R.," 124.

95. Puttkammer, "Forstliche Rechts- und Verwaltungsprobleme der Gegenwart," *Forst- und Holzwirtschaft* 2:3 (1948): 42; SOPADE, "Die Forstwirtschaft in der Sowjetzone," 10.

96. Naimark, *Russians in Germany,* 29.

97. SOPADE, "Die Forstwirtschaft in der Sowjetzone," 16.

98. Naimark, *Russians in Germany,* 98, cites Soviet archival sources. Corruption was almost as hard on the rural population as the Party's functionaries' brutality. Naimark commented: "In 1947 and 1948, a group of officers in a rich Mecklenburg agricultural district purposely underreported grain and meat production to be sure to meet their quotas, and then sold off the excess in private markets. Two majors in the Angermünde region sold ration cards to the Germans in a scam that depended on reporting a higher number of residents than actually lived in the area."

99. Ryle, "Germany: Military Government, C.C.G.," 162.

100. Hansrath,"Was geschiet mit den Revierförster?" *Allgemeine Forstzeitschrift* 3:24 (15 December 1948): 261.

101. Frithjof Paul, interview by author, Tharandt, 26 March 1991.

102. Klaus Weidermann, "Abriß der Geschichte der Fakultät für Forstwirtschaft in Tharandt," *Forstarchiv* 15:11/12 (1966): 1253. Weidermann discussed foresters' "reactionary" image, fueled by forestry students hanging banners and establishing an armory at Tharandt in support of the Kapp Putsch in 1920 and their particular favor with Frederick the Great as army scouts (the *Preußisches Feldjäger*), a relationship Göring enthusiastically revived in the 1930s.

103. SOPADE, "Die Forstwirtschaft in der Sowjetzone," 34.

104. Hildebrandt, "Die Forsteinrichtung in der D.D.R.," 124; *Forestry Abstracts* 10:2 (Oxford: Imperial Forestry Bureau, 1948), 274. The Tharandt and Eberswalde forestry schools reopened in 1948, Eberswalde with five professors and seven lecturers under Dean Dr. K. Hueck. Tharandt's dean, H. Sachse, presided over five professors and one lecturer. The first class of forty ex-soldiers from these schools graduated in 1950.

105. Arnd Bauerkamper, "'Loyale Kader'? Neue Eliten und die SED-Gesellschaftspolitik auf dem Lande von 1945 bis zu den fruhen 1960er Jahren," *Archiv für Sozialgeschichte* 39 (1999): 265.

106. SOPADE, "Die Forstwirtschaft in der Sowjetzone," 33.

107. Ibid., 5, 6; Carl J. Friedrich and Henry Kissinger, eds., *The Soviet zone of Germany* (New Haven: Bechtle, 1956), 373–74. Rau was the former chairman of the State Planning Commission, ex-head of the German Economic Commission (DWK), and full member of the Party's Central Committee and Politburo and an "old-guard Communist."

108. SOPADE, "Die Forstwirtschaft in der Sowjetzone," 5, 6, 26.

109. Albert Milnick, interview by author, Eberswalde, 20 March 1991.

110. Bauerkamper, "'Loyale Kader'?" 265.

111. Our Own Correspondent, "The worst winter," *The Times* (5 February 1947), 4.

112. Our Own Correspondent, "German canals icebound: Threat to Ruhr food," *The Times* (8 January 1947), 4.

113. SOPADE, "Die Forstwirtschaft in der Sowjetzone," 3, 4, 29, 446.

114. SOPADE, "Die Forstwirtschaft in der Sowjetzone," 2.

115. Christian F. Ostermann, "'This is not a Politburo, but a madhouse,' the post-Stalin succession struggle, Soviet Deutschlandpolitik and the SED: New evidence from Russian, German, and Hungarian archives," *Cold War International History Project Bulletin* #10 (March 1998); 108n4; Naimark, *Russians in Germany,* 238–50. The Soviets

established the Wismut uranium mining complex in 1947 as a Soviet stock company (SAG). Naimark reports, "Wismut produced about 215,559 tons of uranium between 1945 and 1990, 13% of the total global uranium production (to 1990)."

116. SOPADE, "Die Forstwirtschaft in der Sowjetzone," 47.

117. Ibid.

118. Ibid., 48.

119. *The Times,* "German timber for Britain: Troops at work in famous forests. 'Operation Woodpecker,'" *The Times* (15 April 1947), 2. In the first quarter of 1947 they sent 40,000 tons of timber to Britain out of 50,000 logged in the Harz Mountains, "sufficient for 7,000 houses."

120. Speer, "Die Forswirtschaft im Wirtschaftsgeschehen des Jahres 1948," 1.

121. Our Berlin Correspondent, "Berlin Claustrophobia," *The Times* (18 March 1948), 5.

122. "Report on reparations," *Telegraf* (Berlin) (7 March 1949), cited in SOPADE, "Raubbau an den Ostzonen-Wäldern" (April 1949), 10.

123. SOPADE, "Der Terror in der Sowjetzone geht weiter," #894 (February 1950), 59, 60. See also numbers 736, 737, 741, 795, and 817; Christian Klemke and Jan Lorenzen, "Die sowjetische Militärherrschaft 1945 bis 1994" (Christoph Links Verlag—Links-Druck GmbH, 2002). From the three-part German television series 21, 26, and 28 April 2002, *Roter Stern über Deutschland*, "Speziallager," http://www.orb.de/roterstern/content/37Speziallager.html (accessed 5 June 2002). Lavrenty Pavlovich Beria (1899–1953) headed the MVD. The camps were closed under order no. 0022, "Zur Liquidierung der Speziallager des MWD der UdSSR in Deutschland" (Moscow, 6 January 1950).

124. "Gulag," Encyclopedia Britannica Online. http://search.eb.com/eb/article?eu=39259 (accessed 26 May 2003). Gulag, the "abbreviation of Glavnoye Upravleniye Ispravitelno-trudovykh Lagerey (Chief Administration of Corrective Labour Camps), the system of Soviet labour camps and accompanying detention and transit camps and prisons that from the 1920s to the mid-1950s housed the political prisoners and criminals of the Soviet Union."

125. Klemke and Lorenzen, "Die sowjetische Militärherrschaft 1945 bis 1994." The camps were established under the Soviet military administration's order no. 15 (18 April 1945) and modeled on the camps of the Soviet gulag.

126. SOPADE, "Der Terror in der Sowjetzone geht weiter," #894 (February 1950). This contemporary report reported that of the 240,000 Germans the Soviets imprisoned, almost 40 percent died. Naimark, *Russians in Germany,* 377. Naimark estimates that 95,643 prisoners died, which fits with contemporary accounts.

127. Rainer Hildebrandt, *Querschnitt durch Politik und Wirtschaft,* "Die Todesmühlen der SS übertroffen" (July 1949): 44.

128. *DPD* (Hamburg) (26 March 1949) reported in SOPADE (April 1949), 52.

129. *Die Neue Zeitung* (Munich) (30 May 1948) reported in SOPADE, "200,000 Deportierte" (June 1948), 98.

130. SOPADE, "Von dem Menschenrechten hat Herr Ulbricht noch nichts gehört" (May 1949), 44. Kurt Schumacher (1895–1952), the leader of the Social Democratic

Party in western Germany, spent ten years in Nazi concentration camps. Konrad Adenauer (1876–1967), a prisoner in a Nazi concentration camp in 1944, was leader of the conservative Christian Democrats and West Germany's first chancellor in 1949.

### *Chapter 4. "A Law Would Be Good"*

1. Charles P. Kindleberger, Letter "Berlin #10" (13 August 1946), 5, in *The German Economy, 1945–1947: Charles P. Kindleberger's letters from the field* (Westport, Conn.: Meckler, 1989), 61.

2. Herbert Hoover, "Text of Hoover mission's findings on the food requirements of Germany," *New York Times* (28 February 1947).

3. *Neues Deutschland* (28 July 1946) editors' comment on a *Hamburger Volkszeitung* (17 July 1946) article on British zone ration cuts. Despite the great agricultural and fisheries wealth of the British zone, the *Neues Deutschland* writer reported, "hunger is turning into starvation." Also, Our Own Correspondent, "£40,750,000 spent on western zone," *The Times* (26 May 1947).

4. The United States, the Soviet Union, and Great Britain agreed at Potsdam (France was not yet included among the victors) that the natural flow of food surpluses from eastern to western Germany would continue as a part of a program to administer Germany as an economic whole.

5. *Neues Deutschland,* "Nur Demokratie kann die Hungerkrise im Westen überwinden," *Neues Deutschland* (26 November 1946).

6. Max Fechner, "Klarheit in der Ostfrage!" *Neues Deutschland* (14 September 1946), 1; Carl J. Friedrich and Henry Kissinger, eds., *The Soviet zone of Germany* (New Haven: Bechtle, 1956), 392–93. Fechner, the Party's top candidate for the October 1946 Berlin elections and fourth man in the SED hierarchy, became minister of justice and a Central Committee member. He was purged after the June Workers' Uprising of 1953.

7. Fechner, "Klarheit in der Ostfrage!" 1.

8. Secretary of State James Francis Byrnes (1879–1972).

9. *Neues Deutschland.* "Die SED zur Grenzfrage," *Neues Deutschland* (19 September 1946), 3, reporting on the Party's Management Committee's (*Vorstand*) meeting on 19 September 1946.

10. Wechselberger Parchmann, "Schluß mit dem Liberalismus in der Forst- und Holzwirtschaft," *Der Deutsche Forstwirt* 16 (1934): 409–11; Wechselberger Parchmann, "Von liberalistische zu nationalsozialistische Forst- und Holzwirtschaft," *Der Deutsche Forstwirt* 16 (1934): 941–44; Erich Wohlfarth, "Zur Waldbaulichen Lage der Gegenwart"; Erich Wohlfarth, "Waldbau oder Waldpflege?" *Allgemeine Forst- und Jagdzeitung* 121 (1950): 115. As late as the early 1950s Erich Wohlfahrt, a leading forest policy scholar in West Germany, attacked clear-cutting and even-aged management as "market-oriented silviculture" (*Marktgemäßen Waldbau*), an invidious reference to capitalism typical of radical conservative foresters and consistent with Marxist-Leninist propaganda.

11. G. B. Ryle, "Forestry in western Germany, 1948," *Forestry* 22:2 (1948): 158.

12. Arbeitsgemeinschaft Naturgemäße Waldwirtschaftswald, "Stellungnahme der Schriftleitung: Erste Norddeutsche Tagung der ANW," *Allgemeine Forstzeitschrift* 5

(1950): 496–99; Ferdinand Beer, "Kahlschlag oder Einzelstammentnahmen," *Forst- und Holzwirtschaft* 3:14 (15 July 1949): 11; Ferdinand Beer, "Naturgemäßer Waldwirtschaft," *Forst- und Holzwirtschaft* 3:5 (15 March 1951): 65; F. Backmund, "Naturgemäße Waldwirtschaft—ein neues Schlagwort?" *Forstwissenschaftliches Centralblatt* 72 (1953): 144; Karl Dannecker, "Um das Plenterprinzip in Sudwestdeutschland," *Schweizerische Zeitschrift für Forstwesen* 100:9 (1949): 414–29; Karl Dannecker, "Vom Naturwald zum naturgemässsen Wirtschaftswald," *Allgemeine Forst- und Jagdzeitung* 121 (1950): 73; Eilhard Wiedemann, "Naturgemäßer Wirtschaftswald und nachhaltige Höchsleistungswirtschaft," *Allgemeine Forstzeitschrift* 5 (1950): 157–62; Witzgall, "Naturgemäßer Waldwirtschaftswald und Holzwirtschaft," *Allgemeine Forstzeitschrift* 5 (1950): 565–66; W. Wobst, "Waldbau—Ein geistloses Handwerk?" *Allgemeine Forstzeitschrift* (1948); E. Wohlfarth, "Waldbau oder Waldpflege?"; E. Wohlfarth, "Natur und Technik im Waldbau," *Allgemeine Forstzeitschrift Wien* (1959); E. Wohlfarth, "Was folgt aus dem Vorstellung vom Wald als Ganzheit für die praktische Waldbehandlung. Vortrag," *Allgemeine Forst- und Jagdzeitung* (1961).

13. Kurt Hueck, "Aktuelle Aufgaben der Forstwirtschaft: Speech by dean of Forstfakultät Eberswald at the Agricultural Science Congress, Berlin, 4 February 1947," *Forst- und Holzwirtschaft* 1:1 (1 April 1947): 6.

14. Friedrich and Kissinger, eds., *Soviet zone of Germany,* 7.

15. Wilhelm Münker, *Gerichtstag im Walde. Die Waldwesen klagen an* (Bielefeld: Deutsche Heimat-Verlag Ernst Gieseking, 1944); Albert Milnick, interview by author, Eberswalde, 20 March 1991. Milnick, a forester active in eastern Germany from 1945 through 1989, particularly singled out Münker's book as reflective of the postwar mood among foresters.

16. The forester is also called *der Forstmann,* an old-fashioned, affectionate title for foresters.

17. Norman M. Naimark, *The Russians in Germany: A history of the Soviet zone of occupation, 1945–1949* (Cambridge: Belknap Press of Harvard University Press, 1995), 144–45. Naimark observed: "Virtually every agricultural expert agreed that breaking up the large estates would hurt rather than help productivity. Expropriating the Junkers and their agents meant destroying the basic economic unit of agricultural production." Also Wolfgang Zank, "Als Stalin Demokratie befahl," *Die Zeit* 25 (23 June 1995): 75; H.-G. Merz, "Bodenreform in der SBZ. Ein Bericht aus dem Jahre 1946," *Deutschland Archiv* 11:24 (1991): 1159. Wilhelm Pieck, Walter Ulbricht, Gustav Sobottka, and Anton Ackermann were at the meeting. Ackermann was responsible for the "special German way toward socialism," which fell out of favor after the Tito crisis in spring 1948. Ulbricht later "purged Ackermann for his 'conciliatory attitude' following the June 1953 workers' revolt."

18. Jochen Laufer, "'Genossen, wie ist das Gesamtbild?' Ackermann, Ulbricht und Sobottka in Moskau im Juni 1945," *Deutschland Archiv* 29:3 (May-June 1996): 355, noted later by Walter Ulbricht, the de facto head of the German Communist Party.

19. Ekkehard Schwartz, "Die demokratische Bodenreform, der Beginn grundlegender Veränderungen der Waldeigentums und der Forstwirtschaft im Gebiet der Deutschen Demokratischen Republik," *Sozialistische Forstwirtschaft* 20:10 (1970): 290; Zank, "Als Stalin Demokratie befahl," 75.

20. Wolfgang Leonhard, *Die Revolution entlässt ihre Kinder* (Cologne: Kiepenheuer and Witsch, 1955); Merz, "Bodenreform in der SBZ," 112. Wolfgang Leonhard was director of the Party School responsible for drafting the first land reform legislation and edicts.

21. Wolfgang Leonhard, "Iron Curtain: Episode 2," interview (4 October 1998), National Security Archive, http://www.gwu.edu/~nsarchiv/coldwar/interviews/episode-2/leonhard2.html (accessed 8 March 2003); Leonhard, *Die Revolution entlässt ihre Kinder*, 348–58, 389–90; Wolfgang Leonhard, "Es muß demokratisch aussehen," *Die Zeit* (7 May 1965): 3, cited in Mark Kramer, "The Soviet Union and the founding of the German Democratic Republic: 50 years later—a review," *Europe-Asia Studies* 51:6 (September 1999): 1093.

22. Leonhard, "Es muß demokratisch aussehen," 3; Leonhard, *Die Revolution entlässt ihre Kinder,* 348–58, 389–90.

23. *Pravda* (15 February 1956), cited in Melvin Croan, "Soviet uses of the doctrine of the 'Parliamentary Road' to Socialism: East Germany 1945–1946," *American Slavic and East European Review* 17:3 (October 1958): 302.

24. Hermann Weber, *Geschichte der DDR Autorreferat* (Munich: Deutscher Taschenbuch Verlag, 1985), 186; Edgar Tümmler, Konrad Merkel, and Georg Blohm, *Die Agrarpolitik im Mitteldeutschland und ihre Auswirkung auf Produktion und Verbrauch landwirtschaftliche Erzeugnisse* (Berlin: Duncker and Humblot, 1969).

25. Merz, "Bodenreform in der SBZ," 1159; Walter Ulbricht, "Der Kampf um Deutschland," *Demokratische Aufbau* 7 (October 1946): 193.

26. Hans Lemmel, "Der deutsche Wald in der Bodenreform," *Allgemeine Forst- und Jagdzeitung* 125:3 (1954): 107; Werner Bröll, "Das sozialistische Wirtschaftssystem," in Heinz Rausch and Theo Stammen, *DDR* (Munich: Verlag C. H. Beck, 1978), 138.

27. Hermann Weber, *Geschichte der DDR Autorreferat* (Munich: Deutscher Taschenbuch Verlag, 1985), 111.

28. Hämmerle, "Das Osthilfegesetz und seine Auswirkungen auf die Forstwirtschaft," *Der Deutsche Forstwirt* 14 (1932): 118; Lemmel, "Der deutsche Wald in der Bodenreform," 106–7. Land reform has a long history in Germany. The Prussian government actively pursued land reform and redistribution programs. The *Ansiedlung* (settlement) program between 1886 and 1914 redistributed 700,000 hectares in Prussia to farm workers and peasants. The *Reichssiedlunggesetz* (Reich Settlement Law) distributed a further 1.3 million hectares in farms over 100 hectares between 1914 and 1941. The *Ostpreußenhilfe* (Support for East Prussia) in 1928–29 and the *Gesetz des Ostpreußenhilfsgesetz* (Law on East Prussian Support, 18 May 1929) furthered these policies. Adolf Damaschke, *Die Bodenreform*, 19th ed. (Jena: Gustav Fischer, 1922). The work of many *völkisch* intellectuals in the 1920s, such as Karl Haushofer and Rudolf Steiner, also reflected land reform's themes. *Die Bodenreform* had run through nineteen editions by 1922.

29. Heinrich Brüning (1885–1970).

30. Hämmerle, "Das Osthilfegesetz," 118. The *Osthilfe* land reform program cost RM2 billion, US$6.1 billion in current value.

31. Dr. Hans Schlange-Schöningen (1886–1960).

32. Francis Ludwig Carsten, *A history of the Prussian Junkers* (Aldershot, Hants, Eng.: Gower, 1989), 165–66.

33. General Kurt von Schleicher (1882–1934), an army major general and political intriguer, manipulated Germany's aging president, Paul von Hindenburg (1847–1934), into dismissing Brüning on 30 May 1932, ending the Weimar Republic. Six months later, on 30 January 1933, Hitler outmaneuvered and seized the chancellorship from von Schleicher in the Nazi *Machtergreifung*. S.S. assassins murdered von Schleicher on 30 January 1934, the "Night of the Long Knives."

34. *Neues Deutschland* is filled in the years between 1945 and 1947 with attacks on Schlange-Schöningen's polices, on his aristocratic heritage, and his protection of "Junker interests." Following are typical: "Herr Schlange will den Spuren verwischen," *Neues Deutschland* (24 October 1946); "Nur Demokratie kann die Hungerkrise im Westen überwinden," *Neues Deutschland* (26 November 1946); "Schlange-Schöningen muß gehen!" *Neues Deutschland* (30 November 1946); "Der Weg Schlange-Schöningen: Zweierlei Maß für Umsiedler," *Neues Deutschland* (29 December 1946).

35. Damaschke, *Die Bodenreform;* Rolf Badstübner and Wilfried Loth, eds., *Wilhelm Pieck—Aufzeichnungen zur Deutschland Politik, 1945–1953,* "Mitteilungen Semjonows vom 25.8.45" ZPA NL 36/734, Bl. 119–20 (Berlin: Akademie Verlag, 1994); Karl Kautsky, *Die Agrarfrage; eine Übersicht über die Tendenzen des modernen Landwirtschaft und die Agrarpolitik der Sozialdemokratie* (Stuttgart: J. H. W. Dietz Nachf. [G.m.b.H.], 1899); Manfred Bensing, "Müntzer, Thomas," from Encyclopedia Britannica Online, http://search.eb.com/eb/article?eu=55671 (accessed 24 March 2003). Müntzer (1490–1525) led the failed Peasants' Revolt (1524–25) in Thuringia during the Reformation. "A controversial figure in life and in death, Müntzer is regarded as a significant force in the religious and social history of modern Europe. Marxists in the twentieth century viewed him as a precursor in the struggle for a classless society."

36. Lenin, at the Second World Congress of Comintern (19 July–7 August 1920). Arnd Bauerkamper, "'Loyale Kader'? Neue Eliten und die SED-Gesellschaftspolitik auf dem Lande von 1945 bis zu den fruhen 1960er Jahren," *Archiv für Sozialgeschichte* 39 (1999): 270, 273; Erich Honecker, "Bündnis war, ist und bleibt Eckpfeiler unserer Politik," *Neues Deutschland* (6–7 September 1975), 3. Naimark, *Russians in Germany,* 155, 163. Erich Honecker, architect of the Wall and East Germany's leader in its final two decades, proudly declared that Soviet experience was of "overwhelming significance," particularly Lenin's 1917–28 agricultural policy and Stalin's forced collectivization and breaking of the kulaks' power (1929–33). In 1917 the bolsheviks purged the largest landowners, the *pomeshchiki,* a class very roughly similar to the Junkers, allowing the large farmers, the kulaks (similar to the German *Großbauern*), to remain to supply food to the cities during Lenin's New Economic Program from the end of the civil war (1918–20) until Lenin's death in 1924.

37. Schwartz, "Die demokratische Bodenreform," 290; Joachim Piskol, "'Junkerland in Bauernhand': Wie deutsche Antifaschisten die demokratische Bodenreform 1945 vorbereiteten," *Neues Deutschland* 198 (24 August 1985): 13.

38. Wilhelm Pieck, *Bodenreform: "Junkerland in Bauernhand,"* (Berlin: Verlag Neuer Weg GmbH, 1945), 3–16. Cited in Naimark, *Russians in Germany,* 156. Wilhelm Pieck, the titular head of the embryonic East German government, disingenuously declared, "the revolution in the countryside was to be a controlled and moderate one, fitting the

needs of the anti-fascist democratic transformation, securing agricultural supplies for the Soviet troops, and avoiding all 'class warfare' in the villages."

39. Transcript of the functionaries' conference (6 January 1946), quoted in Naimark, *Russians in Germany,* 155n60, 512.

40. Croan, "Soviet uses of the doctrine of the 'parliamentary road,' " 303.

41. Bauerkamper, " 'Loyale Kader'?" 273. Land reform "gave the German Communist Party leadership the greatest possibility to extend their influence with the rural population and to create a strong alliance between the workers and peasants."

42. Walter Ulbricht, *Zur Geschichte der neuesten Zeit,* vol. 1 (Berlin: 1955). Fifty-seven percent were independents. Naimark, *Russians in Germany,* 156, also discusses the SED's political motivations in land reform, in particular their competition with the CDU for the support of former Nazi farmers.

43. Bauerkamper, " 'Loyale Kader'?" 270; Arnd Bauerkamper, "Zwangsmodernisierung und Krisenzyklen: Die Bodenreform und Kollektivierung in Brandenburg, 1945–1960/61," *Geschichte und Gesellschaft: Zeitschrift für Historische Sozialwissenschaft* 25:4 (1999): 556–88.

44. Ibid.

45. Badstübner and Loth, eds., *Wilhelm Pieck.* Wilhelm Pieck in an undated speech given at the end of August 1947 before the Second Party Congress.

46. Ibid., 83. Wilhelm Pieck, notes on telephone conversation with V. S. Semyonov (4 October 1946) ZPA NL 36/734, Bl. 213–15.

47. Lemmel, "Der deutsche Wald in der Bodenreform," 89. The "*Provinzalverordnung über die Bodenreform,*" published on 3 September 1945 for Saxony-Anhalt. Essentially the same as Pieck's 4 September 1945 announcement at the pro forma peasants' assemblies in Kyritz.

48. Piskol, " 'Junkerland in Bauernhand!' " 13.

49. Walter Ulbricht, "Die demokratisches Bodenreform—ein rühmreiches Blatt in den deutschen Geschichte," *Einheit* 10 (1955): 849. The "bloc party" politicians protesting land reform expropriation were Waldemar Koch, the Liberal Party chairman, and the conservative Christian Democratic leaders Andreas Hermes and Walther Schreiber. Hermes was a member of the Karl Goerdeler (1884–1945) circle and an active anti-Nazi condemned to death for his part in the 20 July plot against Hitler.

50. Bauerkamper, " 'Loyale Kader'?" 270–75; H. O. Spielke, G. Breithaupt, H. Bruggel, and H. Stand, *Ökonomik der sozialistischen Forstwirtschaft* (Berlin: VEB Deutsche Landwirtschaftsverlag, 1964). Land taken from 13,699 farms and forests and more than 7,000 estates.

51. Lemmel, "Der deutsche Wald in der Bodenreform," 102; Ralf Neubauer, "Rückkehr der Junker?" *Die Zeit* 36 (9 September 1994): 10.

52. Nazis owned 4.3 percent of Soviet zone farm and forest land.

53. Schwartz, "Die demokratische Bodenreform," 291; Naimark, *Russians in Germany,* 156.

54. Anton Hilbert, *Denkschrift über die ostdeutschen Bodenreform.* In H.-G. Merz, ed., "Bodenreform in der SBZ: Ein Bericht aus dem Jahre 1946," *Deutschland Archiv* 11:24 (1991): 1166.

55. Ibid., 1167.

56. Philipp-Christian Wachs, *Die Bodenreform von 1945: die zweite Enteignung der Familie Mendelssohn-Bartholdy* (Baden-Baden: Low & Vorderwulbecke, 1994); Julius Hans Schoeps, ed., *Enteignet durch die Bundesrepublik Deutschland: der Fall Mendelssohn-Bartholdy: eine Dokumentation* (Bodenheim: Philo Verlagsgesellschaft, 1997). The Mendelssohn-Bartholdy estate (*Gut*) was in Börnicke, Kreis Niederbarnim, Brandenburg.

57. S. Duschek, "Wirtschaftspolitische Betrachtungen des deutschen Großgrundbesitzes," *Zeitschrift für Weltforstwirtschaft* 2 (1935): 477. Statistisches Bundesamt, "Bevölkerung und Wirtschaft, 1872–1972" (1973): 152; 200 ha was also the upper end of the size class toward which eastern German farms were equilibrating in the interwar period. The increase in 20–100 ha parcels (*Großbauerlichen Betriebe*) was greatest in East Prussia and Pomerania.

58. Klaus Peter Krause, "Begriffsbewirrungen über die 'Bodenreform' zwischen 1945 und 1949," *Frankfurter Allgemeine Zeitung* (2 September 1994), 8. Only sixty-six expropriated farms in the Soviet zone were larger than one thousand hectares.

59. Beate Ruhm von Oppen, *Documents on Germany under occupation, 1945–1954* (London: Oxford University Press, 1955), 148; Schwartz, "Die demokratische Bodenreform," 292.

60. Von Oppen, *Documents on Germany under occupation,* 148.

61. Friedrich and Kissinger, eds., *Soviet zone of Germany,* 445.

62. Wolfgang Zank, " 'Junkerland in Bauernhand!' 3.3 millionen Hektar Land wurden 1945–49 während der Bodenreform in der Sowjetzone beschlagnahmt," *Die Zeit* 42 (12 October 1990): 49.

63. Werner Klatt, "Food and farming in Germany. II. Farming and land reform," *International Affairs* 26:2 (April 1950): 202; Bundesministerium für gesamtdeutsche Fragen (BGF), *SBZ von A-Z* (Bonn: Deutscher Bundesverlag, 1953); Neubauer, "Rückkehr der Junker?" 10; Naimark, *Russians in Germany,* 157; Horst Kohl, ed., *Ökonomische Geographie der Deutschen Demokratischen Republik,* 3rd ed. (Gotha and Leipzig: VEB Hermann Haack, 1976), 354; Konrad Merkel, *Die Agrarwirtschaft in Mitteldeutschland. "Sozialialisierung" und Produktionsergebnisse* (Bonn: Bundesmininsterium für gesamtdeutsche Fragen, 1963); Gerhard Seidel, Kurt Meiner, Bruno Rausch, and Alfonso Thoms, *Landwirtschaft in der Deutsche Demokratischen Republik,* trans. Gunvor Leeson (Leipzig: VEB Edition, 1962).

64. Seidel, *Landwirtschaft in der Deutsche Demokratischen Republik,* 31; Bundesministerium für gesamtdeutsche Fragen (BGF). *SBZ von A-Z,* 29.

65. Gerd Friedrich, et al., *Die Volkswirtschaft der DDR,* Akademie für Gesellschaftswissenschaften beim ZK der SED (Berlin: Verlag die Wirtschaft, 1979), 156; Gerhard Seidel, *Die Landwirtschaft der Deutschen Demokratischen Republik,* 31; Bauerkamper, " 'Loyale Kader'?" 275; Panorama DDR, *Agriculture in the German Democratic Republic: Some information about the life and work of the cooperative farmers* (Berlin: Panorama DDR, 1979).

66. Similar to conditions in National Socialist agricultural policy and to the Nazi Farm Inheritance law.

67. Ernst Goldenbaum, "Demokratische Bodenreform hat unseren Bauern eine gesicherte Zukunft eröffnet," *Neues Deutschland* 204 (28 August 1975): 3; Schwartz, "Die

demokratische Bodenreform," 291, repeats the myth of "spontaneous action" of peasants, with workers rallying to their comrades in the countryside.

68. "Life in the Soviet zone," *The Times* (27 December 1945), 5.

69. Friedrich and Kissinger, eds., *Soviet zone of Germany,* 237.

70. Erwin Kienitz, *Denkschrift über forstwirtschaftlichsorganisatorischen Reformen, insbesondere des Bauernwäldes der Deutschen Demokratischen Republik: Ein Beitrag zur sozialistischen Umgestaltung der Forstwirtschaft* (Tharandt: Institut für forstliche Wirtschaftslehre, 1958). Erwin Kienitz reported in depth to the Party leadership in 1958 on the growing alienation of the rural population.

71. Dorothea Faber, "Entwicklung und Lage der Wohnungswirtschaft in der sowjetischen Besatzungszone, 1945–1953," *Wirtschafts-Archiv* 8:17 (5 September 1953): 5943.

72. Edwin Hörnle, "Ein Jahr nach der Bodenreform, Materialzussamenstellung für einen Bericht an die Sowjetische Militäradministration" (9 December 1946), Archives of the DDR Agricultural Ministry, quoted by Ulrich Kluge, historian at the Technische Universität Dresden (the TUD) in a letter to the editor, *Frankfurter Allgemeine Zeitung* (13 September 1994), 9, 15. Schwerin's vice-president was Herr Möller.

73. Naimark, *Russians in Germany,* 155. The communists deeply distrusted the traditional cooperatives, the "Raffeisen cooperatives," named after a nineteenth-century farm innovator.

74. Hilbert, "Denkschrift über die ostdeutschen Bodenreform," 1169.

75. Klatt, "Food and farming in Germany: II," 195.

76. Hilbert, "Denkschrift über die ostdeutschen Bodenreform," 1169.

77. SOPADE, "Farmers in Saxony" (August 1947), 5. Reporting on farming in Saxony; SOPADE 7 (October 1947), 21.

8. Hörnle, "Ein Jahr nach der Bodenreform, Materialzussamenstellung," 15; Wolfgang Hassel, "'Junkerland in Bauernhand' war damals die Kampflosung: Dokumente des Staatsarchivs Magdeburg über die Bodenreform," *Neues Deutschland* 255 (27 October 1984): 13. The land reform laws gave the elected Land Reform Commission authority to monitor the land reform processes and responsibility to notify farmers to be expropriated well in advance of any action, to treat expropriated farmers humanely, and to divide expropriated farm inventories and equipment fairly. The expropriation of farms under 100 hectares in area was prohibited but done where "class enemies"—former local government officials and political opponents of the German communists—owned farm and forest land of any area.

79. Hassel, "'Junkerland in Bauernhand' war damals die Kampflosung," 13.

80. Gerhard Grüneberg, "30 Jahre Marxistisch-Leninistische Agrarpolitik—30 Jahre Bündnis der Arbeiterklasse mit den Bauern," *Neues Deutschland* 188 (9 August 1975): 3. Gerhard Grüneberg, hard-line Marxist-Leninist, minister of agriculture and forestry in the 1970s and developer of Industrial Production Methods for farming and forestry, promoted land reform as a spontaneous "anti-imperialist, democratic agrarian revolution, the largest and most comprehensive mass action, a victorious revolution in German history undertaken by the collective action of the workers and peasants. It was wholly carried out by the workers themselves."

81. Naimark, *Russians in Germany,* 85–86.

82. Hilbert, "Denkschrift über die ostdeutschen Bodenreform," 1168.

83. Ibid. Soon after Dr. Kolter's arrest, he "died under mysterious circumstances."

84. Ibid.; Heinz Kuhrig, "Demokratische Bodenreform legte den Grundstein für stetig steigende Agrarproduktion," *Neues Deutschland* 198 (21 August 1975): 3. Kuhrig became agriculture minister after Grüneberg's death in 1981.

85. *Wirtschaft und Arbeit* (Essen) (17 May 1947), cited in SOPADE (Executive Committee, Social Democratic Party, Germany), "Die Agrarsituation in der Ostzone," #3 (June 1948): 17.

86. Naimark, *Russians in Germany,* 161; Kuhrig, "Demokratische Bodenreform legte den Grundstein," 3; Zank, " 'Junkerland in Bauernhand!' " 49–50.

87. Friedrich and Kissinger, eds., *Soviet zone of Germany,* 466; Economic Commission for Europe, *Economic Survey of Europe in 1954* (Geneva, 1955), 49, cited in Joint Session of the Economic Reports, "Trends in economic growth, a comparison of the Western powers and the Soviet bloc" (Washington, D.C., 1955), 292. "The per capita consumption of meat, meat products, and milk remained well below prewar levels as late as the second half of 1953.

88. Friedrich and Kissinger, eds., *Soviet zone of Germany,* 3.

89. Our Diplomatic Correspondent, "Dearth of food in Soviet zone: Demonstrations and arrests: Stocks reduced by requisitioning for Berlin," *The Times* (4 August 1948); Our Diplomatic Correspondent, "Western powers and the Moscow talks: Dearth of food." *The Times* (11 August 1948).

90. Our Diplomatic Correspondent, "Western powers and the Moscow talks: Dearth of food," *The Times* (11 August 1948).

91. *Tägliche Rundschau* (28 September 1947), cited in Our Own Correspondent, "Absenteeism in the Soviet zone," *The Times* (29 September 1947).

92. Dieter Staritz, *Die Gründung der DDR: Von der Sowjetischen Besatzungsherrschaft zum sozialistischen Staat,* 2nd ed. (Munich: Deutscher Taschenbuch Verlag, 1987); Naimark, *Russians in Germany,* 162–63; "Ulbricht, Walter" from Encyclopedia Britannica Online, http://search.eb.com/eb/article?eu=76086 (accessed 7 May 2002). Walter Ulbricht became deputy prime minister on the formation of the German Democratic Republic (11 October 1949), rising to the rank of general secretary of the SED in 1950. When, on the death of President Wilhelm Pieck (1960), the presidency was abolished and a Council of State instituted in its stead, Ulbricht became its chairman, thus formally taking supreme power.

93. R. Reutter, "Volkssolidarität auf dem Lande. Zwei nachahmenswerte Beispiele," *Neues Deutschland* (9 July 1946). Hörnle was also president of the German Central Organization for Agriculture and Forestry.

94. "Mecklenburg baut für die Neubauern," *Demokratische Aufbau* 1 (April 1946): 28.

95. Ibid. The "Verordnung Nr. 69 über die Wohnungen für die Neubauern im Lande Mecklenburg-Vorpommern" (28 February 1946) called for building 12,000 new farmhouses and stables in 1946. Paragraph 6 set the total cost of each farmstead at 3,500 marks, including windows, chimneys, and utilities, not to mention brick and stone, tile roofs, etc., or whatever standard of plumbing the act's authors deemed appropriate.

96. Walter Ulbricht, "Lebendige Demokratie," *Demokratische Aufbau* (November 1947), 321.

97. SOPADE, "Die Forstwirtschaft in der Sowjetzone," 29.

98. Ibid.

99. Bundesministerium für gesamtdeutsche Fragen (BGF), *SBZ von A-Z*, 48.

100. SOPADE. "Die Forstwirtschaft in der Sowjetzone," 10.

101. Schwartz, "Die demokratische Bodenreform," 292; E. Reichenstein, "Die forstwirtschaftliche Lage Deutschland vor und nach dem 2. Weltkrieg," *Forstarchiv* 21:1/3 (1950). In contrast, the British and American zonal governments handed management of former Reich forest, only 30 percent of the western zones' forest area, back to the individual states, the *Länder*. Both Soviet and western zone governments, however, abolished forestry's separate ministry and brought forestry back within the Agriculture Ministry's suzerainty, a reversal of National Socialist reforms that signaled the importance of production and cash flow over capital asset growth and close-to-nature forest management throughout Germany.

102. J. Säglitz, "Die Forstwirtschaft in Ostdeutschland—Stand, Probleme, Ziele," *Forstarchiv* 61:6 (November-December 1990): 226. The Party kept 56 percent (585,000 hectares) of forestland expropriated under the land reform.

103. W. Schindler, "30 Jahre staatliche Forstwirtschaftsbetriebe—30 Jahre sozialistische Entwicklung in der Forstwirtschaft. Staatliche Forstwirtschaftsbetrieb Löbau, Löbau, D.D.R," *Sozialistische Forstwirtschaft* 32:11 (1982): 321–23; Spielke, Breithaput, Bruggel, and Stand, *Ökonomik der sozialistischen Forstwirtschaft*, 60. The Party distributed 500,000 hectares of forestland to new peasants.

104. Hämmerle, "Das Osthilfegesetz," 119.

105. Hilbert, "Denkschrift über die ostdeutschen Bodenreform." The traditional cooperatives were the *Waldbauergenossenschaften*, Peasant Forest Cooperatives.

106. Kurt Mantel, "Forstgeschichte," in "Stand und Ergebnisse der forstlichen Forschung seit 1945," *Schriftenreihe des AID* (1952): 144–53.

107. Kienitz, *Denkschrift*. Farmers' fear of collectivization permeates this confidential study of farmers' attitudes in the mid-1950s.

108. Editorial, *Allgemeine Forstzeitschrift* 2:14 (15 July 1947): 1.

109. Ernst Quadt, "Der Ofen und das Brennmaterial," *Neues Deutschland* (20 December 1946).

110. "One hundred decisive days: How will Berlin get its firewood?" *Neues Deutschland* (27 June 1946); "Nicht nur für den Ofen . . . ," *Neues Deutschland* (23 June 1946).

111. Kienitz, *Denkschrift*, 61.

112. "Nicht nur für den Ofen . . . ," *Neues Deutschland*.

113. Wolfgang F. Stolper and Karl W. Roskamp, *The structure of the East German economy: "German wood production 1935/6–1958"* (Cambridge: Harvard University Press, 1960). Assuming Soviet reparations took 80 percent of the total *Derbholz*/sawlog harvest.

114. Edwin Hörnle, *Volksstimme* (Chemnitz) (3 July 1947) cited in SOPADE, "Landwirtschaft in der Ostzone" (December 1947), 9. Hörnle complained of Soviet requisitions of livestock to haul logs from the forest: "The worst aspect of the shortage of work horses is that all horses are completely overworked skidding wood." The Party's impressments of farm labor for forestry work was a constant source of irritation to farmers struggling to meet the Plan's implacable production quotas.

115. Ibid.

116. Klatt, "Food and farming in Germany: II," 195.

117. "Bloß ein Kolchose," *Telegraf* (Berlin) (8 October 1947) cited in SOPADE, "Landwirtschaft in der Ostzone" (February 1948), 40.

118. Naimark, *Russians in Germany*, 162–63.

119. Lemmel, "Der deutsche Wald in der Bodenreform," 3.

120. Fritz Lange, quoted in Naimark, *Russians in Germany*, 144.

121. Hilbert, "Denkschrift über die ostdeutschen Bodenreform." Hilbert was a land reform administrator in Thuringia from fall 1945 through spring 1946.

122. Schwartz, "Die demokratische Bodenreform," 289–93.

123. Edwin Hörnle, "Wie kann die deutschen Landwirtschaft ihre Aufgabe erfüllen?" *Neues Deutschland* (16 May 1946).

124. *Wirtschaft und Arbeit,* "Die Agrarsituation in der Ostzone," 17.

125. Ibid. "Large farmers" meant independent farmers with twenty to one hundred hectares.

126. Hörnle, "Ein Jahr nach der Bodenreform, Materialzussamenstellung," 15; SOPADE (Executive Committee, Social Democratic Party, Germany), "KZs in der Ostzone" 5 (August 1947), 66.

127. Paul Merker, editorial, *Neues Deutschland* (19 December 1946).

128. *Wirtschaft und Arbeit,* "Die Agrarsituation in der Ostzone," 17. The Party formally allowed peasants to sell surplus produce on the free market "after fulfillment of their delivery obligations," but the Party increased quotas once they were met, constantly absorbing surpluses and killing farmers' incentives to produce more food.

129. Bauerkamper, " 'Loyale Kader'?" 284; Naimark, *Russians in Germany*, 155. "Some 10,000 new peasants, according to Soviet reports, abandoned the land altogether. In the worst case, Mecklenburg, nearly 20 percent of the new peasants left their settlements between 1945 and 1949"; Götz, "Als der Klassenkampf in der DDR begann," 13. By mid-1947, 1,166 new owners in Brandenburg had left their land because of inadequate support and their inability to service their debt to the state; Kienitz, *Denkschrift*. Kienitz also commented upon the great practical difficulties facing small farmers and forest owners in the 1940s and 1950s.

130. Kienitz, *Denkschrift*.

131. "Bauern sichern die Ernährung: Antifascisten aufs Dorf," *Neues Deutschland* (1 June 1946). The 21–22 February 1946 Land Reform Conference in Berlin.

132. Ibid.

133. The Mühleberg concentration camp.

134. SOPADE, "KZs in der Ostzone," 77.

135. Our Special Correspondent, "Arrests of German children: Protest by parents," *The Times* (25 June 1946), 4; Our Special Correspondent "Arrests in the Russian zone: German parents' complaints, re-education motive," *The Times* (19 August 1946), 3; Our Own Correspondent, "Youth group arrests in eastern Berlin: 'Subversive propaganda,' " *The Times* (16 June 1949), 3; Our Own Correspondent, "Arrests by Russians in Germany: Christian youth leaders," *The Times* (24 March 1947), 5; Our Own Correspondent, "Soviet arrests in Berlin: 'Fascist activities' alleged," *The Times* (29 March 1947).

136. Merkel and Wahl, *Das geplünderte Deutschland,* 29; Mike Dennis, *German*

*Democratic Republic: Politics, economics and society* (London: Pinter Publishers, 1931), 23. Ulbricht's speech to the Party's *Hochschule.*

137. Ulbricht's speech at the 16 September 1948 Party conference. Merkel and Wahl, *Das geplünderte Deutschland,* 29.

138. Friedrich and Kissinger, eds., *Soviet zone of Germany.*

139. Hörnle, "Wie kann die deutschen Landwirtschaft ihre Aufgabe erfüllen?"

140. Just as fuelwood harvests slackened.

141. Ferdinand Beer, "Über einige Erfolge und Probleme bei der Aufforstungsarbeiten im Jahre 1949," *Forst- und Holzwirtschaft* 3:21 (1 November 1949): 329; Wolfgang Mühlfriedel, "Der Wirtschaftsplan 1948: Der erste Versuch eines einheitlichen Planes der deutschen Wirtschaftskommission zur ökonomischen Entwicklung der sowjetischen Besatzungszone," *Jahrbuch für Wirtschaftsgeschichte* 3 (1985): 9–26. The First Two-Year Plan was the first comprehensive economic plan for eastern Germany. Its central goals were to increase production by one-third to 80 percent of 1936 levels and to increase work productivity by 30 percent, mostly through increased work norms and *Aktivistenbewegung* (Activists Movement) modeled on the Soviet Stachanov-System, rechristened in East Germany as the *"Hennecke-Bewegung."*

142. Forst- und Holzwirtschaft, "Die Forstwirtschaft in der Deutschen Demokratischen Republik," 225. Ferdinand Beer, "Unser Aufforstungsplan 1949," *Forst- und Holzwirtschaft* 3:6 (5 March 1949): 1. The 1950 goal was increased first by 50 percent and then by 100 percent. Between 1951 and 1955 woodsworkers replanted 320,000 hectares, more than 10 percent of the forest area.

143. Erich Wohlfarth, "Zur waldbaulichen Lage der Gegenwart," *Forstarchiv* 23:4 (1 May 1952). 145,000 hectares were "suitable for afforestation"; Deutsches Institut für Wirtschaftsforschung (DIW), "SBZ 1953 inventory," *(DIW)-Wochenbericht* (1954), 96; SOPADE, "Die Forstwirtschaft in der Sowjetzone."

144. A prescription similar to Dean Acheson's made to State Department officers after he was appointed as secretary of state in January 1949: "Don't just do something, sit there."

145. G. Schröder, "Ökonomische Probleme des zweiten Fünfjahresplans in der Forstwirtschaft," *Forst und Jagd,* part 2 of 4, 7:5 (May 1957): 198.

## *Chapter 5. The Landscape's "Socialist Transformation"*

1. Harald Thomasius, interview by author, Tharandt, 20–26 March 1991. Thomasius knew Krutzsch and described him as dogmatic but also applied and practical.

2. "Vorratspfleglicher Waldwirtschaft, das Gebot der Stunde!" *Forst- und Holzwirtschaft,* 2–part special issue: 3:18 (15 September 1949) and 3:23/24 (December 1949); Johannes Blanckmeister, "Vorratspflegliche Waldwirtschaft. Referat auf der Waldbautagung am 14. Juni in Menz," *Forst- und Holzwirtschaft* 9:5 (15 September 1951): 260; Hermann Krutzsch, "Vorratspflege," *Forst- und Holzwirtschaft* 3:7 99 (1 April 1949). Krutzsch's central principle: "Fell the worst, save the best."

3. Thomasius, interview.

4. Even though *Dauerwald*'s economic advantages were unproved and Blanckmeister's claims contentious.

5. Blanckmeister, "Vorratspflegliche Waldwirtschaft," 260.

6. Ferdinand Beer, "Naturgemäßer Waldwirtschaft," *Forst- und Holzwirtschaft* 3:5 (15 March 1951): 65; Ferdinand Beer,"Unser Aufforstungsplan 1949," *Forst- und Holzwirtschaft* 3:6 (5 March 1949): 1; Ferdinand Beer, "Fortschrittliche Forstwirtschaft—ein Beitrag zur deutsche Demokratisierung," *Forst- und Holzwirtschaft* 2:1 (1 January 1948); Ferdinand Beer, "Kahlschlag oder Einzelstammentnahmen," *Forst- und Holzwirtschaft* 3:14 (15 July 1949): 11; Ferdinand Beer, "Über einige Erfolge und Probleme bei der Aufforstungsarbeiten im Jahre 1949," *Forst- und Holzwirtschaft* 3:21 (1 November 1949): 329; Ferdinand Beer, "Die Forstwirtschaft und unsere junge Deutsche Demokratische Republik," *Forst- und Holzwirtschaft* 3:20 (15 October 1949): 313.

7. Karl Dannecker, "Um das Plenterprinzip in Sudwestdeutschland," *Schweizerische Zeitschrift für Forstwesen* 100:9 (1949): 414–29; Karl Dannecker, "Vom Naturwald zum Naturgemäßsen Wirtschaftswald," *Allgemeine Forst- und Jagdzeitung* 121 (1950): 3; Karl Dannecker, "Beispielbetriebe der Praxis unter Kritik der Wissenschaft," *Forstwissenschaftliches Centralblatt* 69 (1950): 744–64; Editors, "Wiederaufbau der Wälder," *Allgemeine Forst- und Jagdzeitung* 18 (1949): 165. Felix Funke, *Die Bärenfelser naturgemäße Waldwirtschaft; Grundsätze, Ziele und Erfolge* (Berlin: Deutsche Bauernverlag, 1954), 13; W. Wobst, "Waldbau—Ein geistloses Handwerk?" *Allgemeine Forstzeitschrift* 172–74 (1948); E. Wohlfarth, "Zur Waldbaulichen Lage der Gegenwart," *Forstarchiv* 23:4 (1 May 1952). The meeting was held at Schwäbisch-Hall. K. Dannecker who edited five special issues of the *Allgemeine Forstzeitschrift* dedicated to ecological forestry, was less dogmatic than Krutzsch. Dannecker presented a moderate *Dauerwald* manifesto with *Plenterwald* forestry (selection forest management) as the ultimate goal, despite objections that it was not radical enough.

8. A. Richter, "Aufgaben und Methoden gegenwartsnaher Forsteinrichtung," *Archiv für Forstwesen* 1:1/2 (1952): 31–46; A. Richter, "Zur Ethik im Forstberuf," *Beiträge für die Forstwirtschaft und Landschaftsökologie* 27:1 (1993): 14–17; Eilhard Wiedemann, "Naturgemäßer Wirtschaftswald und nachhaltige Höchstleistungswald," *Allgemeine Forstzeitschrift* 5 (1950): 157–62; E. Wagenknecht, Alexis Scamoni, and J. Lehrmann, *Wege zur standortsgerechter Forstwirtschaft* (Radebeul, 1956).

9. Richter, "Aufgaben und Methoden gegenwartsnaher Forsteinrichtung," 31–46.

10. Arbeitsgemeinschaft Naturgemäße Waldwirtschaftswald, "Stellungnahme der Schriftleitung. Erste Norddeutsche Tagung der ANW," *Allgemeine Forstzeitschrift* 5 (1950): 496–99; H. Schoepffer, "Die 'Naturgemäße Waldwirtschaft' und ihre Grundsätze. Darstellung der Entwicklung und Erläuterung des Begriffes," *Forstarchiv* 54:2 (1983): 40.

11. Schoepffer, "Die 'Naturgemäße Waldwirtschaft,'" 42.

12. Ibid.

13. Blanckmeister, "Vorratspflegliche Waldwirtschaft," 260. The Menz Conference was held at Neuglobzow near Menz.

14. Heinz Werner, "Vorratspfleglicher Waldwirtschaft. Die Bedeutung der Menzer Tagung vom 14–5 Juni 1951," *Forst- und Holzwirtschaft* 9:5 (1951): 257; Funke, *Die Bärenfelser naturgemäße Waldwirtschaft*, 13; Ministerium für Land- Forst- und Nahrungsgüterwirtschaft, *Umstellung der Kahlschlagwirtschaft auf vorratspflegliche Waldwirtschaft, Anweisung vom 20 November 1951* (Berlin: Hauptabteilung Forstwirtschaft,

1951); Kohlsdorf, interview. This resolution was formalized on 20 November 1951 in the management instructions "Replacement of Clear-cut Management with Optimal Stocking Forest Management." The foresters who grumbled about the end of "classic forestry" (industrial forest management) were right. If the Menz Resolution had been followed, clear-cutting, Hartig's Eight General Rules, and central decision making for all forest operations would have stopped and decision making devolved to the foresters on the spot.

15. Thomasius, "Waldbauliche Auffassungen," 726, 729.

16. A. Richter, "Aufgaben und Methoden gegenwartsnaher Forsteinrichtung," *Archiv für Forstwesen* 1:1/2 (1952): 31–46.

17. G. Schröder, Ökonomische Probleme des zweiten Fünfjahresplans in der Forstwirtschaft, *Forst und Jagd,* part 1 of 4, 7:4 (April 1957): 145. Ulbricht announced an increase in stocking to 160 cubic meters per hectare by 1956, up from 117 cubic meters per hectare at the Plan's beginning.

18. G. Schröder, "Ökonomische Probleme des zweiten Fünfjahresplans," 145; Horst Ruffer and Ekkehard Schwartz, *Die Forstwirtschaft der Deutschen Demokratischen Republik* (Berlin: VEB Deutscher Landwirtschaftsverlag, 1984), 17; Schindler, "30 Jahre staatliche Forstwirtschaftsbetriebe," 321–23.

19. H. Werner, "Vorratspfleglicher Waldwirtschaft: Die Bedeutung der Menzer Tagung vom 14–15 Juni 1951," *Forst- und Holzwirtschaft* 9:5 (15 September 1951): 257.

20. Ibid.; Thomasius, "Waldbauliche Auffassungen," 726; Felix Funke, *Die Bärenfelser naturgemäße Waldwirtschaft; Grundsätze, Ziele und Erfolge* (Berlin: Deutsche Bauernverlag, 1954), 13.

21. SOPADE (Executive Committee, Social Democratic Party, Germany), "Die Forstwirtschaft in der Sowjetzone," Denkschriften, Sopadeinformationsdienst (Bonn: Vorstand der Sozialdemokratischen Partei Deutschlands, 1955), 28.

22. Hubert Hugo Hilf, "Forstwirtschaft zwischen Gestern und Morgen," *Forstarchiv* 31:3 (15 March 1960): 1.

23. Steven Lukes, "Low Marx," review of *Main currents of Marxism: Its rise, growth and dissolution* by Leszek Kolakowski, trans. P. S. Falla, Oxford University Press, vol. 3: *The breakdown, New York Review of Books* (29 May 1980).

24. G. Schröder, "Ökonomische Probleme des zweiten Fünfjahresplans in der Forstwirtschaft," *Forst und Jagd* 7:4 (April 1957): 145, part 1 of 4; G. Schröder, "Ökonomische Probleme des zweiten Fünfjahresplans in der Forstwirtschaft," *Forst und Jagd* 7:5 (May 1957): 197, part 2 of 4; G. Schröder, "Ökonomische Probleme des zweiten Fünfjahresplans in der Forstwirtschaft," *Forst und Jagd* 7:7 (July 1957): 289, part 3 of 4; G. Schröder, "Ökonomische Probleme des zweiten Fünfjahresplans in der Forstwirtschaft," *Forst und Jagd* 7:8 (August 1957): 338, part 4 of 4.

25. Kurth, interview.

26. SOPADE, "Die Forstwirtschaft in der Sowjetzone," 25.

27. Harvest quotas were only 89 percent filled.

28. Kaiser, "Report on annual plan performance as of 30 April 1953" (in German), *Der Wald* 3:6 (1953).

29. Barth, interview.

30. VEB Forstprojektierung Potsdam, *Forsterhebung 1949. Flächen- und Vorratsgliederung nach Besitzverhältnissen*, v. 5–161–2 FE-Unterlagen (Potsdam, 1949). Inventory data for 1949 for *Volkseigener Wald* ("People's Forest" or nationalized) only, without privately held forest (*Privatwald*). *Übersicht der Holzbodenfläche und Holzvorrat des Volkswaldes der Deutschen Demokratischen Republik* (VEB Forstprojektierung Potsdam, 1 January 1952), typewritten summary. Data for 1952 from VEB Forstprojektierung Potsdam; *Flächen und Vorratsgliederung für den Waldzustand vom 1.1.54* (*Volkswald* [nationalized forest] only) (1954). Data for 1954 from Forstwirtschaftliches Institut Potsdam.

31. Food and Agriculture Organization (FAO) of the United Nations, European Forest Commission, *Forest policy—national progress reports*, 8th session, Rome, no. FAO/EFC/81–H-J, pt. 7, West Germany (1955). State Forest Districts had an average stocking of 93 cubic meters per hectare (Vfm/ha) compared to an average of 75 Vfm/ha in private forests. People's Forest growth averaged between 2.0 and 2.5 cubic meters per hectare; Haden-Guest, Wright, and Teclaff, *World geography of forest resources*, 289; Kienitz, *Denkschrift*.

32. Schmidt-Renner, *Wirtschaftsterritorium Deutsche Demokratische Republik*, 193.

33. *Forestry Abstracts* 7 (1945–46) (Oxford: Imperial Forestry Bureau, 1946).

34. Friedrich Wilhelm Leopold Pfeil (1783–1859); Friedrich Wilhelm Leopold Pfeil, *Neue vollständige Anleitung zur Behandlung, Benutzung und Schätzung der Forsten: ein Handbuch für Forstbesitzer und Forstbeamte* (Berlin: Veit, 1821).

35. Georg Ludwig Hartig (1764–1837); Georg Ludwig Hartig, *Lehrbuch für Jäger und die es werden wollen* (Tübingen: In der J. G. Cotta'schen Buchhandlung, 1811); Georg Ludwig Hartig, *Neue Instructionen für die königlich-preussischen Forst-Geometer und Forst-Taxatoren: Durch Beispiele erklärt mit einem Karten-Schema und einer illuminierten Forst-Karte* (Berlin: In Commission bei der Kummerischen Buchhandlung zu Leipzig, 1819); Georg Ludwig Hartig, *Lehrbuch für Förster und für die, welche es werden wollen*, 10th ed., ed, Dr. Theodor Hartig (Stuttgart, 1861); C. Wiebecke, "Wiedemanns 'Eisernes Gesetz des Örtlichen,'" *Forstarchiv* 61:5 (September-October 1990): 183; Lemmel, "Der Dauerwaldgedanke und 'das eiserne Gesetz des Örtlichen,'"; Ekkehard Schwartz, interview by author, Eberswalde, 13 March 1991. Schwartz observed that there is a direct line from Pfeil through Alfred Möller to Albert Richter.

36. Some of Wiedemann's important articles and criticisms of *Dauerwald* and Krutzsch were: Eilhard Wiedemann, "Der laufender Zuwachs 1913–1924 in Bärenthoren," *Zeitschrift für Forst- und Jagdwesen* 58 (1926): 717–56; Eilhard Wiedemann,"Friedrich von Kalitsch," *Zeitschrift für Forst- und Jagdwesen* 60:1 (1928): 1; Eilhard Wiedemann, "Über die Beziehung des forstlichen Standortes zu dem Wachstum und dem Wirtschaftserfolg im Wald," *Deutsche Forstbeamten Zeitung* 34:5 (1934): 103; Wiedemann, "Review of Krutzsch-Weck's 'Bärenthoren 1934' at the request of Hermann Göring" (in German) (1936): 1157; Wiedemann, "Review of Krutzsch-Weck's 'Bärenthoren 1934'" (1937); Wiedemann, "Bärenthoren 1934" (1937); Wiedemann, "Das Ergebnis von 'Bärenthoren 1934,'" 1065–69; Wiedemann,"Grundsätzliche Fragen des Dauerwaldes," 265–64, 281–82; Wiedemann,"Naturgemäßer Wirtschaftswald und nachhaltige Höchsleistungswirtschaft," 157–62; von Dieterich, "Wiedemann's research on pine *Dauerwald*"; Alfred

Dengler, "Die Untersuchung Professor Dr. Wiedemann's zur Kieferndauerwaldfrage," *Zeitschrift für Forst- und Jagdwesen* 58 (July 1926): 431; Wiebecke, "Wiedemanns 'Eisernes Gesetz des Örtlichen,'" 182.

37. Wagenknecht, Scamoni, and Lehrmann, *Wege zur standortsgerechter Forstwirtschaft.* At the same time East German forests were split into formal management groups. Despite the formal division of forests into management categories, and the further identification of 22.6 percent of total forest area as nature preserves and landscape protection and nature reserve forests (*Landschaftsschutzgebiete*), virtually all the forests in the GDR were managed as production forests.

38. G. Schröder, "Einige Probleme der Auswertung der Grundfragen des zweiten Fünfjahresplans für die StFB," *Forst und Jagd* 6:4 (April 1956): 145; Kienitz, *Denkschrift,* 242. The Third Party Conference program of 1956 also demanded "Modernization, Mechanization, Automation" to raise industrial production by 55 percent.

39. Kienitz, *Denkschrift;* Hermann Weber, *Geschichte der DDR. Autorreferat* (Munich: Deutscher Taschenbuch Verlag, 1999), 297.

40. Thomasius, "Waldbauliche Auffassungen," 727.

41. Mann, *Prinzipien der Preisbildung,* 46.

42. Ministerium für Land- Forst- und Nahrungsgüterwirtschaft, *Grundsätze zur waldbaulichen Behandlung der Forstwirtschaft in der Deutschen Demokratischen Republik,* Anweisung vom 18 Oktober 1961, Abteilung Forstwirtschaft (Berlin, 1961). The guidelines were: "Grundsätze zur waldbaulichen Behandlung der Forstwirtschaft in der Deutschen Demokratischen Republik" (Principles of Silviculture for Forestry in the German Democratic Republic).

43. Helmuth Schrötter, "Zum Begriff der Nachhaltigkeit," *Archiv für Forstwesen* 13:12 (1964): 1280–81; Kurth, interview; Hildebrandt, "Die Forsteinrichtung in der DDR," 130. The formula decided upon was: sustained yield harvest (*Hiebsatz*) equals production quota (*Nutzungssoll*).

44. Schrötter, "Zum Begriff der Nachhaltigkeit," 1281. See also Hildebrandt.

45. G. Laßmann, "Die Rolle und Bedeutung der Forstwirtschaft in System der Volkswirtschaft der Deutschen Demokratischen Republik," *Shriftenreihe für Forstökonomie,* vol. 5 (Berlin: VEB Deutsche Landwirtschaftsverlag, 1960), 14.

46. Ibid.

47. Ibid., 29.

48. Wiebecke, "Zum Stand der deutschen Forststatistik," 6; Bundesministerium für Ernährung, Landwirtschaft und Forsten, *Statistisches Jahrbuch über Ernährung, Landwirtschaft und Forsten der Bundesrepublik Deutschland.* Second forest inventory results, 1948 (Münster-Hiltrup: Landwirtschaftsverlag, 1955); Haden-Guest, Wright, and Teclaff, *World geography of forest resources,* 289. In contrast, West German forests in the mid-1950s grew at 3.7 cubic meters per hectare and harvests limited to 82 percent of growth.

49. Haden-Guest, Wright, and Teclaff, *World geography of forest resources,* 289. Wiebecke, "Zum Stand der deutschen Forststatistik," 6.

50. Through the *Verordung über die Bildung von staatlichen Forstwirtschaftsbetrieben* of 14 February 1952, effective as of 1 January 1952. W. Schindler, "30 Jahre staatliche

Forstwirtschaftsbetriebe—30 Jahre sozialistische Entwicklung in der Forstwirtschaft. Staatliche Forstwirtschaftsbetrieb Löbau, Löbau, DDR," *Sozialistische Forstwirtschaft* 32:11 (1982): 321–23; Kohl, *Ökonomische Geographie der Deutschen Demokratischen Republik,* 461. State Forest Districts averaged 40,500 hectares in the lowlands and 29,000 in the uplands and were built up from four to five old County Forest Districts (renamed *Oberförstereien*) of roughly 7,000 hectares each. The final organizational structure remained constant, *Revieren* of 1,300 hectares each. The number of State Forest Districts shrank further to seventy-two in the 1970s as large area management intensified.

51. Horst Kurth, "Die Entwicklung der Forstwirtschaft in der DDR," *Allgemeine Forstzeitschrift* 35 (1990): 894; Bundesministerium für gesamtdeutschen Fragen (BGF), *SBZ von A-Z,* 4th ed. (Bonn: Deutscher Bundesverlag, 48/9 1958), 103.

52. Ruffer and Schwartz, *Die Forstwirtschaft der Deutschen Demokratischen Republik,* 17; Schindler, "30 Jahre staatliche Forstwirtschaftsbetriebe," 321–23.

53. Kurth, "Die Entwicklung der Forstwirtschaft in der DDR," 894.

54. Ruffer and Schwartz, *Die Forstwirtschaft der Deutschen Demokratischen Republik,* 42.

55. Weber, *Geschichte der DDR,* 226; Hildebrandt, "Die Forsteinrichtung in der DDR," 124; Mike Dennis, *German Democratic Republic: Politics, economics and society* (London: Pinter Publishers, 1988), 22.

56. H. Heidrich, "Die Aufgaben des Betriebsleiters bei der Lösung der ökonomischen Beiträge," *Forst und Jagd* (September 1960): 2.

57. Horst Paucke, "The German Democratic Republic"; Gyorgy Eneyedi, August J. Gijawijt, and Barbara Rhode, eds., *Environmental policies in East and West* (London: Taylor Graham, 1987), 155. The new laws replaced the Nazis' 1935 *Reichsnaturschutzgesetz* and the 1936 *Naturschutzverordnung;* Nikola Knoth, "Die Naturschutzgesetzgebung der DDR von 1954," *Zeitschrift für Geschichtswissenschaft* 39:2 (1991): 163.

58. Hildebrandt, "Die Forsteinrichtung in der DDR," 124. The comment on "cameralist thinking" comes from Paul, interview. Professor Frithjof Paul was East Germany's preeminent forest policy scientist at Tharandt and worked at the core of GDR forest policy formation from the late 1950s until reunification.

59. Bundesministerium für gesamtdeutsche Fragen (BGF), *SBZ von A-Z,* 4th ed. 102/3. The order, *Status der Staatliche Forstwirtschaftsbetriebe,* also mandated that individual districts "meet their own production responsibilities with their own resources." Bundesministerium für gesamtdeutsche Fragen (BGF), *SBZ von A-Z,* 1st ed. (Bonn: Deutscher Bundesverlag, 1953), 48–49.

60. Paul, interview.

61. Bruno Gleitze, *Die Industrie der Sowjetzone unter dem gescheiterten Siebensjahrplan* (Berlin: Duncker und Humblot, 1964), 15; Stolper and Roskamp, *Structure of the East German economy,* 6. Losses, profits, or supports were termed "Fonds."

62. Stolper and Roskamp, *Structure of the East German economy,* 6.

63. Schröder, "Ökonomische Probleme des zweiten Fünfjahresplans," 197.

64. Wolfgang F. Stolper, "The labor force and industrial development in Soviet Germany," *Quarterly Journal of Economics* 71:4 (November 1957): 533.

65. Kienitz, *Denkschrift,* 5.
66. Ibid., 67.
67. Ibid.
68. Ibid., 47.
69. Ibid., 55. Seventy-five cubic meters per hectare versus a state stocking of ninety-three cubic meters per hectare.
70. Bundesministerium für Ernährung, *Statistisches Jahrbuch,* sec. 14, Land- und Forstwirtschaft—Pflanzliche Produktion.
71. Ibid.; Deutsches Institut für Zeitgeschichte, *Jahrbuch der Deutschen Demokratischen Republik* (Berlin: Verlag der Wirtschaft, 1959), 174.
72. Private owners did respond well to such incentives as free delivery of seedlings and favorable financing in 1956 after which private forest owners overfilled the afforestation quota by 143 percent with fast-growing hardwoods.
73. Kienitz, *Denkschrift,* 57.
74. Ibid., 60.
75. SOPADE, "Die Forstwirtschaft in der Sowjetzone," 11, 12. The *Forstwirtschaftliche Arbeitsrechtlinien für Waldgemeinschaften und Bauernförster* (Forest Management Guidelines for Cooperative and Peasant Forests).
76. Schröder, "Ökonomische Probleme des zweiten Fünfjahresplans," 289; Kienitz, *Denkschrift,* 63. Repayment was required in ten years and interest rates were unattractive; thus, in Bezirk Neubrandenburg peasants drew down on only DM 16,000 in credit out of DM 180,000 available.
77. Kienitz, *Denkschrift.*
78. Ibid. Kienitz reports that "Parzellenweise" management represented 88 percent of the total "nonsocialist" forest holdings (754,748 hectares). Eighty-eight percent of peasant owners with 754,748 hectares chose communal management in 1,151 private cooperatives.
79. Ibid., 61. Valuable *Derbholz* (over seven centimeters in diameter) taken for fuelwood accounted for 2,779,800 cubic meters, 42 percent of the total 1956 harvest on 30 percent of the total forest area. Peasant harvest in 1956 = 3.25 cubic meters per hectare.
80. Ibid., 16, 68. The loss was 1,334,304 cubic meters *of high-quality Derbholz* valued at DM 48.55 per cubic meter.
81. Ibid., 56. Private forest growth was 0.8 cubic meters per hectare lower than *Volkswald* growth, 12 percent of the People's Forest harvest of 6.6 million cubic meters. Poor private forest growth "cost" the economy 800,000 cubic meters annually.
82. Ibid., 3, 6.
83. Ibid., 6.
84. Ibid., 5.
85. Ibid., 55.
86. Ibid.
87. Ibid., 3.
88. Laßmann, "Die Rolle und Bedeutung der Forstwirtschaft," 27.
89. Deutschen Institut für Zeitgeschichte, *Jahrbuch der Deutschen Demokratischen Republik* (Berlin: Verlag der Wirtschaft, 1959), 173. The Party leadership dubbed this forestland *Betreuungswald* (Trustee Forest); Ruffer and Schwartz, *Die Forstwirtschaft*

*der Deutschen Demokratischen Republik;* Spielke, Breithaupt, Bruggel, and Stand, *Ökonomik der sozialistischen Forstwirtschaft.*

90. Paul, interview.

91. Kienitz, *Denkschrift,* 17; SOPADE, "Die Forstwirtschaft in der Sowjetzone," 4, 10, 11, 12. Peasant forests were brought into the *Revierförstereien,* where they were managed by so-called *Bauernförstern* under the *Kreisforstämtern* (Country Forest Administration). After 1952 the state set all harvests on peasant forests under five hectares. Even peasant harvests for their own use had to be approved by the *Kreisforstamt* and the State Forest Service. At the end of 1951, the peasant forest area totaled 974,608 hectares, of which only 378,586 remained under traditional forest cooperatives. The County Councils were responsible for private forests through the *Sachgebiete Forstwirtschaft bei den Räte der Kreise;* Kienitz, *Denkschrift,* 58.

92. Bundesministerium für gesamtdeutschen Fragen (BGF), *SBZ von A-Z,* 6th ed. (Bonn: Deutscher Bundesverlag, 1960), 127.

93. Kienitz, *Denkschrift,* 17. Forty percent (with 353,668 hectares) were new peasants from the land reform and 60 percent (with 530,516 hectares) existing small private farm owners.

94. M. C. Kaser, ed., "Institutional change within a planned economy," in *The economic history of Eastern Europe, 1919–1975,* vol. 3 (Oxford: Clarendon Press, 1986), 79. Kaser saw "a decisive turn from private to collectivized farming" between 1957 and 1965 throughout the Soviet bloc.

95. Kohl, *Ökonomische Geographie der Deutschen Demokratischen Republik,* 357. Between 1959 and 1960 the number of Type III LPG forests (fully collectivized) increased from 285,000 hectares to 447,755 hectares. By 1963 socialist farm collectives (LPGs) accounted for 86 percent of total GDR farmland; Deutschen Institut für Zeitgeschichte (DIZ), *Jahrbuch der Deutschen Demokratischen Republik* (Berlin: Verlag der Wirtschaft, 1961), 221.

96. Grüneberg, "30 Jahre Marxistisch-Leninistische Agrarpolitik," 3.

97. Friedrich, *Die Volkswirtschaft der DDR,* 50.

98. Ernst Reuter, "Zur Öffnung der Sektorengrenzen und zur Bedeutung des 17. Juni 1953, 9 July 1953 broadcast (RIAS Berlin), http://www.17juni53.de/chronik/5307_1.html (accessed 22 March 2004); "Nachrichtenmeldung über die Aufhebung der Sperren an den Sektorengrenzen," 8 July 1953 broadcast (DDR-Rundfunk, East Berlin), http://www.17juni53.de/chronik/5307_1.html (accessed 22 March 2004).

99. Roesler, "Rise and fall of the planned economy," 52; Weber, *Geschichte der DDR,* 297; Frowen, "Economy of the German Democratic Republic," 37.

100. The First Seven-Year Plan was designed at the Fifth Party Central Committee Conference in July 1958. N. F. R. Crafts, "The golden age of economic growth in western Europe, 1950–1973," *Economic History Review* 48:3 (1995): 429.

101. Wolfgang F. Stolper, "The labor force and industrial development in Soviet Germany," *Quarterly Journal of Economics* 71:4 (November 1957): 539. Stolper, a leading West German economist, noted, "I conclude that Soviet Germany lags substantially in overall industrial production behind the Federal Republic [West Germany]," and the gap between East and West Germany increased after the East Germans launched the First Five-Year Plan (1951–55).

102. Ibid., 538. Despite East German claims to have reached 1936 levels of industrial production in 1949.

103. Ibid., 541.

104. Hardach, *Political economy of Germany,* 125.

105. Ibid., 126. Apel's plans were set at the Fifth SED Central Committee Conference and made law by the *Volkskammer* (the People's Chamber) in October 1959. Erich Apel, "Durch sozialistische Rekonstruktion und Erhöhung der Arbeitsproduktivität zur Erfüllung des Siebenjahrplans," *Decision of the Fifth SED Central Committee Conference* (Berlin: Dietz Verlag, 1959): 83. Erich Apel (1917–65) had been raised to the Central Committee in 1961. Ackermann promoted the "special German path to socialism," an idea which fell out of favor after the Tito crisis in spring 1948 and Stalin's suspicion of political initiative in his satellites. Ulbricht purged Ackermann for his conciliatory attitude following the June 1953 Workers' Uprising. Apel and Ackermann both wanted greater independence in East German economic policy vis-à-vis the Soviet Union, but both were disappointed as East Germany aligned itself ever more tightly with the Soviet economy.

106. Roesler, "The rise and fall of the planned economy," 55; People's Chamber of the German Democratic Republic, *Law of the Seven-Year Plan for the development of the national economy of the German Democratic Republic from 1959 to 1965,* passed *Volkskammer* 1 October 1959, photocopy (1960). The goal was to pass West Germany in per capita consumption of the "most important industrial consumer goods and foodstuffs."

107. G. Schröder, "Das bedeutendste Gesetzwerk," 483. Taking its cue from the Sixth Plenum of the Party's Central Committee's "Plan of the Victory of Peace and Socialism" (6 May 1959) and echoing the *Volkskammer*'s 1 October 1959 declaration, "The economic preconditions for the victory of socialism in the German Democratic Republic are clearly laid out."

108. People's Chamber of the German Democratic Republic, *Law of the Seven-Year Plan,* 4; Apel, "Durch Sozialistische Rekonstruktion und Erhöhung."

109. Abteilung Forstwirtschaft, Ministerium für Landwirtschaft, Erfassung und Forsten, "Die sozialistische Rekonstrution in der Forstwirtschaft," *Forst und Jagd* 10:6 (June 1960): 248.

110. At the Fifth SED Central Committee Conference.

111. Apel, "Durch sozialistische Rekonstruktion und Erhöhung," 12. Apel ruled against increasing forest stocking: forest capital: "The most important results we will achieve with Socialist Reconstruction are the highest possible increase in workers' productivity, rapid increases in the production and manufacture of products at the lowest manufacturing cost and best quality. [Through scientific and technical knowledge we will achieve] more and better products at lower costs under better working conditions with less labor. [Socialist Reconstruction is] the most rational organization of production on the basis of the highest level of science and technology."

112. Ibid., 10.

113. Abteilung Forstwirtschaft, "Die sozialistische Rekonstrution in der Forstwirtschaft," 247; Marx (*Das Kapital,* vol. 1, p. 44) is cited as the authority for this statement.

114. H.-F. Joachim, interview by author, Eberswalde, 11 June 1991; Weber, *Ge-*

*schichte der DDR,* 299–300; Hardach, *Political economy of Germany,* 125; Apel, "Durch sozialistische Rekonstruktion und Erhöhung," 6, 83–84. Apel noted Soviet primacy in the *Decision of the Fifth SED Central Committee Conference:* "Step by step we demonstrate the superiority of our economy over the capitalist economy of West Germany. With this we achieve our contribution to the goal set at the Twenty-first Party Conference of the Communist Party of the Soviet Union, for the socialist camp to produce half the world's industrial production by 1965."

115. Säglitz, "Die Forstwirtschaft in Ostdeutschland," 226. East Germany imported an average of 3.5–4.0 thousand cubic meters in raw wood from the Soviet Union annually. Soviet imports, principally pulp from Siberia, began in earnest after 1957; Schröder, "Ökonomische Probleme des zweiten Fünfjahresplans," 146.

116. Frowen, "Economy of the German Democratic Republic," 37.

117. Barciok, interview.

118. Säglitz, "Die Forstwirtschaft in Ostdeutschland," 226; Deutschen Institut für Zeitgeschichte, *Jahrbuch der Deutschen Demokratischen Republik* (Berlin: Verlag der Wirtschaft, 1959), 174; Ruffer and Schwartz, *Die Forstwirtschaft der Deutschen Demokratischen Republik,* 18; Friedrich, *Die Volkswirtschaft der DDR,* 182.

119. Deutschen Institut für Zeitgeschichte, *Jahrbuch der Deutschen Demokratischen Republik* (Berlin: Verlag der Wirtschaft, 1959), 174; G. Schröder, "Das bedeutendste Gesetzwerk in der Geschichte Deutschlands," *Forst und Jagd* 9:11 (November 1959): 483; Schröder, "Ökonomische Probleme des zweiten Fünfjahresplans," 146. East bloc imports allowed the Party to reduce the 1956 harvest 47 percent from 1947 levels to 6.5 million cubic meters. In 1965 imports were planned to double to more than 4 million cubic meters, over 40 percent of East Germany's wood supply, whereas in 1955 Soviet exports had accounted for only 20 percent of total RGW imports.

GDR Wood Imports (Tfm Rohholz)

| | 1955 | 1956 | Plan 1957 |
|---|---|---|---|
| USSR | 221.4 | 334.1 | 1,100.0 |
| Other RGW | 704.7 | 781.9 | 700.0 |
| TOTAL | 926.1 | 1,116.0 | 1,800.0 |

120. Deutschen Institut für Zeitgeschichte, *Jahrbuch der Deutschen Demokratischen Republik* (Berlin: Verlag der Wirtschaft, 1959), 174.

121. SOPADE, "Die Forstwirtschaft in der Sowjetzone," 4, 10.

122. G. Hildebrandt, "Die Forsteinrichtung in der DDR 1950 bis 1965. Ein Beitrag zur jüngeren deutschen Forsteinrichtungsgeschichte aus Anlaß des 80. Geburtstages von Albert Richter in Eberswalde," *Allgemeine Forst- und Jagdzeitung* 160:6 (1989): 123; Milnick, interview; A. Richter, "Vom Ende der Forstfakultät Eberswalde 1963–Ein persönlicher Bericht," *Allgemeine Forst- und Jagdzeitung* 11/12:162 (1991): 229.

123. Joachim, interview.

124. Kohlsdorf, interview; Peter Allrich, interview by author, Eberswalde, 25 March 1991; Rüffler, interview.

125. Rudolf Rüffler, "Zur Geschichte des Instituts für Forstwissenschaften Eberswalde," *Beiträge für die Forstwirtschaft* 14:3/4 (1980): 94; G. Schröder, "Zielsetzung und Methode" *Forst und Jagd* 10:1 (January 1960); G. Schröder, "Die sozialistischen Rekonstruktion erfordert, im größeren Zusammenhängen zu denken," *Forst und Jagd,* part 1 of 2, 10:1 (January 1960); G. Schröder, "Die sozialistischen Rekonstruktion erfordert, im größeren Zusammenhängen zu denken," *Forst und Jagd,* part 2 of 2, 10:5 (May 1960): 193; G. Laßmannn, "Über die Durchführung der sozialistischen Rekonstruktion in der Forstwirtschaft," *Forst und Jagd* 10:3 (March 1960): 97.

126. Schröder, "Zielsetzung und Methode," part 1 of 2.

127. Schröder, "Die sozialistischen Rekonstruktion erfordert, im größeren Zusammen hängen," 193.

128. Laßmannn, "Über die Durchführung der sozialistischen Rekonstruktion," 98–99.

129. H. Heidrich, "Sozialismus und Kommunismus werden siegen," *Sozialistische Forstwirtschaft* 12:1 (January 1962): 3; Abteilung Forstwirtschaft, "Die sozialistische Rekonstrution in der Forstwirtschaft," 247.

130. Abteilung Forstwirtschaft, "Die sozialistische Rekonstrution in der Forstwirtschaft," 248.

131. People's Chamber of the German Democratic Republic, *Law of the Seven-Year Plan,* 17; Schröder, "Zielsetzung und Methode," part 1 of 2.

132. Schröder, "Zielsetzung und Methode," part 1 of 2, 524.

133. Ibid.

134. People's Chamber of the German Democratic Republic, *Law of the Seven-Year Plan,* 17.

135. Abteilung Forstwirtschaft, "Die sozialistische Rekonstrution in der Forstwirtschaft," 247.

136. Roesler, "Rise and fall of the planned economy," 53; Frowen, "Economy of the German Democratic Republic," 37.

137. Jeffries and Melzer, eds., *East German economy,* 26–27.

138. Bruno Leuschner, *Ökonomie und Klassenkampf,* 411. Cited in Roesler, "Rise and fall of the planned economy," 46–61.

139. Smith, *Germany beyond the Wall,* 96–97.

## *Chapter 6. The Landscape Transformed*

1. Gert Ritter and Joseph G. Hajdu, "The East-West German Boundary," *Geographical Review* 79:3 (July 1989): 326, 333; Timothy Garton Ash, "Big brother isn't watching anymore," *Guardian* (13 March 1999), 1.

2. The width of the Wall system is 5.75 kilometers, its length 1,386 kilometers. Total area equals 7,970 sq. km. Connecticut's area is 8,922 sq. km, Delaware's is 4,006 sq. km.

3. Alexander Fischer, ed., *Ploetz, die Deutsche Demokratische Republik: Daten, Fakten, Analysen* (Freiburg: Ploetz, 1988), 36. Inge Bennewitz and Rainer Potratz, *Zwangsaussiedlungen an der innerdeutschen Grenze: Analysen und Dokumente* (Berlin: Ch. Links, 1994) "Forschungen zur DDR-Geschichte," Bd. 4, pp. 16, 18. The Soviet Military Administration unilaterally declared its intention to defend the demarcation line

in December 1946; Thomas Flemming, *Die Berliner Mauer: Geschichte eines politischen Bauwerks,* Dokumentation Berliner Mauer-Archiv, Hagen Koch (Berlin: Bebra Verlag, 1999); Wayne C. Thompson, Susan L. Thompson, and Juliet S. Thompson, *Historical dictionary of Germany* (Metuchen, N.J.: Scarecrow Press, 1994).

4. *New York Times,* "Soviet zone adds border guards," *New York Times* (18 June 1949).

5. Ritter, "East-West German boundary," 326, 333; Bennewitz and Potratz, *Zwangsaussiedlungen an der innerdeutschen Grenze,* 16, 18; Flemming, *Die Berliner Mauer.*

6. West Germany, the "Bonn Republic," claimed the exclusive right to represent the entire German people as the legitimate successor to Germany's last democratic government, the Weimar Republic. This policy was codified in mid-1955 and named after Konrad Adenauer's state secretary in the Foreign Ministry, Walter Hallstein. It led to West Germany's breaking of relations with Yugoslavia in 1957 (19 October) and with Cuba in 1963 but was watered down by Adenauer's opening of relations with the Soviet Union in 1956 in exchange for the return of German POWs and finally ended with Willy Brandt's *Ostpolitik* in the 1960s.

7. Christian F. Ostermann, "New research on the GDR," *Cold War International History Project Bulletin* 4:34 (Washington, D.C.: Woodrow Wilson International Center for Scholars, 1994): 48, for an English translation of both Russian and German versions of the 7 April 1952 conversation; Stefan Creuzberger, "Abschirmungspolitik gegenüber dem westlichen Deutschland im Jahre 1952," in *Die sowjetische Deutschland-Politik in der Ära Adenauer,* ed. Gerhard Wettig (Bonn, 1997), 12–36. Soviet notes from the meeting show that Stalin saw British and American policy as defensive and not hostile to the Soviet Union, even though Ulbricht later praised the Wall as the "antifascist protection wall."

8. "Minutes of conversation between Stalin and leaders of the SED (East German Communist Party)" (7 April 1952), http://members.fortunecity.com/stalinmao/DDR/dokument/sed.html (accessed 26 February 2005). Stalin, angry at the West's refusal to accept his suggestions for reunifying Germany (the "Stalin Note" of 10 March 1952), angrily instructed the East German Party leadership in April 1952 to "organize your own state" on the demarcation line, East Germany's "dangerous" frontier with the West. Stalin was not concerned that the United States and Britain were gathering strength to confront the Soviet Union, but that they were consolidating their power through building an economically powerful and independent state in Central Europe, West Germany.

9. German Historical Museum, http://www.dhm.de/~roehrig/ws9596/texte/kk/dhm/zeit.html (accessed 30 June 2003). The 26 May 1952 "Verordnung über Maßnahmen an der Demarkationslinie zwischen der Deutsche Demokratischen Republik und der westlichen Besatzungszonen Deutschlands"; Bennewitz and Potratz, *Zwangsaussiedlungen an der innerdeutschen Grenze,* 16, 18; Flemming, *Die Berliner Mauer;* Thompson, *Historical dictionary of Germany.*

10. Our Own Correspondent, "Border zone in east Germany: Stricter system of passes, cuts in telephones," *The Times* (28 May 1952), 6.

11. Our Own Correspondent, "Cutting Germany in two," *The Times* (4 June 1952), 6.

12. Our Own Correspondent, "Deportations in east Germany," *The Times* (14 June

1952), 6; Jack Raymond, "Allies to tighten patrolling on East Germany's border," *New York Times* (28 June 1952).

13. The VoPos marched to celebrate the third anniversary of the founding of the East German republic.

14. Walter Ulbricht, "Open Letter" *Junge Welt,* cited in *The Times*, "More work for same pay in East Germany: Legitimate to kill" (8 September 1961), 11.

15. Dr. Konrad Adenauer (1876–1967) signed the Moscow Agreement on 13 September 1955. Board of Editors, "The Trojan barge?" *The Times* (9 December 1955), 11.

16. Our Diplomatic Correspondent, "Moscow," *The Times* (17 September 1955), 5.

17. Board of Editors, "Realities," *The Times* (21 September 1955), 9. The Moscow Treaty was signed on 20 September 1955.

18. Our Correspondent, "Soviet aims in East German pact," *The Times* (21 September 1955), 8.

19. Ibid. Granted through an exchange of notes between the East German and Soviet foreign ministries (Dr. Lothar Bolz, East German foreign minister, and Soviet Deputy Foreign Minister Valerian A. Zorin).

20. Our Correspondent, "Right to intervene retained by Soviet government," *The Times* (23 September 1955), 8.

21. Warsaw Pact Treaty signed 14 May 1955. The Moscow Treaty was signed on 20 September 1955.

22. Our Correspondent, "Soviet aims," 8.

23. Ibid.

24. 27 November 1955.

25. Our Correspondent, "East Berlin as 'capital,'" *The Times* (30 November 1955), 6.

26. Ulbricht spoke in a *Volkskammer* (People's Chamber) speech. SED Agitation Department, "Wer die Deutsche Demokratische Republik verläßt, stellt sich auf die Seite der Kriegstreiber," Notizbuch des Agitators (Berlin: SED Agitation Department, 1955), trans. in "German Propaganda Archive," http://www.calvin.edu/academic/CAS/gpa/notiz3.htm (accessed 2003.08.05).

27. Board of Editors, "East Germany," *The Times* (3 August 1955), 9.

28. New York Times, "Berlin depots to be renamed," *New York Times* (8 November 1950).

29. The Stettin Station's name was changed to "North Station" in December 1950 and the Silesia Station's name to "East Station" in May 1951. The Party tore down the North, Stettin Station, a nineteenth-century architectural treasure rebuilt in the 1870s according to Theodore August Stein's designs, after the Berlin Wall severed its direct lines to the east and Stettin/Szezecin. The East, Silesia Station, the station through which Soviet leaders arrived in Berlin, was consolidated within the Hauptbahnhof in 1987 but took back the name "Am Ostbahnhof" (East Railway Station) in 1998, nine years after reunification, the name "Silesia Station" lost for good.

30. Jörg Roesler, "The rise and fall of the planned economy in the German Democratic Republic, 1945–1989," *German History* 9:1 (February 1991): 46–61; Bruno Leuschner, *Ökonomie und Klassenkampf: Ausgewählte Reden und Aufsätze, 1945–1965*, Institut für Marxismus-Leninismus beim ZK der SED (Berlin: Dietz, 1984), 411.

31. Jeffrey Kopstein, "Ulbricht embattled: the quest for socialist modernity in the light

of new sources," *Europe-Asia Studies* 46:4 (Soviet and East European History) (1994): 602, 610.

32. Martin S. Ochs, "German red purge sweeps out books," *New York Times* (10 February 1951), 1.

33. Martin S. Ochs, "Marx and Engels works revised under East Germany's book purge," *New York Times* (15 March 1952).

34. Alfred Dengler, Ernst Röhrig, and A. Gussone, *Waldbau auf ökologischer Grundlage: Baumartenwahl, Bestandesbegründung und Bestandespflege,* Bd. 2 (Ulmer [Eugen], 1990); Alfred Dengler, Ernst Röhrig, and Norbert Bartsch, *Waldbau auf ökologischer Grundlage: Der Wald als Vegetationsform und seine Bedeutung für den Menschen,* Bd.1 (Ulmer [Eugen], 1992).

35. Albert Richter, "Vom Ende der Forstfakultät Eberswalde 1963—Ein persönlicher Bericht," *Allgemeine Forst- und Jagdzeitung* 11/12 162 (1991): 229; Joachim, interview.

36. At a conference in West Germany in 1957. Richter, "Vom Ende der Forstfakultät Eberswalde," 229.

37. As a *Revierförster* at StFB Weißwasser.

38. G. Laßmann, "Die Rolle und Bedeutung der Forstwirtschaft," 44–45. The dominance of Marxist theory in the Party's forest policy is shown in Laßmann's 1960 statement: "Forest economics are conditioned by class relations. Therefore bourgeois forest economics cannot serve as a theoretical and ideological foundation. But bourgeois forest economics are still widespread and many still cling to bourgeois forest management teaching."

39. H. Heidrich, "Die Aufgaben des Betriebsleiters," 1; Laßmann, "Die Rolle und Bedeutung der Forstwirtschaft," 44–45.

40. Laßmann, "Die Rolle und Bedeutung der Forstwirtschaft," 46.

41. Heidrich, "Die Aufgaben des Betriebsleiters," 1.

42. *Sozialistische Forstwirtschaft,* 1962. "Was wir heute pflanzen ernten wir unter Kommunismus!" *Sozialistische Forstwirtschaft* 1:1.

43. At the Party's Thirty-third Central Committee meeting in October 1957.

44. Our Correspondent, "Political rumblings in East Germany: Leaders' concern at criticism," *The Times* (19 February 1957), 6.

45. Department of Propaganda-Agitation of the East German Communist Party, "Einzelbauer Arnold und sein Verhältnis zum Sozialismus: Ein Wort an die Einzelbauern—vor allem an jene, die es bleiben wollen" (Bezirk Karl-Marx-Stadt: Department for Propaganda-Agitation, 1960), "German Propaganda Archive," trans. Randall Bytwerk, http://www.calvin.edu/academic/cas/gpa/arnold.htm (accessed 11 July 2003).

46. Ibid.

47. *Neues Deutschland* (19 April 1960); Our Correspondent (Bonn), "Mass flight of peasants from East Germany awaited," *The Times* (20 April 1960), 10.

48. Our Correspondent, "Brain washing down on the farm: Recent east German advances in collectivization by consent," *The Times* (30 March 1961), 10.

49. Rolf Badstubner and Wilfried Loth, eds., *Wilhelm Pieck—Aufzeichnungen zur Deutschlandpolitik, 1945–1953,* trans. Stephen Connors (Berlin: Akademie Verlag, 1994), 396–97; Walter Ulbricht, "2. Gespräch des Staatsratsvorsitzenden Ulbricht mit Präsident Nasser am 28. Februar 1965 von 18.00 Uhr bis 19.40 Uhr," *Vierteljahrshefte für Zeitgeschichte* 46:4 (October 1998): 800.

50. Our Correspondent (Bonn), "Mass flight of peasants from east Germany awaited," *The Times* (20 April 1960), 10.

51. Our Correspondent (Bonn), "Easter exodus from east Germany," *The Times* (26 April 1960), 8.

52. Vladislav M. Zubok's interview of Troyanovsky, 23 March 1993, Washington, D.C., cited in Vladislav M. Zubok, "Khrushchev and the Berlin Crisis (1958–1962)," *Cold War International History Project Working Paper*, #6 (Washington, D.C.: Woodrow Wilson International Center for Scholars, 1993): 24.

53. Peter Przybylski, *Tatort Politbüro: die Akte Honecker* (Berlin: Rowohlt, 1991), vol. 2, 351–52; Dölling, Ambassador in Moscow, "Note of a Discussion on 27 February 1962," 5 March 1962, marked "for personal use only," trans. in Douglas Selvage, "The end of the Berlin Crisis: New evidence from the Polish and East German archives," *Cold War International History Project Bulletin* 11 (Washington, D.C.: Woodrow Wilson International Center for Scholars, 1999): 222.

54. Our Correspondent, "Maize sowing lags in East Germany," *The Times* (6 May 1960), 12.

55. Our Special Correspondent in East Germany, "Living with Pankow and the Wall," *The Times* (21 September 1962), 13.

56. *Neue Justiz* (15 August 1960); Board of Editors, "All out of step," *The Times* (15 August 1960), 9.

57. Our Correspondent, "Exodus from East Germany," *The Times* (27 July 1961), 8.

58. Our Correspondent, "More work for same pay in East Germany," *The Times* (8 September 1961), 11.

59. Our Own Correspondent, "200,000 a year in migration from East Germany: West Berlin seeks to retain a higher proportion of influx," *The Times* (6 March 1961), 10.

60. Our Correspondent (Berlin), "Youth bored with indoctrination," *The Times* (10 March 1961), 9.

61. Vladislav M. Zubok, "Khrushchev and the Berlin Crisis (1958–1962)," 31–32.

62. Selvage, "End of the Berlin Crisis," 219; Hope Millard Harrison, "Ulbricht and the concrete rose: New archival evidence on the dynamics of Soviet-East German relations and the Berlin crisis, 1958–1961," *Cold War International History Project Working Paper*, #5 (Washington, D.C.: Woodrow Wilson International Center for Scholars, 1993).

63. Goronwy Rees, "From Berlin to Munich," *Encounter* 22:4 (April 1964): 3.

64. Dölling, "Note of a Discussion on 26 February 1962," 218.

65. Adzhubei quoted in Selvage, "End of the Berlin Crisis."

66. Dölling, "Note of a Discussion on 26 February 1962," 226–27.

67. Ibid., 222; Vladislav M. Zubok and Constantine V. Pleshakov, *Inside the Kremlin's Cold War: From Stalin to Khrushchev* (Cambridge: Harvard University Press, 1997), 249.

### *Chapter 7. Cybermarxism and Innovation*

1. N. F. R. Crafts, "The Golden Age of economic growth in Western Europe, 1950–1973," *Economic History Review* 48:3 (1995): 429; Gianni Toniolo, "Europe's Golden Age, 1950–1973: Speculations from a long-run perspective," *Economic History Review*

51:2 (1998): 252; Alan S. Milward, *The European rescue of the nation state* (London: Routledge, 1992); Angus Maddison, *The world economy: A millennial perspective*, OECD Development Centre; Martin Wolf, "Bright spots amid the gloom," *Financial Times* (14 March 2001).

2. Nikita Khrushchev (1894–1971) in a "remark to Western diplomatists" at the Kremlin (18 November 1956). Our Special Correspondent, "Western ambassadors walk out," *The Times* (19 November 1956), 8; Benjamin Welles, "Khrushchev bangs his shoe on desk: Khrushchev adds shoe-waving to his heckling antics at U.N.," *New York Times* (13 October 1960), 1; Our United Nations Correspondent, "If I go to the bottom I shall drag you down too," *The Times* (14 October 1960), 10; Our Diplomatic Correspondent, "Mr. Khrushchev's crescendo of publicity," *The Times* (17 October 1960), 10; William Taubman, "Nikita Khrushchev and the shoe," *International Herald Tribune* (26 July 2003), http://www.iht.com/articles/103353.html (accessed 4 September 2004).

A *New York Times* writer noted Khrushchev's threatening tone: "To Khrushchev, peaceful coexistence does not mean reconciliation with the West; it means war, to be fought with every possible means short of a nuclear conflict." Philip E. Mosely, "Is it 'peaceful' or 'coexistence'?" *New York Times Magazine* (7 May 1961).

3. Soviet bloc computer scientists used the term "cybernetics" to refer to a broad range of information technology: artificial intelligence, systems analysis, linear and nonlinear programming, operations research, and game theory. Western scientists differentiated between these fields, and cybernetics' founder, Norbert Wiener, flavored his thinking with humanism and philosophy and rejected the literal applications of his scholarship by Soviet bloc "cyberneticians." Norbert Wiener, *Cybernetics* (New York: John Wiley, 1949), 19; Norbert Wiener, *The human use of human beings: Cybernetics and society* (Boston: Houghton Mifflin, 1954).

4. The Soviets tested a fifty-eight-megaton hydrogen bomb in the atmosphere on 30 October 1961.

5. Khrushchev spoke at the village of Odnova on 18 May 1962. *Pravda* (19 May 1962) and *Izvestiya* (20 May 1962), cited in Albert Parry, "Science and technology versus communism," *Russian Review* 25:3 (July 1966): 231.

6. Jacques Barzun, *From dawn to decadence: 500 years of cultural life, 1500 to the present* (New York: HarperCollins, 2000).

7. Khrushchev spoke at the Seventeenth Soviet Communist Party (KPSU) Central Committee Meeting in fall 1962. Stephen F. Frowen, "The economy of the German Democratic Republic," in David Childs, *Honecker's Germany* (London: Allen and Unwin, 1985), 38; Gert Leptin, "The GDR," in *The new economic systems of Eastern Europe*, ed. Hans-Hermann Höhmann, Michael Kaser, and Karl C. Thalheim (London: C. Hurst, 1975), 47; Karl Willy Hardach, *The political economy of Germany in the twentieth century*, trans. by author (Berkeley: University of California Press, 1980), 132.

8. Michael Keren, "The New Economic System in the GDR: An obituary," *Soviet Studies* 24:4 (April 1973): 556.

9. Uwe-Jens Heuer, *Demokratie und Recht im neuen ökonomischen System der Planung und Leitung der Volkswirtschaft* (Berlin, 1965), cited in Baylis, "Economic reform as ideology," 219. Uwe-Jens Heuer was a leading East German legal theorist.

10. L. Smolinski and P. Wiles, "The Soviet planning pendulum," *Problems of Communism*

(November-December 1963): 23, cited in Egon Neuberger, "Libermanism, computopia, and visible hand: The question of informational efficiency" (in *Knowledge, information, and innovation in the Soviet economy*), *American Economic Review* 56:1/2 (March 1966): 131.

11. Board of Editors, "Wanted: A new man," *The Times* (3 May 1963), 15.

12. Adam Bruno Ulam, *Unfinished revolution: An essay on the sources of influence of Marxism and communism* (New York: Random House, 1960); Thomas Arthur Baylis, "Economic reform as ideology: East Germany's New Economic System," *Comparative Politics* 3:2 (January 1971): 212.

13. Anton Ackermann, "Gilt es einen besonderen deutschen Weg zum Sozialismus?" *Einheit: Zeitschrift für Theorie und Praxis des Wissenschaftlichen Sozialismus* (February 1946) Sonderheft; Melvin Croan, "Soviet uses of the doctrine of the 'Parliamentary Road' to socialism, East Germany, 1945–46," *American Slavic and East European Review* 17:3 (October 1958): 303.

14. Hardach, *Political economy of Germany,* 132.

15. The NÖS was developed further at the June 1963 Economic Conference of the Central Committee. On 11 July 1963 the Ministerrat published a detailed draft of the NÖS and the planning phase was complete. Dennis, *German Democratic Republic,* 33; Hermann Weber, *Geschichte der D.D.R. Autorreferat* (Munich: Deutscher Taschenbuch Verlag, 1985), 350; Leptin, "GDR," 47.

16. As documented in the 11 February 1963 Erlaß "Hauptverwaltung Forstwirtschaft bei der Produktionsleitung des Landwirtschaftsrates beim Ministerrat." Lehmann, "Gedanken zur Anwendung des Produktionsprinzips," 294; Horst Ruffer and Ekkehard Schwartz, *Die Forstwirtschaft der Deutschen Demokratischen Republik* (Berlin: VEB Deutscher Landwirtschaftsverlag, 1984), 18; Frowen, "Economy of the German Democratic Republic," 39.

17. Bundesministerium für gesamtdeutsche Fragen (BGF), *SBZ von A-Z,* 3rd ed. (Bonn: Deutscher Bundesverlag, 1956), 287; Bundesministerium für gesamtdeutsche Fragen (BGF), *SBZ von A-Z,* 11th ed., 218; Horst Kohl, ed., *Ökonomische Geographie der Deutschen Demokratischen Republik,* 3rd ed. (Gotha and Leipzig: VEB Hermann Haack, 1976), 761; Joachim Lehmann, "Gedanken zur Anwendung des Produktionsprinzips bei der Leitung unserer staatliche Forstwirtschaftsbetriebe," *Sozialistische Forstwirtschaft* 14:10 (October 1964): 295; Edgar Tümmler, Konrad Merkel, and Georg Blohm, *Die Agrarpolitik im Mitteldeutschland und ihre Auswirkung auf Produktion und Verbrauch landwirtschaftliche Erzeugnisse* (Berlin: Duncker and Humblot, 1969); Leptin, "GDR," 47; Hardach, *Political economy of Germany,* 132. State Forest District directors now reported directly to one of the five VVB Forest Management, *Oberlandforstmeister*, directors located in: Waren, Potsdam, Cottbus, Karl-Marx-Stadt, and Suhl. The *Vereinigungen Volkseigene Güter* (VVG) operated like VEB in the rural sector to take over management of agricultural land acquired through land reform.

18. Fritz Behrens, "Zum Problem der Ausnutzung ökonomische Gesetze in der Übergangsperiode," *Zur ökonomische Theorie and Politik in der Übergangsperiode*, 3. Sonderheft, *Wirtschaftswissenchaft* 5 (1957): 105–40; Herbert S. Levine, "Economics," in George Fischer, ed. *Science and ideology in Soviet society* (New York, 1967), 107–38.

19. G. Laßmann, *Die Rolle und Bedeutung der Forstwirtschaft in System der Volks-*

*wirtschaft der Deutschen Demokratischen Republik* Schriftenreihe für Forstökonomie (Berlin: VEB Deutsche Landwirtschaftsverlag, 1960), 46.

20. E. Wagenknecht, "Der Waldbau zwischen Heute und Morgen," *Archiv für Forstwesen* 10:4/6 (1961): 366.

21. Johannes Blanckmeister, "Kurswechsel im Waldbau," *Sozialistische Forstwirtschaft* 12:1 (January 1962): 7. *Dauerwald* had been selected as the principal silvicultural method at the Menz Conference (Menzer Tagung, 14–15 June 1951).

22. Rudolf Rüthnick, *Erste Konferenz der VVB Forstwirtschaft Potsdam vom 15.–17: Juni 1964 in Leipzig-Markkleeburg* (Potsdam, 1964); Tümmler, Merkel, and Blohm, *Die Agrarpolitik im Mitteldeutschland,* 102; Lehmann, "Gedanken zur Anwendung des Produktionsprinzips," 294; W. Schult, *Bedeutung und Inhalts eines Zweigsprogrammes und wissenschaftlich-technischen Konzeptionen für die perspektivische Planung in der Forstwirtschaft* (Leipzig-Markkleeburg: Landwirtschaftsaustellung der D.D.R., 1965).

23. Rüthnick, *Erste Konferenz der VVB Forstwirtschaft Potsdam.*

24. Kurth, "Die Entwicklung der Forstwirtschaft in der D.D.R.," 894.

25. Walter Ulbricht, *Dem VI Parteitag entgegen. Referat auf der 17. Tagung des ZK der SED* (Berlin: Dietz Verlag, 1962), 43, introduced at the Seventeenth Plenum of the SED Central Committee in 1962.

26. "Neue Waldbauliche und Holzartenrichlinien." Johannes Blanckmeister, *Wege und Irrwege des Waldbaus in der letzten 150 Jahre,* in *Kolloquium anläßlich des 75 Geburtstag von Nationalpreisträger Prof. (em.) Dr. ing. Habil. Johannes Blanckmeister* (Tharandt: Technische Universität Dresden, 1973), 12.

27. Klaus Höppner and Paul Hauenschild, interview by author, Eberswalde, 13 March 1991.

28. Spielke, Breithaupt, Bruggel, and Stand, *Ökonomik der sozialistischen Forstwirtschaft,* 53; Lehmann, "Gedanken zur Anwendung des Produktionsprinzips," 294.

29. G. Hildebrandt, "Die Forsteinrichtung in der DDR 1950 bis 1965. Ein Beitrag zur jüngeren deutschen Forsteinrichtungsgeschichte aus Anlaß des 80. Geburtstages von Albert Richter in Eberswalde," *Allgemeine Forst- und Jagdzeitung* 160:6 (1989): 130.

30. Thalheim, *Stagnation or change in Communist economies?* 16–17. Thalheim sees price reform at the core of the NÖS, not empowerment of managers through decentralization; Ian Jeffries and Manfred Melzer, "The New Economic System of planning and management, 1963–1970, and recentralization in the 1970s," in Jeffries and Melzer, eds. *East German economy,* 34.

31. Jeffries and Melzer, "New Economic System of planning," 34; Hardach, *Political economy of Germany,* 133; Leptin, "GDR," 47.

32. Wolfgang F. Stolper, "The Labor force and industrial development in Soviet Germany," *Quarterly Journal of Economics* 71:4 (November 1957): 534.

33. Smith, *Germany beyond the Wall,* 97.

34. Frowen, "Economy of the German Democratic Republic," 39.

35. Smith, *Germany beyond the Wall,* 97.

36. Walter Ulbricht, "Schlußwort zur Wirtschaftskonferenz der SED 1961," *Die Wirtschaft* (special edition) (18 October 1961), 3, cited in F. Walter, "Möglichkeiten einer vertieften wirtschaftlichen Rechnungsführung der Forstwirtschaft mit Hilfe veränderter Finanzierungsmethoden," *Sozialistische Forstwirtschaft* 5 (1965): 1.

37. Frithjof Paul, "Die Wirkungsweise ökonomischer Gesetze unter den gegenwärtigen Bedingungen der Forstwirtschaft der Deutschen Demokratischen Republik." In *Beiträge zum Neuen Ökonomischen System der Planung und Leitung der Volkswirtschaft in der sozialistischen Forstwirtschaft der D.D.R.* 62 (1963): 70.

38. Smith, *Germany beyond the Wall,* 97; Hardach, *Political economy of Germany,* 134; Thalheim, *Stagnation or change in Communist economies?* 16–17. Revaluation of capital assets by 52 percent on 30 June 1964 technically allowed managers to calculate depreciation costs for the first time. Of course, depreciation was only as good as the original cost estimate fixed by the Plan. This massive revaluation was closely related to increased prices of raw material. Realistic direct and indirect costs were key to establishing profitability as the criterion of performance under the NÖS and to reducing waste. The NÖS also placed a 6 percent levy on long-term and working capital to encourage conservation of capital and raised interest rates to keep inventory levels down and encourage efficiency.

39. Christine L. Zvosec, "Environmental deterioration in East Europe," *Survey* 84:28/4 (1984): 103.

40. Mann, *Prinzipien der Preisbildung,* 37.

41. Paul, interview; Josef Füllenbach, *European environmental policy: East and West,* trans. Frank Carter and John Manton (London: Butterworths, 1981), 85.

42. Alfred Zauberman, "Liberman's rules of the game for Soviet industry" (in Notes and Comment), *Slavic Review* 22:4 (December 1963): 734.

43. G. Schröder, "Neue Maßtäbe für die Planung und Leitung der Forstwirtschaft. Rückblick und Ausblick nach dem 11. Plenum des ZK der SED," *Sozialistische Forstwirtschaft* 16:3 (March 1965): 67.

44. Hildebrandt, "Die Forsteinrichtung in der D.D.R.," 130.

45. The forest inventories were conducted through H. Grossmann's "Grossrauminventur." Some of his important articles are: "Present position and possibilities in the information supplied by continuous large-scale inventory," *Sozialistische Forstwirtschaft* 13:2 (1963): 43–45; "Toward improving the growing stock and increment inventory of the E. German Republic," *Sozialistische Forstwirtschaft* 14:6 (1964): 174–75; "Forest inventories as a basis for planning, appraisal of Plan fulfillment, and permanent verification of the condition of the forest," *Archiv für Forstwesen* 18:2 (1969): 211–33; "Ten years permanent large-area inventory in the German Democratic Republic," *Sozialistische Forstwirtschaft* 22:3 (March 1972): 74–76.

46. Barth, interview.

47. R. J. Boys, D. P. Forster, and P. Jozan, "Mortality from causes amenable and nonamenable to medical care: The experience of eastern Europe," *British Medical Journal* 303:6807 (1991): 879–83.

48. Goronwy Rees, "From Berlin to Munich," *Encounter* 22:4 (April 1964): 3.

49. Erich Apel, "Einige Grundfragen der Leitung unserer sozialistischen Volkswirtschaft," *Einheit: Zeitschrift für Theorie und Praxis des Wissenschaftlichen Sozialismus* 11/12 (1961): 1632, cited in Paul W. Sanderson, "Scientific-technical innovation in East Germany," *Political Science Quarterly* 96:4 (winter, 1981–82): 580.

50. Keren, "New Economic System in the GDR," 572.

51. Erich Apel and Günther Mittag, *Ökonomische Gesetze des Sozialismus und neues ökonomisches System der Planung und Leitung der Volkswirtschaft,* 2nd ed. (Berlin: Dietz, 1964); Erich Apel and Günther Mittag, *Wissenschaftlich Führungstätigkeit: Neue Rolle der VVB* (Berlin, 1964), cited in Baylis, "Economic reform as ideology," 216–17.

52. Erich Apel to Walter Ulbricht (12 November 1964), Institut für Geschichte der Arbeiterbewegung (IdGA), Zentrales Parteiarchiv (ZPA), NL 182/971, cited in Jeffrey Kopstein, "Ulbricht embattled: The quest for socialist modernity in the light of new sources," *Europe-Asia Studies* 46:4 (1994): 600, 604.

53. Howard Swearer in Roy D. Laird, ed., *Soviet agriculture and peasant affairs* (Lawrence: Kansas University Press, 1963), 14; Jerzy F. Karcz, "The new Soviet agricultural programme," *Soviet Studies* 17:2 (October 1965).

54. Paul W. Sanderson, "Scientific-technical innovation in East Germany," *Political Science Quarterly* 96:4 (winter 1981–82): 576.

55. Parry, "Science and technology versus Communism," 227

56. *Neues Deutschland* (15 November 1970), 3, cited in Michael J. Sodaro, "Ulbricht's grand design: Economics, ideology, and the GDR's response to détente—1967–1971," *World Affairs* 142:3 (winter 1979–80): 167n30. Ulbricht needed to deflect criticism that the new elites usurped the working class and that cybernetics contradicted the primacy of the economic struggle and Marx's Labor Theory of Value.

57. Peter Christian Ludz, *The changing party elite in East Germany* (Cambridge: MIT Press, 1973), cited in John M. Starrels, "Comparative and elite politics," *World Politics* 29:1 (October 1976): 130; Baylis, "Economic reform as ideology"; Daniel Bell, "The dispossessed—1962," *Columbia University Forum* 5:4 (fall 1962): 5–6.

58. Baylis, "Economic reform as ideology," 215–16.

59. Laßmann, "Die Rolle und Bedeutung der Forstwirtschaft," 46.

60. Ibid., 31.

61. Kosing and Heinrich, "Natur-Mensch-Gesellschaft," 3; Alfred Kosing, "Natur und Gesellschaft," *Einheit: Zeitschrift für Theorie und Praxis des Wissenschaftlichen Sozialismus* 39:11 (1984): 1020.

62. G. Zillmann, "Fragen des Überganges zu industriegemäßigen Produktionsmethoden in der Forstwirtschaft," *Sozialistische Forstwirtschaft* 14:2 (February 1964): 35; Spielke, Breithaupt, Bruggel, and Stand, *Ökonomik der sozialistischen Forstwirtschaft,* 67.

63. Friedrich, *Die Volkswirtschaft der D.D.R.,* 28; Hoffman and Paul, interviews; Gerhard Breithaupt, "Die landeskulturellen Leistungen der Forstwirtschaft und die Problematik ihrer Bewertung und ökonomische Erfassung," *Wissenschaftliches Zeitschrift der Technischen Universität Dresden* 16:5 (1967): 1589.

64. Friedrich, *Die Volkswirtschaft der D.D.R.,* 181.

65. Schult, *Bedeutung und Inhalts eines Zweigsprogrammes,* 6.

66. Ziegler, "Die wirkung der Industrie-Rauchschäden," 777–87.

67. Deutschen Institut für Zeitgeschichte (DIZ), *Jahrbuch der Deutschen Demokratischen Republik* (Berlin: Verlag der Wirtschaft, 1958), 204.

68. *Liberal-Demokratische Zeitung,* "Wie steht's mit unserer Luft?" *Liberal-Demokratische Zeitung* (Halle, 1964).

69. Kohlsdorf, Erich, "Denkschrift über der Rauchschaden Situation im Bereich des mittlerer und östlicher Erzgebirges," typewritten manuscript in Tharandt archives (1964), and Kohlsdorf, interview.

70. Rudolf Rüthnick, "Die weiteren Aufgaben bei der Verwirklichen des Neuen Ökonomischen System der Planung und Leitung der Volkswirtschaft im Bereich der VVB Forstwirtschaft Potsdam," in *Die Anwendung des Neuen Ökonomischen Systems im Bereich der VVB Forstwirtschaft Potsdam,* ed. R. Rüthnick (Potsdam-Babelsberg: VVB Forstwirtschaft Potsdam, 1966), 7, 8. Rüthnick asserted (at the Seventh Party Congress in April 1967 when the Economic System of Socialism [ÖSS] replaced the NÖS) an "indissoluble link between the national policy for the maintenance of peace and the exemplary fulfillment of our complex economic tasks."

71. Spielke, Breithaupt, Bruggel, and Stand, *Ökonomik der sozialistischen Forstwirtschaft,* 53. Heidrich, perhaps as a consequence of his opposition to closer economic ties with the Soviet Union, was cashiered in 1971 and replaced by the more reliable director of VVB Potsdam, Rudolf Rüthnick. Senior levels of the Forest Service bureaucracy were now filled with Horst Kurth and Roland Barth in Forest Management (*Forsteinrichtung*) and Blanckmeister's student Harald Thomasius as de facto head of silviculture in Tharandt. By the end of the decade the Forest Service was purged of anyone who might question Marxist-Leninist dogma.

72. Along with the Third Agricultural Reform, the final stage of socialist farm and forest policy after land reform and the Socialist Spring in the Countryside. Ulbricht, *Dem VI Parteitag entgegen,* 43; Karl Eckart, "Veränderungen in der Landwirtschaft der D.D.R. seit Anfang der siebziger Jahre," *Deutschland Archiv* 18:4 (1985): 396; G. Zillmann, "Fragen des Überganges zu industriegemäßigen Produktionsmethoden in der Forstwirtschaft," *Sozialistische Forstwirtschaft* 14:2 (February 1964): 35, 37.

73. Frithjof Paul, "Beiträge zu den Grundlagen der Forstökonomik," *Schriftenreihe für Forstökonomie* 1 (1960): 158; G. Laßmann, *Die Rolle und Bedeutung der Forstwirtschaft in System der Volkswirtschaft der Deutschen Demokratischen Republik* (Berlin: VEB Deutsche Landwirtschaftsverlag, 1960), 10.

74. Hans Immler, *Agrarpolitik in der D.D.R.* (Köln: Verlag Wissenschaft und Politik, 1971), 172.

75. Rüthnick, *Die weiteren Aufgaben,* 18.

76. H. Heidrich, "Die nächsten Aufgaben bei der Entwicklung und Festigung der sozialistischen Forstwirtschaft in der Deutschen Demokratischen Republik nach dem VII. Parteitag der SED," *Sozialistische Forstwirtschaft* 17:9 (September 1967): 273.

77. Rüthnick, *Die weiteren Aufgaben,* 18.

78. H. Heidrich, *Schlußwort,* in "Die Anwendung des Neuen Ökonomischen Systems im Bereich der VVB Forstwirtschaft Potsdam," ed. R. Rüthnick (Potsdam-Babelsberg: VVB Forstwirtschaft Potsdam, 1966), 176–77; Heidrich, "Die nächsten Aufgaben," 260.

79. G. Schröder, "Das bedeutendste Gesetzwerk in der Geschichte Deutschlands," *Forst und Jagd* 9:11 (November 1959), part 1 of 2, 483.

80. G. Schröder, "Zielsetzung und Methode der Sozialistischen Rekonstruktion in der Forstwirtschaft," *Forst und Jagd* 9:12 (December 1959), part 2 of 1, 524. Investments in plywood and oriented strand-board production doubled to absorb pulp imports from the Soviet far east.

Pulp imports from the Soviet Union cost DM 200 million annually and amounted to 40 percent of East Germany's wood consumption by 1965. Imports rose from 1,980 million cubic meters in 1958 to 4,135 million cubic meters in 1965 (in raw wood equivalent). VEB Forstprojektierung, *Der Waldfonds in der Deutsche Demokratischen Republik,* 129, 131, sec. 5.–5.1.2.

81. Heidrich, *Schlußwort,* 172.

82. Ivan T. Berend, *The Hungarian economic reforms, 1953–1988* (Cambridge: Cambridge University Press, 1990).

83. Keren, "New Economic System in the GDR," 557, 564.

84. Lehmann, Joachim, "Gedanken zur Anwendung des Produktionsprinzips bei der Leitung unserer staatliche Forstwirtschaftsbetriebe," *Sozialistische Forstwirtschaft* 14:10 (October 1964): 294. Lehmann's statement tracks well with Spielke's, author of the standard East German textbook on forest economics, definition of Democratic Socialism, "directing all business from the center out and ranking the hierarchy under the center through voting of the people for all power centers and the acceptance of a personal duty for accountability to higher levels as well as widespread acceptance by the masses of State leadership."

H. O. Spielke, G. Breithaupt, H. Bruggel, and H. Stand, *Ökonomik der sozialistischen Forstwirtschaft* (Berlin: VEB Deutsche Landwirtschaftsverlag, 1964), 46. Spielke cites *Grundlagen des Marxismus-Leninismus,* chap. 21, 607.

85. Kohlsdorf, interview.

86. Vladimir Kontorovich, "Lessons of the 1965 Soviet economic reform," *Soviet Studies* 40:2 (April 1988): 308.

87. At the Seventh Party Congress (17–22 April 1967). Douglas Selvage, "The end of the Berlin Crisis: New evidence from the Polish and East German archives," *Cold War International History Project Bulletin* #11 (Washington, D.C.: Woodrow Wilson International Center for Scholars, 1999); Michael J. Sodaro, "Ulbricht's grand design: Economics, ideology, and the GDR's response to détente—1967–1971," *World Affairs* 142:3 (winter 1979–80): 160–62.

88. Theo Stammen, *Von der SBZ zur DDR,* in "DDR: Das politische, wirtschaftliche und soziale System," ed. Heinz Rausch and Theo Stammen, 4th ed. (Munich: Verlag C. H. Beck, 1978), 48.

89. The constitution's Article 15 mandated environmental protection.

90. Horst Paucke, "Soziologie und Sozialpolitik," *Soziologie und Sozialpolitik* 1:87 (Berlin: Akademie der Wissenschaften der D.D.R., 1987), 155. The *Landeskulturgesetz,* 4 May 1970.

91. Bröll, *Das sozialistische Wirtschaftssystem,* 143.

92. Christian F. Ostermann, " 'This is not a Politbüro, but a madhouse,' the post Stalin succession struggle, Soviet Deutschlandpolitik and the SED: New evidence from Russian, German, and Hungarian Archives," *Cold War International History Project Bulletin* #10 (Washington, D.C.: Woodrow Wilson International Center for Scholars, March 1998).

93. *Neues Deutschland* (12 November 1961); Our Correspondent, "East Germany 'unaffected by attack on personality cult,' " *The Times* (13 November 1961).

94. Record of conversation between Honecker and Brezhnev, 20 August 1970, cited in Grieder, "Overthrow of Ulbricht," 38n212. Brezhnev and Ulbricht met in October 1964

at Ulbricht's villa northeast of Berlin in the resort Werbellinsee. Peter Przybylski, *Tatort Politbüro: Die Akte Honecker* (Berlin: Rowohlt, 1991): 287.

95. Record of private meeting between Ulbricht and Brezhnev (28 July 1970). Przybylski, *Tatort Politbüro,* cited in, Kopstein, "Ulbricht embattled," 611. Peter Grieder, "The overthrow of Ulbricht in East Germany," *Debatte* 6:1 (1998): 1.

96. Ulbricht's speech to East German and Soviet Communist Party delegates (21 August 1970) cited in Peter Grieder, "The overthrow of Ulbricht in East Germany," *Debatte* 6:1 (1998): 17nn61,65.

97. On 12 April 1971. Norbert Podewin "Global denken, lokal handeln. Walter Ulbrichts Modell des Sozialismus. Eine Würdigung," (30 June 2003), http://www.jungewelt.de/2003/06–30/002.php (accessed 21 September 2003).

98. Wolfgang Leonhard, *Das kurze Leben der DDR. Bericht und Kommentare aus vier Jahrzehnten* (Stuttgart, 1990), 143–43, cited in Grieder, "Overthrow of Ulbricht," 44.

99. David Childs, "Marxism-Leninism in the German Democratic Republic: The Socialist Unity Party (Party)," *Soviet Studies* (April 1981): 317; Michael J. Sodaro, "Ulbricht's grand design: Economics, ideology, and the GDR's response to détente, 1967–1971," *World Affairs* 142:3 (winter 1979–80): 149–50. Ulbricht tried to distance East Germany from the Soviet Union. He "wanted to drive a harder bargain with the West over Berlin" than the Soviets. This "ultimately destroyed him."

100. At the Fifteenth Central Committee meeting in May 1971, cited in Peter Grieder, "The overthrow of Ulbricht in East Germany," *Debatte* 6:1 (1998): 18n81; Theo Stammen, "Von der SBZ zur D.D.R," in *DDR. Das politische, wirtschaftliche und soziale System,* ed. Heinz Rausch and Theo Stammen, 4th ed. (Munich: Verlag C. H. Beck, 1978); Dennis, *German Democratic Republic,* 37.

101. Editorial, "The two Germanies," *The Times* (20 January 1970), 9; Roger Berthoud and Gretel Spitzer, "Warsaw Pact leaders seek Ulbricht compromise on Berlin," *The Times* (3 December 1970), 5.

102. Editorial, "Busy days in Bonn," *The Times* (15 January 1970), 11; Brandt's policy of *Ostpolitik* had already reoriented West German foreign policy toward cooperation with the Soviet bloc. He dropped the Hallstein Doctrine, granting West German diplomatic recognition to countries with relations with East Germany, accepted the postwar boundaries as final, and negotiated the 12 August 1970 Moscow-Bonn Treaty. These actions robbed Ulbricht of any excuse for rigidity or opposition to rapprochement.

103. Hans Immler, *Agrarpolitik in der DDR* (Köln: Verlag Wissenschaft und Politik, 1971).

104. Ibid., 172.

105. Some "Soviet and East German scientists even demanded a reinterpretation of dialectical materialism in the light of the discoveries of modern physics." Meyer argued that "self-legitimation" was Soviet ideology's primary function. Alfred E. Meyer, "The functions of ideology in the Soviet political system," *Soviet Studies* 17 (January 1966): 280; Rüdiger, Thomas, *Modell D.D.R. Die kalkulierte Emanzipation*, 2nd ed. (Munich, 1977), 25–26.

106. These comments are from Kurt Hager (1912–98), Politburo member and chief (*Chefideologe*) of the Politburo's Ideology Commission. Kurt Hager, "Die entwickelte sozialistische Gesellschaft," *Einheit: Zeitschrift für Theorie und Praxis des Wissenschaft-*

*lichen Sozialismus* 11 (1971): 1214, cited in Steffen Werner, *Kybernetik statt Marx?: Politische Ökonomie und marxistische Philosophie in der DDR unter dem Einfluss der elektronischen Datenverarbeitung* 39 (Stuttgart: Verlag Bonn Aktuell, 1977): 103; Franz Loeser, "Sind die formalisierten Methoden des marxistischen Gesellschaft Wissenschaften Klassenindifferent?" *Staat und Recht* 3 (1969): 467; Georg Klaus, "Kybernetik und ideologischen Klassenkampf," *Einheit: Zeitschrift für Theorie und Praxis des Wissenschaftlichen Sozialismus* 9 (1970): 1180.

107. Eric Honecker, "Fragen von Wissenschaft und Politik," *Einheit: Zeitschrift für Theorie und Praxis des Wissenschaftlichen Sozialismus* 1 (1972): 8, cited in Werner, *Kybernetik statt Marx?* 103.

108. Herbert Hörz, "Die Wirksamkeit der ideologischen Arbeit erhöhen! Diskussion auf Einladung der Einheit," *Einheit: Zeitschrift für Theorie und Praxis des Wissenschaftlichen Sozialismus* 1 (1972): 21, cited in Werner, *"Kybernetik statt Marx?* 103.

109. Ulbricht's dismissal came at the 3 May 1971 Sixteenth Central Committee meeting. Important scholars have seen Honecker's coup as a *Zäsur,* a dramatic break with the past when the Party's Central Committee abandoned Ulbricht's "reform concept" and reintroduced central planning. Jeffries and Melzer, "New Economic System of planning," 35; Phillip J. Bryson and Manfred Melzer, *The end of the East German economy: From Honecker to reunification* (New York: Macmillan, 1991), 29; Mike Dennis, *German Democratic Republic: Politics, economics and society* (London: Pinter Publishers, 1931), 35; Hermann Weber, *Geschichte der D.D.R. Autorreferat* (Munich: Deutscher Taschenbuch Verlag, 1985), 404.

Weber called the *Zäsur* the triumph of the *Primat der Politik* and emphasized Honecker's more pragmatic and flexible style. Yet it is difficult to think of Honecker as "pragmatic." As reflected in East Germany's forests, rather than a *Zäsur,* Honecker's 1971 coup reimposed the orthodoxy from which Ulbricht seemed to be drifting.

110. Letter cited in Grieder, "Overthrow of Ulbricht," 43n245.

111. The East German Politburo vote was on 27 April 1971. Leonhard, *Das kurze Leben der DDR*, 44; Harry Nick, "Sozialistische Rationalisierung, wissenschaftlich-technisch Revolution und Effektivität," *Einheit: Zeitschrift für Theorie und Praxis des Wissenschaftlichen Sozialismus* 2 (1971): 169.

112. Roger Berthoud, "Mr. Brezhnev says time is ripe for speedy settlement on Berlin," *The Times* (17 June 1971).

113. *Neues Deutschland* (20 June 1971).

114. Gretel Spitzer, "Hint from East Germany of new line on détente," *The Times* (21 June 1971), 6; Deutsch-sowjetischen Vertrag vom 12. August 1970 ("Moskauer Vertrag"), http://www.auswaertiges-amt.de/www/de/infoservice/politik/dokumente_html (accessed 23 September 2003).

115. Ulbricht's dismissal came at the 3 May 1971 Sixteenth Central Committee meeting. Jeffries and Melzer, "New Economic System of planning," 35; Phillip J. Bryson and Manfred Melzer, *The end of the East German economy: From Honecker to reunification* (New York: Macmillan, 1991), 29; Mike Dennis, *German Democratic Republic: Politics, economics and society* (London: Pinter Publishers, 1931), 35; Hermann Weber, *Geschichte der D.D.R. Autorreferat* (Munich: Deutscher Taschenbuch Verlag, 1985), 404.

116. Ruffer and Schwartz, *Die Forstwirtschaft der Deutschen Demokratischen*

*Republik,* 21. The continuity with Eighth Party Congress policies is also discussed in Alfred Kosing and Richard Heinrich, "Natur-Mensch-Gesellschaft: Das Verhältnis der sozialistischen Gesellschaft zur Natur," *Neues Deutschland* 127 (1 June 1987): 33–34.

117. B. Fahner, G. Weiß, F. Ullmann, and G. Ervert, "Parteitagsdelegierte aus der Forstwirtschaft berichten vom 10. Parteitag der Party: Unter der Führung der Party weiter voran auf den bewährten Kurs der Hauptaufgabe, der Einheit von Wirtschafts- und Sozialpolitik," *Sozialistische Forstwirtschaft* 31:6 (1981): 161; Frowen, "Economy of the German Democratic Republic," 42; Jeffries and Melzer, *East German economy,* 44; Erich Honecker, *Die Aufgaben der Partei bei der weiteren Verwirklichung der Beschlüße des IX Parteitages der Partei* (Berlin: Dietz Verlag, 1978).

118. Weber, *Geschichte der DDR* (Munich: Deutscher Taschenbuch Verlag, 1985), 260; Erich Honecker, "Program of the Party's Tenth Party Congress," *Neues Deutschland* (12 April 1981), 3; Günther Mittag, "Tenth Party Congress demands," *Einheit: Zeitschrift für Theorie und Praxis des Wissenschaftlichen Sozialismus,* vol. 5; Frowen, "Economy of the German Democratic Republic," 42; Jeffries and Melzer, *East German economy,* 47; Wilma Merkel and Stephanie Wahl, *Das geplünderte Deutschland: Die wirtschaftliche Entwicklung im östlichen Teil Deutschlands von, 1949–1989,* 2nd ed. (Bonn: Instituts für Wirtschaft und Gesellschaft, 1991), 26; Arthur A. Stahnke, "GDR economic strategy in the 1980s: The 1981–1985 Plan," in *Studies in GDR culture and society: Nine selected papers from the Fourteenth New Hampshire Symposium on the German Democratic Republic,* ed. Margy Gerber (New York: University Press of America, 1983), 1; Karl Christian Thalheim, *Stagnation or change in Communist economies?* (London: Center for Research into Communist Economies, 1986), 29; Werner Bröll, "Das sozialistische Wirtschaftssystem," in Heinz Rausch and Theo Stammen, *D.D.R.* (Munich: Verlag C. H. Beck, 1978), 138.

119. Horst Ruffer and Ekkehard Schwartz, *Die Forstwirtschaft der Deutschen Demokratischen Republik* (Berlin: VEB Deutscher Landwirtschaftsverlag, 1984), 18, 21; Kurth, interview.

120. Jimmy Carter, "Town meeting remarks, Bardstown, Kentucky" (31 July 1979), *Public papers of the presidents of the United States: Jimmy Carter,* book 2 (Washington, D.C.: Superintendent of Documents, 1979), 1340; Jimmy Carter, "Speech to the nation" (15 July 1979), http://www.pbs.org/wgbh/amex/carter/filmmore/ps_crisis.html (accessed 28 February 2005).

### *Chapter 8. The Grüneberg Era*

1. East Germany mined more than 300 million metric tons of lignite in 1990 on a 1987–installed capacity of 23,596 megawatts. Economist Intelligence Unit (Great Britain), *EIU country report, East Germany* (London: Economist Intelligence Unit, 1989), 28; Rita Ökten, *Die Bedeutung des Umweltschutzes für die Wirtschaft der Deutsche Demokratischen Republik* (Berlin: A. Spitz, 1986), 44; Carl Graf Hohenthal, "Die Umwelt-Last der D.D.R.," *Frankfurter Allgemeine Zeitung* 18 (22 January 1990): 12.

2. House Committee on Energy and Commerce, "Acid rain in Europe," 8; National Foreign Assessment Center, *Estimating Soviet and East European hard currency debt,* USSR: Hard Currency Debt, table A–1 (Washington, D.C.: Central Intelligence Agency,

1980), 15; Manfred Schmidtz, "Der ungeteilte Dreck: Saubere Luft braucht die Kooperation von Bundesrepublik und DDR," *Die Zeit* (1987): 42. Schmidtz cites a 1987 Deutsches Institut für Wirtschaftsforschung (DIW) study; John Ardagh, *Germany and the Germans* (New York: Harper and Row, 1987), 121; David Goodheart, "Man with the power to light up the east," *Financial Times* (19 April 1990), 25.

3. Gerhard Grüneberg, "30 Jahre Marxistisch-Leninistische Agrarpolitik—30 Jahre Bündnis der Arbeiterklasse mit den Bauern," *Neues Deutschland* 188 (9 August 1975): 3.

4. H. O. Spielke, G. Breithaupt, H. Bruggel, and H. Stand, *Ökonomik der sozialistischen Forstwirtschaft* (Berlin: VEB Deutsche Landwirtschaftsverlag, 1964), 34; Meinhard Ott, interview by author, Eberswalde, 18 March 1991.

5. Political forest boundaries were reimposed on 1 July 1975. Horst Kohl, ed., *Ökonomische Geographie der Deutschen Demokratischen Republik,* 3rd ed. (Gotha and Leipzig: VEB Hermann Haack, 1976), 461; Ian Jeffries and Manfred Melzer, "The New Economic System of planning and management, 1963–1970, and recentralization in the 1970s," in Jeffries and Melzer, eds., *East German economy* (1987), 26–27.

6. Alfred Kosing and Richard Heinrich, "Natur-Mensch-Gesellschaft: Das Verhältnis der sozialistischen Gesellschaft zur Natur," *Neues Deutschland* 127 (1 June 1987): 3–4; Alfred Kosing et al., *Sozialistische Gesellschaft und Natur: Wissenschaftlichen Rates für Marxistisch-Leninistisch Philosophie der DDR* (Berlin: Dietz, 1989); Alfred Kosing, "Natur und Gesellschaft," *Einheit: Zeitschrift für Theorie und Praxis des Wissenschaftlichen Sozialismus* 39:11 (1984): 1018–23; K. Müller, R. Budzin, and H. Trinks, "Die Berechnung der Lehre von Karl Marx für die gesellschaftliche Nutzung der Naturkräfte des Waldes," *Sozialistische Forstwirtschaft* 18:5 (May 1968): 129; Rolf Steffens, *Wald, Landeskultur und Gesellschaft,* 2nd ed. (Jena: VEB Gustav Fischer Verlag, 1978).

7. Frithjof Paul, interview by author, Tharandt, 26 March 1991. Professor Paul was East Germany's senior forest policy expert and professor of forest policy at Tharandt. Although a communist, Paul fought with the cadres for a rational and scientific forest policy.

8. Hans Herbert Götz, "Als der Klassenkampf in der DDR begann: Die Bodenreform vor 40 Jahren," *Frankfurter Allgemeine Zeitung* 206 (6 September 1985): 13.

9. Harald Thomasius, "Waldbauliche Auffassungen, Probleme und Wege in der DDR," *Allgemeine Forstzeitschrift* 28–29 (14 July 1990): 726, and Harald Thomasius, interview by author, Tharandt, 20, 26 March 1991.

10. Thomasius, "Waldbauliche Auffassungen," 727; Hans Herbert Götz, "Eine Landwirtschaft mit Mammut-Betrieben: 30 Jahre Agrarpolitik in der D.D.R," *Frankfurter Allgemeine Zeitung* 269 (17 November 1979): 15.

11. Erich Honecker and Irma Brandt, "Arbeiter und Bauern schreiben neues Kapitel der Geschichte," *Neues Deutschland* 312 (6 September 1975): 1; Staatliches Komitee für Forstwirtschaft, "Diskussionsmaterial"—Die Wege zu Intensivierung des forstlichen Reproduktionsprozess und zur Erhöhung seiner Effektivität im sein, typewritten (Berlin, 1972), 3, 5.

12. Honecker and Brandt, "Arbeiter und Bauern schreiben neues Kapitel," 1; Harald Thomasius, "Gesetzmäßigkeiten in der historischen Entwicklung des Waldbaus," in *Kolloquium anläßlich des 75 Geburtstag von Nationalpreisträger Prof. (em.) Dr.-ing. habil. Johannes Blanckmeister* (Tharandt: Technische University Dresden, 1973), 3;

F. Fischer, ed., *Industriemäßige Produktionsmethoden der Rohholzbereitstellung aus Fichtenvornutzung. Wissenschaftliche Tagung 13. bis 25. Oktober 1974* (Freiberg: Zentralen Druckerei der Bergakademie Freiberg, 1974), 2; Horst Kurth, "Entwicklungstendenzen der Forsteinrichtung als Ergebnis des internationalen Forsteinrichtungssymposium," in *Kolloquium anläßlich des 75 Geburtstag von Nationalpreisträger Prof. (em.) Dr.-ing. habil Johannes Blanckmeister* (Tharandt: Technische Universität Dresden, 1973).

13. Honecker, *Die Aufgaben der Partei,* 4; Heinz Kuhrig, "Demokratische Bodenreform legte den Grundstein für stetig steigende Agrarproduktion," *Neues Deutschland* 198 (21 August 1975): 3; Horst Kurth, ed., "Wissenschaftliche-technischen Darlegung zur Intensivierung der Holzproduktion und zur komplexen und volkswirtschaftlichen Effektiven Holzverwertung," typewritten and marked "VVS" and "VD" (1979).

14. Grüneberg, "30 Jahre Marxistisch-Leninistische Agrarpolitik," 3. The Party leadership based their farming policies on the model of Soviet *Kolchoze* and the massive April 1975 Soviet agrarian reform program. Soviet agricultural expenditures in 1975 doubled 1974 levels, to DM 130 billion (37.1 billion rubles), half to finance the collectives (*Sowchosen,* or *Kolchoze*) with the remainder financing price supports.

15. Erich Honecker, "Bündnis war, ist und bleibt Eckpfeiler unserer Politik," *Neues Deutschland* (6–7 September 1975): 3.

16. Marion, Gräfin Dönhoff, "Dogma oder Weizen?" *Die Zeit* 37 (5 September 1975): 6.

17. Rudolf Rüffler, interview by author, Eberswalde, 20 May 1991.

18. Gerhard Hofmann, interview by author, Eberswalde, 7 March 1991.

19. Gerd Friedrich, et al., *Die Volkswirtschaft der DDR*. Akademie für Gesellschaftswissenschaten beim ZK der SED (Berlin: Verlag Die Wirtschaft, 1979).

20. Müller, Budzin, and Trinks, "Die Berechnung der Lehre von Karl Marx," 129.

21. Karl Marx ("A Rhinelander"), "Remarks on debates on the law on thefts of wood and the Proceedings of the Sixth Rhine Provincial Assembly," *Rheinische Zeitung* 88 (October 1842). First published in the supplement to the *Rheinische Zeitung,* nos. 298, 300, 303, 305, and 307 (25, 27, and 30 October, 1 and 3 November 1842), transcribed by director@marx.org, November 1996. http://www.marxists.org/archive/marx/works/1842/10/25.htm (accessed 6 July 2003).

22. Müller, Budzin, and Trinks, "Die Berechnung der Lehre von Karl Marx," 132.

23. G. Zillmann, "Fragen des Überganges zu industriegemäßigen Produktionsmethoden in der Forstwirtschaft" *Sozialistische Forstwirtschaft* 14:2 (February 1964): 35, 37; W. Schult, *Bedeutung und Inhalts eines Zweigsprogrammes und wissenschaftlichtechnischen Konzeptionen für die perspektivische Planung in der Forstwirtschaft* (Leipzig-Markkleeburg: Landwirtschaftsaustellung der DDR, 1965), 46–47.

24. G. Schröder, "Zu einigen Problemen der Forstwissenschaft und Praxis in Prozeß der wissenschaftlich-technischen Revolution," *Sozialistische Forstwirtschaft* 15:11 (1965): 323.

25. H. Heidrich, "Schlußwort," in *Die Anwendung des Neuen Ökonomischen Systems im Bereich der VVB Forstwirtschaft Potsdam*, ed. Rudolf Rüthnick (Potsdam-Babelsberg: VVB Forstwirtschaft Potsdam, 1966), 172, 175–76.

26. Schult, *Bedeutung und Inhalts eines Zweigsprogrammes,* 5; Rudolf Rüthnick, ed., *Die Anwendung des Neuen Ökonomischen Systems im Bereich der VVB Forstwirtschaft Potsdam* (Potsdam-Babelsberg: VVB Forstwirtschaft Potsdam, 1966).

27. Karl Marx, "Lohnarbeit und Kapital," *Werke,* vol. 6 (Berlin: Dietz Verlag, 1959), 407.

28. Dieter Sachse, "Einst Objekt der Ausbeutung, heute Quelle des Wohlstands," *Neues Deutschland* 288 (6 December 1980): 10; Jeffries and Melzer, "New Economic System," 37.

29. Karl Marx and Friedrich Engels, *The Communist manifesto* (*Manifest der Kommunistischen Partei*), ed. David McLellan (New York: Oxford University Press, 1992).

30. Karl Marx and Friedrich Engels, *Marx-Engels Werke,* vol. 4 (Berlin: Dietz, 1964), 481; Committee for the World Atlas of Agriculture, *World atlas of agriculture,* 148; Goldenbaum, "Demokratische Bodenreform," 3. The purpose of the socialist collectives (LPGs) was not first food production but "to reduce the difference between urban and rural life, i.e., between modern, large-scale farms and widely scattered peasant farming," a prominent feature of *The Communist manifesto.*

31. Anna Bramwell, *Ecology in the twentieth century: A history* (New Haven: Yale University Press, 1989), 34. Anna Bramwell picked up the religious timbre in Marx's and Engels' thought, observing that they call "for the abolition of the countryside rather in the way that Old Testament biblical prophesy promises that there will be no more sea." Professor Bramwell cites Karl Marx and Friedrich Engels, *Basic writings on politics and philosophy* (Garden City, N.Y.: Doubleday, 1959), 70.

32. Konrad Merkel, *Die Forst- und Holzwirtschaft in Mitteldeutschland.* Forschungsbeirat für Fragen der Wiedervereinigung Deutschlands beim Bundesminister für gesamtdeutsche Frage (Berlin, 1974), cited in *Frankfurter Allgemeine Zeitung* 218 (20 September 1974): 14; Ernst Goldenbaum, "Demokratische Bodenreform hat unseren Bauern eine gesicherte Zukunft eröffnet," *Neues Deutschland* 204 (28 August 1975): 3. Goldenbaum, head of the Farmers' Union, echoed the idea of choosing policy to meet the conditions of Marx's "prophesy."

33. Gertraud Seidenstecher, "DDR," in *Umweltschutz und ökonomisches System in Oste: Drei Beispiele Sowjetunion, DDR, Ungarn,* ed. Hans-Hermann Höhmann and Gertraud Seidenstecher (Stuttgart: Kohlhammer, 1973), 91; B. Bittighöfer, H. Edeling, H. Kulow, "Theoretische und politisch-ideologische Fragen der Beziehung von Mensch und Umwelt," *Deutsche Zeitschrift für Philosophie* 1 (1972): 68. The causes of pollution were "social in nature, the forms of capitalism methods of production." Technology and economic growth were merely "second order phenomena," pollution's origins were social in nature, not economic. Pollution was "an aspect of the general crisis of capitalism in the world historical epoch of the transition to socialism," a "catastrophic remnant of German imperialism caused by the nature of capitalist production relations and monopoly capital," a consequence of presocialist class relations.

34. Kosing and Heinrich, "Natur-Mensch-Gesellschaft," 3, 10–20; Höppner and Hauenschild, interview by author, Eberswalde, 13 March 1991; Seidenstecher, "DDR," 85, 91.

35. Kosing and Heinrich, "Natur-Mensch-Gesellschaft," 3.

36. Evgeny N. Lisitzin, "Collaborative arrangements for environmental protection in European socialist countries," in *Environmental policies in East and West,* ed. Gyorgy Eneyedi, August J. Gijawijt, and Barbara Rhode (London: Taylor Graham, 1987), 352.

37. Gertraud Seidenstecher, *Umweltschutz in der DDR,* "Berichte des Bundesinstituts für ostwissenschaftliche und internationale Studien" (Cologne: Bundesinstitut für ostwissenschaftliche und internationale Studien, 1973), 90–91.

38. Seidenstecher, "DDR," 85.

39. Bittighöfer, Edeling, Kulow, "Theoretische und politisch-ideologische Fragen," 68.

40. Kosing, "Natur und Gesellschaft," 1023.

41. Kurt Voge, et al., " 'Unser Wald soll gesund, sauber, ertragreich sein': Forstwirtschaftsbetrieb Oranienburg an das ZK der Party," *Neues Deutschland* 68 (21 March 1986): 4, 6.

42. Peter Allrich, interview by author, Eberswalde, 25 March 1991.

43. Conference on Security and Co-operation in Europe, Final Act Helsinki 1975, sec. 5, "Environment." Text at Hellenic Resources Network (HR-Net) at http://www.hri.org/docs/Helsinki75.html (accessed 21 September 2003).

44. Ibid.

45. Seidenstecher, "DDR," 91.

46. Rudolf Rüthnick, "Unser Wald im guten Händen. Wie wir ein Stück Verantwortung für heutige und künftige Generationen wahrnehmen," *Neues Deutschland* 225 (23 September 1989): 9.

47. Seidenstecher, "DDR," 85. The secrecy level was *Vertrauliche Verschlußsache* (VVS) under the *Anordnung zur Sicherung des Geheimschutzes auf dem Gebiet der Umwelt,* the Council of Ministers' 16 November 1982 "Law for Securing the Protection of Secrecy in the Environment."

48. Manfred Ronzheimer, "Umweltforschung in der DDR—eine Bilanz. Besondere aktuelle Bedeutung. Wichtige Erkentnisse trotz politischer Restriktionen," *Der Tagesspiegel* 13:679 (22 September 1990): 18; Seidenstecher, "DDR," 85. The exclusion of foresters began with the inventory BRA IV, the Waldfonds, and the Waldschäden Berichte. Michael von Berg, "Umweltschutz in Deutschland: Verwirklichung einer deutschen Umweltunion," *Deutschland Archiv* 23:6 (June 1990): 897; Klaus Schikora and H.-E. Wünsche, *Der Waldfonds in der Deutschen Demokratischen Republik* (Potsdam: VEB Forstprojektierung Potsdam, 1977), 132–33; E. Klein, "Forsteinrichtung nach der Wiedervereinigung," *Der Wald,* 41:2 (1991): 60.

49. Horst Kurth, interview by author, Tharandt, 25 March 1991.

50. Frithjof Paul, quoted in *Der Tagesspiegel,* "Sorgen um den Wald in der D.D.R," *Der Tagesspiegel* 11:408 (6 April 1983): 13.

51. Merrill E. Jones, "Origins of the East German environmental movement," *German Studies Review* (1993); Meredith Kirkpatrick, *Environmental problems and policies in East Europe and the USSR* (Monticello, Ill.: Council of Planning Librarians, 1978); Seidenstecher, *Umweltschutz in der DDR;* UNO, *Probleme der Umwelt, Deutsche Demokratische Republik: Nationaler Bericht der Deutsche Demokratischen Republik für die UNO-Konferenz über menschliche Umwelt Bedingungen in Stockholm 1972* (Berlin, 1971).

52. Seidenstecher, "DDR," 91; Seidenstecher, *Umweltschutz in der DDR,* 83–84, 88.

53. Duncan Fisher, "The emergence of the environmental movement in eastern Europe and its role in the revolution of 1989," in *Environmental Action in Eastern Europe,* ed. Barbara Jancar-Webster (Armonk, N.Y.: M. E. Sharpe, 1993), 96; Ökten, *Die Bedeutung des Umweltschutzes für die Wirtschaft,* 65.

54. Christine L. Zvosec, "Environmental deterioration in East Europe," *World Affairs* 147:2 (fall 1984): 117.

55. Staatliches Komitee für Forstwirtschaft, "Forstressourcen-Umfrage," typewritten (Berlin, 1980), 8; Erwin Kienitz, *Denkschrift über forstwirtschaftlichsorganisatorischen Reformen, insbesondere des Bauernwäldes der Deutschen Demokratischen Republik: Eine Beitrag zur sozialistischen Umgestaltung der Forstwirtschaft* (Tharandt: Institut für Forstliche Wirtschaftslehre, 1958), 10.

56. Staatliches Komitee für Forstwirtschaft, "Forstressourcen-Umfrage."

57. Theologischen Studienabteilung beim DDR-Kirchenbund, "Ökonomie, Leistung, Persönlichkeit," *Deutschland Archiv* 15:1 (January 1982): 68; Peter Schmalz, "Das Umweltpflänzchen beginnt zu grünen. Sie galten und sie gelten noch immer als Außenseiter der sozialistischen Gesellschaft: Die Öko-Gruppen in der 'DDR,'" *Die Welt* (24 May 1984); Michael Mara, "Wachsendes Umweltbewußtsein in der D.D.R.," *Der Tagesspiegel* 11:808 (27 July 1984): 3; Michael Mara, "Partei befurchtet eine 'grüne Unterwanderung,'" *Der Tagesspiegel* 12:107 (23 July 1985): 3; Albrecht Hinze, "Exportierte Schäden stören die sozialistische Eintracht. Über die Umwelt in der DDR gibt es nur dürre Nachrichten. 'Menschen nicht verunsichern,'" *Süddeutsche Zeitung* (31 December 1986): 1; Dankwart Guratzsch, "Der Schweigebann bricht," *Die Welt* (20 February 1986).

58. Jones, "Origins of the East German environmental movement."

59. Werner Felfe, "Alles mit den Menschen—Alles für die Menschen. Ausgewählte Reden und Aufsätze," speech welcoming the Cuban delegation headed by Fidel Castro, 16 June 1972 at the Leuna chemical works (Berlin: Dietz Verlag, 1987), 48.

60. Guratzsch, "Der Schweigebann bricht."

61. Erich Honecker, "Gespräch über aktuelle Fragen der Friedenssicherung und Umweltpolitik. Erich Honecker empfing Abordnung des Bundesverbandes Bürgerinitiativen Umweltschutz aus der BRD. Erhalt natürlicher Lebensbedingungen erfordert die Verhinderung eines atomaren Krieges. DDR tut alles, um die Gefahr zu bannen, die durch USA-Erstschlagswaffen in Westeuropa, besonders in der BRD, entstanden ist. Leistungen der DDR beim Umweltschutz gewürdigt," *Neues Deutschland* (6 September 1984): 1; Kosing, "Natur und Gesellschaft," 1023. The West German environmentalists represented the *Bundesverbandes Bürgerinitiativen Umwelts* (Federal Coalition of the Citizens Environmental Initiative).

The British novelist Kingsley Amis has a misogynist character parody the timbre of such rhetoric in his 1984 novel *Stanley and the women*: "Women were like the Russians —if you did exactly what they wanted all the time you were being realistic and constructive and promoting the cause of peace, and if you ever stood up to them you were resorting to cold-war tactics and pursuing imperialistic designs and interfering in their internal affairs. And by the way of course peace was more peaceful, but if you went on promoting its cause long enough you ended up Finlandized at best."

62. Honecker, "Meeting with Bundesverbandes Bürgerinitiativen Umwelts," 1.

63. I rely heavily on the following two articles in my discussion of Honecker's relations with the West German press: Theo Sommer, "Ein deutscher Kommunist, ein deutscher Realist. Anmerkungen zu einem ZEIT-Interview mit Erich Honecker," *Die Zeit* 6 (31 January 1986): 1; Marlies Menge and Theo Sommer, "Miteinander leben, gut miteinander auskommen" Ein ZEIT-Gespräch mit Erich Honecker, *Die Zeit* 6 (31 January 1986): 3.

64. Manfred Grote, "The Party under Honecker," *East European Quarterly* 21:1 (1987): 67–78; Werner Felfe, "Aus dem Bericht des Politbüros an der 5. Tagung des Zentralkomitees der Party," *Neues Deutschland* (17 December 1987): 6.

65. This 1984 agreement came in effect with an addendum to Article 7 of the Basic Treaty. Kanzleramts-Papier 1 June 1984. Karutz, "Beteiligt sich Bonn an den Kosten, dann zeigt D.D.R," 4.

66. Thomas Szasz, *The second sin* (1974), quoted in *The Oxford book of aphorisms*, ed. John Gross (New York: Oxford University Press, 1987), 328.

67. Felfe, "Aus dem Bericht des Politbüros," 6; Bundesministerium für innerdeutsche Beziehungen, ed., *Informationen* 24:4 (1988): 4.

68. Hans Barciok, interview by author, Potsdam, 14 March 1991. Barciok conceded only to earlier knowledge of "localized damage to spruce, two-thirds caused by Czechoslovakia." Barciok was a prisoner of war in an American camp in the Vosges Mountains at the end of the Second World War. He reported that camp conditions became inhuman after the German capitulation and that most of his fellow POWs died from exposure and starvation. When he was finally released he "walked straight to the Soviet zone of occupation" and joined the Communist Party. Fifty years later his hatred of Americans still burned fiercely, almost stronger than his Marxist faith.

69. In response, a forest conference was convened in 1985 (the Bitterfeld Conference) to discuss forest death in spruce. Barciok, interview.

70. J. Säglitz, "Die Forstwirtschaft in Ostdeutschland—Stand, Probleme, Ziele," *Forstarchiv* 61:6 (November-December 1990): 226.

71. Felfe, "Aus dem Bericht des Politbüros," 6. This was a substantial increase from Felfe's earlier estimates.

72. Ulrich Petschow, Jürgen Meyerhoff, and Claus Thomasberger, *Umweltreport DDR: Bilanz der Zerstörung, Kosten der Sanierung, Strategien für den ökologischen Umbau*, Eine Studie des Instituts für Ökologische Wirtschaftsforschung (Frankfurt am Main: S. Fischer, 1990); Runder Tisch, *Information zur Analyse der Umweltbedingungen in der DDR und zu weiteren Maßnahmen* (Berlin, 1990).

73. Ökten, *Die Bedeutung des Umweltschutzes*, 55. West Germany had a net influx of 125,000 tons of sulfurous oxides; House Committee on Energy and Commerce, Subcommittee on Health and the Environment, "Acid rain in Europe: A report, March 1985, on the fact-finding excursion," 99th Cong., 1st sess. Com print 99–F (SD ct. no. Y 4. En 2/3–99F) (LC 85–16244) (Washington, D.C., 1985), 8; John Ardagh, *Germany and the Germans* (New York: Harper and Row, 1987), 121; Christine L. Zvosec, "Environmental deterioration in East Europe," *World Affairs* 147:2 (fall 1984): 101.

74. Hans Richter, ed. *Nutzung und Veränderung der Natur* (Leipzig: Geographisches Gesellschaft der Deutschen Demokratischen Republik, 1981).

75. Kohl, ed., *Ökonomische Geographie der Deutschen Demokratischen Republik,* 455.

76. Autorenkollektiv, *Richtlinie für die Bewirtschaftung,* "Nur für Dienstgebrauch" (Berlin: SKF [Staatliches Komitee für Forstwirtschaft], 1979), 1.

77. Horst Paucke cited in Hans Richter, ed., *Nutzung und Veränderung der Natur* (Leipzig: Geographisches Gesellschaft der Deutschen Demokratischen Republik, 1981), 155.

78. This built on the 1979 initiative of the State Forestry Committee, an agreement between industrial ministries and forestry to reduce pollution. Autorenkollektiv, *Richtlinie für die Bewirtschaftung.*

79. In the 1960s this plan was formulated as "Tree Species Optimization" *(Baumartenoptimierung).* E. Ziegler, "Die Wirkung der Industrie-Rauchschäden auf den Wald, ihre Berüchtsichtigung bei der Raumplanung und die Notwendigkeit ihrer gesetzlichen Regelung," *Wissenschaftliches Zeitschrift der Technischen Universität Dresden* 6 (1956–57): 777–787; Rüthnick, "Unser Wald im guten Händen," 9; Rudolf Rüthnick et al., "40 Jahre DDR 40 Jahre Entwicklung zu einer sozialistischen Forstwirtschaft," *Sozialistische Forstwirtschaft* 39:9 (1989): 257–88; Hinze, "Exportierte Schäden stören die sozialistische Eintracht," 1.

80. Paul, interview.

81. Höppner and Hauenschild, interview. Eberswalde forest policy scholars were tasked with drafting a national forest law in the late 1960s, anticipating the Soviet national forest legislation published in *Pravda* in the 1970s and West Germany's Federal Forest Law (*Bundesforstgesetz*) of 1975.

82. Horst Ruffer, interview by author, Eberswalde, March 1991.

83. Höppner and Hauenschild, interview.

84. Rolf Bartonek, "Umweltkriminalität—in der DDR bislang ein Kavaliersdelikt," *Neues Deutschland* 101 (2 May 1990): 8; Horst Kurth, "Die Entwicklung der Forstwirtschaft in der DDR," *Allgemeine Forstzeitschrift* 35 (1990): 892; Säglitz, "Die Forstwirtschaft in Ostdeutschland," 225.

85. Rüthnick, "Unser Wald im guten Händen," 9.

86. Horst Paucke, "German Democratic Republic," in *Environmental policies in East and West,* ed. Gyorgy Eneyedi, August J. Gijawijt, and Barbara Rhode (London: Taylor Graham, 1987), 148; Ronzheimer, "Umweltforschung in der DDR," 19; Runder Tisch, *Information zur Analyse*; Thomasius, "Waldbauliche Auffassungen," 727. Fertilization began heavily in the late 1960s. Later forest decline was formally acknowledged through the 1983 "Resolution on Measures for the Protection of the Forests of the GDR." This resolution called for increased fertilization and "tree species optimization." Foresters were ordered to supervise "forest cleanliness" more vigorously and to ensure "a clean woods."

87. Doris Cornelson, "Umweltprobleme und Umweltbewußtsein in der DDR," *Gegenwartskunde* 1 (1989): 48.

88. Friedrich-Karl Helmholz and Adolf Sturzbecher, "Wie bei uns Abgeordnete für die Gesunderhaltung des Waldes sorgen," *Neues Deutschland* 116 (19 May 1989): 3. The posted east mark to deutsche mark exchange rate (OM:DM) was 0.56 in 1989; Wilma Merkel and Stephanie Wahl, *Das geplünderte Deutschland. Die wirtschaftliche*

*Entwicklung im östlichen Teil Deutschlands von 1949–1989*, 2nd ed. (Bonn: Instituts für Wirtschaft und Gesellschaft, 1991), table 7, p. 48.

89. Paul, interview.

90. Kurt Sontheimer and Wilhelm Bleek, *The government and politics of East Germany,* trans. Ursula Price (New York: St. Martins Press, 1976), 159.

91. Richard Schröder, "Dreimal den Krieg verloren? Gibt die Bundesregierung dem Druck der Alteigentümer nach, fördert sie den Glauben, in der vereinten Republik seien die Ostdeutschen immer die Dummen," *Die Zeit* 15 (11 April 1997): 5.

92. W. Titel, "Umweltschäden in der DDR dient dem Wohl des Menschen," *Marxistische Blätter* 1 (1972): 49.

93. Staatlichen Zentralverwaltung für Statistik (SZWS), ed., *Statistisches Jahrbuch der Deutsche Demokratischen Republik.* "Geschädigte Waldfläche," sec. 7—Umweltschutz. Quelle: Ministerium für Ernährung, Land- und Forstwirtschaft (Berlin: Staatsverlag der Deutschen Demokratischen Republik, 1990), 149.

94. Majunke, "Tätigkeitsbericht 1990"; Majunke, interview.

95. J. H. Bergmann and W. Flohr, "Zur Wirkung von Fremdstoffen in den Wäldern der DDR unter besonderer Berücksichtigung einer Veränderung der Bodenflora," *Sozialistische Forstwirtschaft* 38:6 (1988): 164–66; Hofmann, interview.

96. Staatlichen Zentralverwaltung für Statistik (SZWS), ed., *Statistisches Jahrbuch der Deutsche Demokratischen Republik.* "Geschädigte Waldfläche," sec. 7—Umweltschutz (Berlin: Staatsverlag der Deutschen Demokratischen Republik, 1990), 148–49; Gerhard Hofmann and D. Heinsdorf, "Depositionsgeschen und Waldbewirtschaftung," *Der Wald* 40:7 (1990): 208.

97. House Committee on Energy and Commerce, "Acid rain in Europe," 8.

98. Fritz Richard Stern, *The politics of cultural despair: A study in the rise of the Germanic ideology,* 2nd ed. (Berkeley: University of California Press, 1974), 15.

99. Arnulf Baring and Volker Zastrow, *Unser neuer Größenwahn. Deutschland zwischen Ost und West,* 3d ed. (Stuttgart: Deutsche Verlags-Anstalt, 1989), 86.

100. Günther Haaf, "Noch zwanzig Jahre deutscher Wald?" *Die Zeit* 2 (7 January 1983): 1.

101. *Die Zeit,* "Special section, letters to the editor in response to 'Noch zwanzig Jahre deutscher Wald?'" *Die Zeit* 6 (4 February 1983): 11.

102. Hans Magnus Enzensberger, "The state of Europe," *Granta* 30 (winter 1990): 136; Kurt Sontheimer, "Intellectuals and politics in Western Germany," *West European Politics* 1 (February 1978): 30. Sontheimer discussed the role of intellectuals and university professors in radical politics and their affinity for radical politics and countercultural movements.

103. Ökten, *Die Bedeutung des Umweltschutzes für die Wirtschaft,* 65.

104. East German government declaration. Seidenstecher, *Umweltschutz in der DDR*, 92. Rudolf Rüthnick, chief of the State Forest Service, defended his record in 1989: "The East German Communist Party and the government have taken effective steps to protect and preserve the forests through intensive forest management, introduction of smoke-tolerant species, fertilization, and stabilization."

105. Kuhrig, "Demokratische Bodenreform legte den Grundstein," 3; Horst Bitschkowski and Franz Krahn, "Wirksamer Beitrag zu der höherer Leistungskraft. Gerhard

Grüneberg bei Genossen in Schwinkendorf," *Neues Deutschland* 258 (1 November 1980): 3.

106. "One of the most deeply reaching political and economic policies in the history of East Germany." Karl Eckart, "Veränderungen in der Landwirtschaft der DDR seit Anfang der siebziger Jahre," *Deutschland Archiv* 18:4 (1985): 409; B. Lietz, "Steigerung der Produktion und Effektivität bei Erhöhung der lebender Holzvorräte—Schwerpunkt der Forstwirtschaft in den 80er Jahre," *Sozialistische Forstwirtschaft* 33:9 (1983): 266–71; 274–75.

107. Karl Christian Thalheim, *Stagnation or change in Communist economies?* (London: Center for Research into Communist Economies, 1986), 31.

108. Honecker and Brandt, "Arbeiter und Bauern schreiben," 4.

109. Honecker, "Bündnis war, ist und bleibt," 3; Jeffries and Melzer, "New Economic System," 37. The Eighth Party Congress published three formal goals for the economy: (1) qualitative and quantitative improvements in the supply of consumer goods; (2) intensified production using the existing means of production through increased application of advanced technology; and (3) increased social benefits, such as better and more plentiful housing, improved pensions, and a shorter work week.

110. Menge and Sommer, "Miteinander leben," 3.

111. The "Reform of the Structure of Economic Management" called for a "unification of planning and execution." The State Forest Districts got additional reform tasks in the "Order on the Principles for Management of East German Forests" ("Verfügung über die Grundsätze für die Bewirtschaftung der Wälder der Deutschen Demokratischen Republik," 10 June 1986). H.-E. Wünsche and K. Schikora, "Der Waldfonds der DDR—ausgewählte Fakten über Entwicklung und Zustand," *Sozialistische Forstwirtschaft* 40:3 (1990): 74.

112. Höppner and Hauenschild, interview.

113. G. Hildebrandt, "Die Forsteinrichtung in der DDR 1950 bis 1965. Ein Beitrag zur jüngeren deutschen Forsteinrichtungsgeschichte aus Anlaß des 80. Geburtstages von Albert Richter in Eberswalde," *Allgemeine Forst- und Jagdzeitung* 160:6 (1989): 124; Klaus Wenske, interview by author, Marienburg, 18 March 1991; H. Barciok, interview by author, Potsdam, March 1991; D. Bieberstein, *Haupttendenzen und Probleme der Waldfondsentwicklung in der Deutsche Demokratischenn Republik, 1944–1950*, in "Forestry in East Germany: A special number on the 150th anniversary of the initiation of forestry research at Eberswalde" (in German) *Beiträge für die Forstwirtschaft* 15:3–4 (1981): 144–50; Horst Kurth, "Die Forsteinrichtung in der Deutschen Demokratischen Republik," IUFRO S.4.04 Conference Report, Bukarest (1983), 81–96.

114. Höppner and Hauenschild, interview.

115. Barciok, interview.

116. Kosing and Heinrich, "Natur-Mensch-Gesellschaft," 3; Kurth, *Wissenschaftliche-technische Darlegung;* Kurth, interview.

117. Despite the increased harvest volume, stocking was to increase from 190 cubic meters per hectare to 195 cubic meters per hectare. J. Piesnack, "Zur Leitung und Planung der Bestandespflege," *Sozialistische Forstwirtschaft* 7:2 (1986): 193–224.

118. Kurth, "Die Entwicklung der Forstwirtschaft in der DDR," 892.

119. Staatliches Komitee für Forstwirtschaft, "Forstressourcen-Umfrage," 8; Piesnack,

"Zur Leitung und Planung der Bestandespflege"; Wünsche and Schikora, "Der Waldfonds der DDR," 75; Bryson and Melzer, *End of the East German economy*, 33.

120. Staatlichen Zentralverwaltung für Statistik (SZWS), ed., *Statistisches Jahrbuch der Deutsche Demokratischen Republik*, sec. 10, "Land- und Forstwirtschaft, Nutzholzeinschlag nach Holzsorten" (Berlin: Staatsverlag der Deutschen Demokratischen Republik, 1989), 34.

121. Merkel and Wahl, *Das geplünderte Deutschland*, 27. The authors commented on "[the] lack of economic and social policy conception and perspective."

122. Paul, interview.

123. The GDR's huge paper mills had to deliver products despite Soviet short-shipments in the late 1980s.

124. Paul, interview.

125. Klaus Schikora and H.-E. Wünsche, *Der Waldfonds in der Deutschen Demokratischen Republik* (Potsdam: VEB Forstprojektierung Potsdam, 1973), 54. In 1972, 12 million cubic meters were destroyed in the Harz and the lowlands, almost twice the normal harvest; Säglitz, "Die Forstwirtschaft in Ostdeutschland," 225.

126. Klaus Schikora and H.-E. Wünsche, *Der Waldfonds in der Deutschen Demokratischen Republik* (Potsdam: VEB Forstprojektierung Potsdam, 1986); Klaus Schikora and H.-E. Wünsche, *Der Waldfonds in der Deutschen Demokratischen Republik*, "KOWA" lists for 1986–89 (Potsdam: VEB Forstprojektierung Potsdam, 1989); Wünsche and Schikora, "Der Waldfonds der DDR—ausgewählte Fakten," 74; Forschungsanstalt für Holz- und Forstwirtschaft, "Bericht über den Gesundheitszustand der Wälder der ostdeutscher Bundesländer (DDR) der Bundesrepublik Deutschland," typewritten (Eberswalde-Finow: IFE, 1991); Barciok, interview; Curt Majunke, ed., "Tätigkeitsbericht 1990," Forschungsanstalt für Holz- und Forstwirtschaft Eberswalde. Abteilung Forstschutz/Hauptstelle für Forstpflanzenschutz, typewritten (Eberswalde-Finow: IFE, 1990). Storms destroyed thirteen thousand cubic meters in 1981 and 1982 in the Southern Uplands, the Vogtland of the Erzgebirge, and the Thüringer Wald. A "storm catastrophe" followed in 1985, extensively damaging forests on the Rostocker Heide in the North German Plain. The serious drought of 1986–87 accelerated pine forest decline. Finally, in January and February of 1990 a severe windstorm in the Harz caused massive breakage damage amplified in 1991 by *Nonne* (nun moth) infestation (*Lymantria monacha L.*).

127. Salvage harvests were the official reason for the increase in spruce's share of the 1981–85 harvests to 38 percent (despite representing only 20 percent of all conifers). Wünsche and Schikora, "Der Waldfonds der DDR—ausgewählte Fakten," 74; Horst Kurth, "Bestandesinventur in der Deutschen Demokratischen Republik," *Sozialistische Forstwirtschaft* 38:4 (1988): 141; Kurth, "Die Entwicklung der Forstwirtschaft in der DDR," 89; Plochmann, "Forestry in the Federal Republic of Germany," 453.

128. Barciok, interview.

129. West German stocking increased from 130 cubic meters in 1949 to 364 cubic meters per hectare in 1987. Ignaz Kiechele, "Bundeswaldinventur," *Forstarchiv* 61:5 (September-October 1990): 204–5; Bundesministerium für Ernährung, Landwirtschaft und Forsten, "Bundeswaldinventur 1987. Übersichtstabellen," Entwurf auf den Ergebnissen vom 23.9.90 basierend. *Schriftenreihe des Bundesministerium für Ernährung,*

*Landwirtschaft und Forsten* (1987); VEB Forstprojektierung Potsdam, *Flächen- und Vorratsgliederung nach Besitzverhältnissen,* vol. 5–161–2 FE-Unterlagen (1949).

130. Stocking levels in 1945 are not known precisely but were probably marginally better than levels in West Germany.

131. Hildebrandt, "Die Forsteinrichtung in der DDR," 123.

132. O. Kandler, "Epidemiologische Bewertung der Waldschädenserhebungen 1983 bis 1987 in der Bundesrepublik Deutschland," *Allgemeine Forst- und Jagdzeitung* 159:9/10 (1988): 179–94.

133. Horst Kurth, "Der Wald im 21. Jahrhundert—eine ökologische Entwicklungsaufgabe im Stoffwechsel der Menschens mit der Natur," *Natur und Umwelt* 2:1 (1986): 5; R. Barth, R. Kallweit, K. Schikora, and G. Schübel, *Der Wald im Land Brandenburg* (Potsdam: Forsteinrichtungsamtes, 1990), 32.

134. Kurth, "Der Wald im 21. Jahrhundert," 5, 22; Säglitz, "Die Forstwirtschaft in Ostdeutschland," 226; Wünsche and Schikora, "Der Waldfonds der DDR—ausgewählte Fakten," 74.

135. Staatlichen Zentralverwaltung für Statistik (SZWS), ed., *Statistisches Jahrbuch der Deutsche Demokratischen Republik*, "Holzeinschlag" (Berlin: Staatsverlag der Deutschen Demokratischen Republik, 1989), 34.

136. Johannes Leithäuser, "'Aber laßt nicht die Massen dran.' Für die DDR-Funktionäre Jagdreviere so groß wie das Saarland," *Frankfurter Allgemeine Zeitung* 269 (17 November 1990): 7–8.

137. Caroline Möhring, "Bleibt Fontanes Märchenplatz am Ufer des Werbellin?" *Frankfurter Allgemeine Zeitung* 37 (13 February 1990): 3.

138. Michael Mara, "DDR Prominenz entspannt sich bei der Jagd," *Der Tagesspiegel* 12:447 (4 September 1986): 12.

139. Staatliches Komitee für Forstwirtschaft, "Forstressourcen-Umfrage," 8.

### *Chapter 9. Reunification*

1. Göring reconciled his passion for hunting with *Dauerwald* only though a prodigious feat of cognitive dissonance; Nazi and Marxist game wildlife management alike made it impossible to restore a close-to-nature forest due to the overburden of game wildlife. Göring was also *Reichsjägermeister* (Reich Master of the Hunt), *Preußischer Landforstmeister* (Prussian Chief of Forestry), and *Landesjägermeister von Preußen* (Prussian Master of the Hunt), with appropriate uniforms for each title. Institut Deutsche Adelsforschung, "Edelleute im preußischen Staatsforstdienst 1936–1945," http://home.foni.net/~herumstreifer/forstoo.htm (accessed 1 April 2005); Heinrich Rubner, *Deutsche Forstgeschichte—1933–1945: Forstwissenschaft, Jagd und Umwelt im NS-Staat* (St. Katharinen: Scriptae Mercaturae Verlag, 1985).

2. Alexander Riedel, "Chancen und Probleme der sachsischen Forstwirtschaft," lecture to graduating students of the Tharandt Forestry School (26 March 1991); Alexander Riedel, "Probleme der Umgestaltung der Forstwirtschaft in Sachsen," *Forstarchiv* 62:3 (1990): 91–94.

3. Horst Kurth, "Die Entwicklung der Forstwirtschaft in der DDR," *Allgemeine*

*Forstzeitschrift* 35 (1990): 896. Putting the State Forest Districts' harvests at a sustained yield level required 350 DM per hectare in supports.

4. HA Forstwirtschaft des Ministeriums für Land-, Forst- und Nahrungsgüterswirtschaft, "Statistisches Übersichtsmaterial 1990," 41, 43; Kurth, "Die Entwicklung der Forstwirtschaft in der DDR," 894. East German State Forest Districts averaged 700 employees compared to 350 in comparable West German forest districts. The employment level in the West also would soon be cut drastically and many forest districts closed.

5. Kurth, "Die Entwicklung der Forstwirtschaft in der DDR," 894.

6. Immediately after the Wall fell 50 percent of the State Forest Districts' staff were dismissed. In Brandenburg alone, the largest of Germany's five new states, 60 percent were furloughed. Roland Barth, interview by author, Potsdam, 15 and 21 March 1991. Barth headed GDR Forest Management. Höppner and Hauenschild, interview; Alexander Riedel, letter to the editor, "Es liegt nicht an der Forstbürokraten der neuen Länder," *Frankfurter Allgemeine Zeitung* (26 February 1993), 8; H. O. Spielke, G. Breithaupt, H. Bruggel, and H. Stand, *Ökonomik der sozialistischen Forstwirtschaft* (Berlin: VEB Deutsche Landwirtschaftsverlag, 1964), 53.

7. Andreas Oldag, "Bauern Opfer im Osten," *Süddeutsche Zeitung* (30 June 1994), 4.

8. Thuringia and Saxony replaced its State Forest Districts of 30,000–33,000 hectares with *Forstämter,* traditional management units of 6,000–7,000 hectares. Sixty *Forstämter* replaced Saxony's huge fifteen State Forest Districts. The new *Forstämter* in Brandenburg and Mecklenburg-Vorpommern were larger, 12,000–15,000 hectares, reflecting the poorer ecological conditions in the northern lowlands. Forest administration returned to traditional management units and to ecological boundaries. Kurth, "Die Entwicklung der Forstwirtschaft in der DDR," 894; Barth, interview; Riedel, *Chancen und Probleme.*

9. Lothar de Maizière, *Neues Deutschland* (20 April 1990).

10. Kurth, interview. The 40 percent share for hardwoods was still far short of the 80 percent share for hardwoods in the close-to-nature forest under "ideal" conditions.

11. S. Kempe and Rudolf Rüffler, "Analyse aus gewahlter Rahmenbedingungen beim Übergang zur eigenverantwortlichen Bewirtschaftung nichtstaatlichen Waldeigentums im Land Brandenburg," *Berichte aus Forschung und Entwicklung* 23:49 (Eberswalde: Forschungsanstalt für Forst- und Holz, 1991); J. Säglitz, "Die Forstwirtschaft in Ostdeutschland—Stand, Probleme, Ziele," *Forstarchiv* 61:6 (November-December 1990): 226; R. Beyse, "DDR-Bodenreform vor dem Bundesverfassungsgericht," *Holz-Zentralblatt* 117:20 (1991): 305; Judy Dempsey, "Eastern Germany in turmoil over agricultural counter-revolution," *Financial Times* (7 October 1994).

12. Richard Schröder, "Dreimal den Krieg verloren? Gibt die Bundesregierung dem Druck der Alteigentümer nach, fördert sie den Glauben, in der vereinten Republik seien die Ostdeutschen immer die Dummen," *Die Zeit* 15 (11 April 1997): 5.

13. Beyse, "DDR-Bodenreform vor dem Bundesverfassungsgericht," 306. Lothar de Maizière, press statement, 16 June 1990.

14. Conversation with Eduard Shevardnadze, Tiblisi, Georgia, December 1993; Norman Stone, "Property values," review of *Schicksalsbuch des Sächsisch-Thüringischen Adels 1945, Restitutionsverbot,* ed., Christoph Rechberg, and *Dokumentation zum neuen Entschädigungsgesetz EALG* by Udo Madaus, *Times Literary Supplement* (10 May 1996), 9. Gorbachev told Stone in 1994 that the Soviets made "no such precondi-

tions. Property transactions in Germany were a matter for the Germans, who were, after all, getting back full sovereignty."

15. Beyse, "DDR-Bodenreform vor dem Bundesverfassungsgericht," 305.

16. Stone, "Property values," 9. Upholding land reform also lowered the financial burden of reunification. By sustaining land reform, the Kohl government saddled former collective farmers with the socialist collectives' debt of 8.6 billion DM. The DG Bank in Frankfurt bought the loan assets (8 billion DM) of the former East German agricultural bank from the Treuhand for 3 billion DM, at book value, 3 percent of DG Bank's assets. Edmund L. Andrews, "Privatized German farms liable for Communist's debt: Ruling is newest burden on burdened east," *New York Times* (9 April 1997), D–2.

17. Friedrich Heitmann, "International Sammlung von Forststatistiken," *Zeitschrift für Weltforstwirtschaft* 10:7/10 (1944): 473; Joseph C. Kircher, "The forests of the U.S. zone of Germany," *Journal of Forestry* 45:4 (1947): 249. Kircher reported that Germany "lost" more than 2,666,936 ha of forest in the Oder-Neiße territories.

18. E. Reichenstein, "Die forstwirtschaftliche Lage Deutschland vor und nach dem 2. Weltkrieg," *Forstarchiv* 21:1/3 (1950): 30.

19. Alexis de Tocqueville, *Selected letters on politics and society,* ed. Roger Boesche, trans. James Toupin and Roger Boesche (Berkeley: University of California Press, 1985), 366–68, 373, http://chnm.gmu.edu/revolution/d/590/ (accessed 18 October 2003).

20. Alexis de Tocqueville, *L'Ancien Régime et la Révolution,* ed. François Furet and Françoise Mélonio, trans. Alan S. Kahan (Chicago: University of Chicago Press, 1998).

21. David Remnick, "Apocalyptic critics offer scenarios for the collapse of perestroika," *Washington Post* (4 November 1989), A–17.

22. *TASS*, "Pravda reports Mikhail Gorbachev's meetings" (8 October 1989); Xinhua News Agency, "Military parade reviewed in Berlin," Xinhua News Agency (7 October 1989), item 1007188; Serge Schmemann, "Gorbachev lends Honecker a hand," *New York Times* (7 October 1989), 1. In addition to the Party bosses and dictators allied with the Soviet Union, the Soviet army's top commanders in the West, all close to Honecker, joined him on the reviewing stand: General Boris Snetkov, commander in chief of the Western Group of Soviet Forces of the USSR; Lt-General Nikolay Moiseyev, member of the Military Council and chief of the Political Directorate of the Western Group; and General Vladimir Shuralov, representative of the commander in chief of the Joint Armed Forces of the Warsaw Treaty member states with East Germany's NVA.

23. Erich Honecker, *Neues Deutschland* (10 October 1989), cited in Serge Schmemann, "German morning after," *New York Times* (11 October 1989), A-1; Craig R. Whitney, "Party coup turned East German tide," *New York Times* (19 November 1989), 1.

24. Erich Honecker, ADN in German (11 October 1989) 1830 GMT and Radio GDR Home 2110 GMT, "SED Politbüro statement on disturbances and emigration," cited in BBC Summary of World Broadcasts," part 2, East Europe; B. Internal Affairs; EE/0586/B/1.

25. Mikhail Gorbachev, 7 October 1989, as reported by Horst Sindermann, president of the *Volkskammer* (People's Chamber) and number three in the Party hierarchy, to the West German news magazine *Der Spiegel* shortly before his death in April 1990. Gorbachev, like Khrushchev, found it easier to encourage reform in East Germany than it was to enact reforms in the Soviet Union. "Unification survey — how it all started," part 1 of 2,

*Economist* (30 June 1990): 4; Federal Research Division of the Library of Congress, Germany, Country Studies/Area Handbook Series sponsored by the U.S. Department of the Army http://countrystudies.us/germany/72.htm (accessed 18 October 2003).

26. Anna Tomforde and Michael Simmons, "Communist Party allies shun Honecker's strategy," *Guardian* (13 October 1989).

27. Mark A. Uhlig, "Ortega's foray: A stunning misstep," *New York Times* (30 October 1989), A–8.

28. John Tagliabue, "Prague seizes dissidents on eve of anniversary," *New York Times* (28 October 1989), A–6. The Czech secret police also arrested Vaclav Maly, a Catholic priest whom the government banned from pastoral duties and his ministry, and Jaroslav Sabata, a lapsed Marxist-Leninist theoretician.

29. Richard Pipes, "The Russians are still coming: Moscow is building up its clients, and winning," *New York Times* (9 October 1989), A–17.

30. Brzezinski's speech at the Soviet Foreign Ministry's Diplomatic Academy (27 October 1989) was reprinted in Zbigniew Brzezinski, "A common house, a common home," *New York Times* (15 November 1989), A–29; Mikhail Ivanov, "Brzezinski welcomed to Soviet Diplomatic Academy," *TASS* (1 November 1989); *TASS*, "Zbigniew Brzezinski on perestroika—*TASS* interview" (3 November 1989); *TASS*, "Common European home, as seen by Brzezinski" *TASS* (9 November 1989); Bill Keller, "Making policy of the inevitable: Gorbachev accepts wide change," *New York Times* (12 November 1989), 1.

31. Murray Feshbach's work on Soviet destruction of the environment remains the best source available. Murray Feshbach, *Ecological disaster: Cleaning up the hidden legacy of the Soviet regime* (New York: Twentieth Century Fund Press, 1995); Murray Feshbach, *Economics of health and environment in the USSR* (Washington, D.C.: Office of Net Assessment, U.S. Department of Defense, forthcoming); Murray Feshbach, "Russia's population meltdown," *Wilson Quarterly* 25:1 (winter 2001); Murray Feshbach, ed., *Environmental and health atlas of Russia* (Moscow: PAIMS Publishing House, 1995); Murray Feshbach and Alfred Friendly, Jr., *Ecocide in the USSR: Health and nature under siege* (New York: Basic Books, 1992); Joan DeBaredeleben, review of M. Feshbach's *Ecocide in the USSR: Health and nature under siege*, M. Turnbull's *Soviet environmental policies and practices: The most critical investment,* and Philip R. Pryde's *Environmental management in the Soviet Union, Slavic Review* 52:3 (fall 1993): 593; David Holloway, "The politics of catastrophe," review of Murray Feshbach's *Ecocide in the USSR*, Grigori Medvedev's *The Truth about Chernobyl* and *No breathing room: The aftermath of Chernobyl*, and Piers Paul Read's *Ablaze: The story of the heroes and victims of Chernobyl, New York Review of Books* 40:11 (16 October 1993): 36; John Massey Stewart, "No place to hide," review of Murray Feshbach's *Environmental and health atlas of Russia* and *Ecological disaster, Nature* 381:6579 (16 May 1996): 203.

32. Lincoln C. Chen, Friederike Wittgenstein, and Elizabeth McKeon, "The upsurge of mortality in Russia: Causes and policy implications" (in Notes and Commentary), *Population and Development Review* 22:3 (September 1996): 517.

33. Dr. Marin Strmecki, interview by author, Westport, Conn., June 2000. Dr. Strmecki was in the audience as Dr. Brzezinski spoke.

34. Tadeusz Mazowiecki, a close ally of Lech Walêsa, became prime minister of a

Solidarity-Communist Party coalition government on 24 August 1989. Brzezinski also warned that there had not yet been a successful "transition from a Communist, dictatorial system to a pluralistic, democratic system."

35. Bill Keller, "Clamor in the East: The Kremlin; Making policy of the inevitable, Gorbachev accepts wide change," *New York Times* (12 November 1989), 1.

36. Fewer than 5,000 bodies have been found in the mass grave at Katyn, however.

37. Theo Waigel, the West German finance minister, made this report to the Budget Committee early in 1990. David Goodheart, "East German industry 'more inefficient than thought,'" *Financial Times* (29 March 1990). The East German Finance minister's report is reported in Economist Intelligence Unit (Great Britain), *EIU country report, East Germany* (London: Economist Intelligence Unit, 1989), 37. East Germany's net external debt equaled $20 billion in current value; David Binder, "Grim state of East Germany's economy is disclosed to Parliament," *New York Times* (16 November 1989), A–20.

38. Andrew Fisher, "Bundesbank warns of tax rise to prop east German economy," *Financial Times* (19 March 1992), 1. Eighty-five billion DM flowed into private households, financing 40 percent of private consumption and perpetuating the Marxist-Leninist culture of dependence, passivity, and depression.

39. Judy Dempsey, "Treuhand's final debt to total DM 270 bn," *Financial Times* (26 July 1994), 2.

40. Hella Pick, "West gets a warning to stay out of reform," *Guardian* (7 October 1989).

41. Nevil Johnson, review of Mary Fulbrook's *Anatomy of a dictatorship: Inside the GDR, 1949–1989* (Oxford: Oxford University Press, 1995), *English History Review* 112:448 (September 1997): 1030–31.

42. Louis Joachim Edinger, *West German politics* (New York: Columbia University Press, 1986), 66.

43. Roy E. H. Mellor, "The German Democratic Republic," in *Planning in eastern Europe,* ed. Andrew H. Dawson (London: Croon Helm, 1987): 139.

44. Henry Kamm, "A riddle for Communists: Why does the East German economy prosper?" *New York Times* (13 March 1989), A–3.

45. Craig Whitney, "German trade bond," *New York Times* (17 November 1989).

46. Terence Roth, "West German firms poised to head East," *Wall Street Journal* (8 November 1989).

47. Adrian Hyde-Price, *European security beyond the cold war* (London: Sage Publications for the Royal Institute for International Affairs, 1991), 14.

48. Simon Levin, *Fragile dominion: Complexity and the commons* (Reading, Mass.: Helix Books, 1999), 13. Levin, a distinguished Princeton biologist, discussed the application of ecosystem analysis to policy analysis. Other useful titles are John Henry Holland, *Hidden order: How adaptation builds complexity* (Reading, Mass.: Addison-Wesley, 1996), and John Henry Holland, *Emergence: From chaos to order* (Reading, Mass.: Addison-Wesley, 1998).

49. Norbert Wiener, *The human use of human beings; Cybernetics and society* (Boston: Houghton Mifflin, 1954), 12, 24–25, 95.

50. Thomas Jefferson's "syntax of the middle landscape" is discussed fully in Leo Marx, *The machine and the garden: Technology and the pastoral ideal in America* (New York: Oxford University Press, 2000), 120–22.

51. Isaiah Berlin wrote of the "barbarous consequences" of faith in absolute truth: "The notion that there must exist final objective answers to normative questions, truths that can be demonstrated or directly intuited, that it is in principle possible to discover a harmonious pattern in which all values are reconciled, and that it is towards this unique goal that we must make; that we can uncover some single central principle that shapes this vision, a principle which, once found, will govern our lives—this ancient and almost universal belief, on which so much traditional thought and action and philosophical doctrine rests, seems to me invalid, and at times to have led (and still to lead) to absurdities in theory and barbarous consequences in practice." Isaiah Berlin, *Five essays on liberty,* 2nd ed. (Oxford: Oxford University Press, 1969), 15–16. Berlin developed his ideas about the plurality of truths in Isaiah Berlin, *Two concepts of liberty, an inaugural lecture delivered before the University of Oxford on 31 October 1958* (Oxford: Clarendon Press, 1961); Henry Hardy, "Isaiah Berlin's key idea," *Philosopher's Magazine* 11 (summer 2000): 15–16.

52. Aldo Leopold discussed such anomalies in *Dauerwald* philosophy in his fascinating record of the long excursion he made to the German forest in the years just before the Second World War. Aldo Leopold, "Deer and Dauerwald in Germany: I. History," *Journal of Forestry* 34:4 (April 1936): 366–75; Aldo Leopold, "Deer and Dauerwald in Germany: II. Ecology and policy," *Journal of Forestry* 34:5 (May 1936): 460–66.

53. George F. Kennan, *Democracy and the student left* (New York, 1968), 9–10.

54. Stefan Heym, quoted in, Timothy Garton Ash, "East Germany: The solution," *New York Review of Books* 37:7 (26 April 1990): 14.

# *Glossary*

*Ära Grüneberg*. The "Grüneberg era" (1971–81), named after Gerhard Grüneberg, the harsh Central Committee secretary for agriculture. The eponymous era was characterized by intensive application of Industrial Production Methods.

*Arbeitsgemeinschaft Naturgemäße Waldwirtschaft* (ANW). Working Group for Close-to-Nature Forestry, the West German forest society started by K. Dannecker in 1950 dedicated to *Dauerwald*, close-to-nature forestry, and *Naturschutz*, nature protection.

*Ausgleich*. The balance of prices and costs under the *Brutto* system of accounts.

*Baumartenoptimierung*. Tree Species Optimization program of 1966 designed to move the species composition of the East German forest to pollution-tolerant species grown in even-aged plantations.

*Bezirke*. Districts established in 1952 to replace the five states (the *Länder*) of eastern Germany.

*Bodenreinertragslehre*. Soil rent theory. Used to set harvests based on present value (the "money yield of financial rotation") and the "normal forest model." Also known as "classic forest management."

*Bodenfonds*. Land account set up in 1945 to hold farm and forest land expropriated under the land reform.

*Brutto* Accounting. The Soviet method of gross accounting employed in East Germany to police Plan tasks.

*Bündnispolitik*. Lenin's "Union of Workers and Peasants" policy which peaked after 1971 during the Ära Grüneberg. Erich Honecker used it to meld Engels' "backward" peasantry into the advanced class of urban workers and farms and forests into industry.

Cadre. Party members advancing the interests of the Communist Party. The cadres, who worked mostly in the districts and counties, were orthodox Marxist-Leninists and poorly educated. The cadres faithfully carried out the leadership's policies under Lenin's principle of "democratic socialism."

Close-to-Nature Forestry. See *Naturgemäße Waldwirtschaft* below. Forest management based on ecosystem management emphasizing long-term capital appreciation, stability, and multiple use.

Coppice. A low forest regenerated by stump-sprouting. Harvests from coppiced woods supply small dimension lumber for charcoal and fuelwood and for fence and barrier material. Coppicing of mixed hardwood forests was the principal method of forest management in Central Europe until the Industrial Revolution.

*Dauerwald.* "Permanent forest" forest management developed by Alfred Möller in the early 1920s. *Dauerwald* remains the ultimate expression of ecological forestry, applied today as *Plenterwald* forest management. *Dauerwald* foresters approached the forest as a complex organism. It is based on an absolute proscription on clear-cutting, focusing on single tree, uneven-aged management; natural regeneration; and a natural species composition. As a silvicultural system it is often summarized as "cut the worst, leave the best."

*Dauerwaldstreit.* The "*Dauerwald* debate" between *Dauerwald* and industrial, "classic forest management" has raged in Germany in various forms since the mid-nineteenth century.

*Demontage.* Soviet reparations taken by dismantling factories and fixed assets for shipment to the Soviet Union.

*Derbholz.* Above-ground portion of trees over seven cm in diameter at breast height. Standing volume is estimated either without bark (o.R.) or with bark (m.R.).

*Deutsche Demokratische Republik* (DDR). East Germany, the German Democratic Republic (GDR), founded in 1949 as the successor government to the Soviet Military Administration in the Soviet zone of occupation.

*Deutsche Wirtschaftskommission.* The German Economic Commission established by the Soviets to coordinate economic policy in the Soviet zone.

*Devisen.* The East bloc term for hard currencies, usually U.S. dollars, earned from trade with the West. East bloc countries regularly "sold" their own soft currency balances earned under trade agreements with second- and third-world economies at discounts in exchange for *Devisen.*

Dirigisme. The policy of state direction and control in economic and social matters.

*Eiserne Gesetz des Örtlichen* (EGÖ). Wilhelm Pfeil's "Iron Law of Site Conditions," which taught that silvicultural planning must follow from local site conditions. The EGÖ stood in direct contrast to G. Hartig's Eight General Rules, which guided even-aged management and industrial forest management.

Even-Aged Management. A system of plantation forest management, usually associated with monocultures, where stands are harvested all at once through clear-cutting and then replanted.

*Festmeter* (Fm). Measure of wood volume equivalent to a cubic meter of wood without air spaces, estimated either without bark (Fm o.R.) or with bark (Fm m.R.). For standing volume calculations, this measure includes only trees with a diameter at breast height of seven cm and greater (*Derbholz*).

Forest Management. The German word *Forstwirtschaft* approaches closest to the meaning of this management system. Forest management is the practical application of silviculture and applied ecology to generate a suite of benefits from forested ecosystems, typically stressing forest products.

*Forsteinrichtung.* Forest management regulation, an extension of forest mensuration and inventory work, which implements periodic comprehensive analysis of mid- and long-term forest management goals and forest conditions. In West Germany it does not include such short-term management tasks as regeneration, harvest, and stand maintenance. In East Germany *Forsteinrichtung* specialists ensured the maximum continuous supply of annually increasing volumes of forest products.

*Hiebsatz.* Annual sustained yield harvest.

Highgrading. Destructive harvesting taking only valuable trees. The resulting forest stand's economic value is ruined as the expense of clearing the inferior wood and replanting often exceeds any potential revenue from a second harvest. Highgrading also leaves inferior phenotypes in its wake and a poor seed source for a future forest. Highgrading is the opposite of Alfred Möller's *Dauerwald* prescription "cut the worst, leave the best."

Industrial Production Methods (*Industriegemäße Produktionsmethoden,* IPM). An extreme variant of industrial forest management ("classic forest management") traditionally associated in National Socialist and romantic conservative environmental literature with "Manchester-school economic liberalism" and "capitalist" forestry. IPM furthered the *Bündnispolitik* and the socialist industrial model in agriculture and forestry and forced constantly rising farm and forest production. It reached its maximum influence during the Ära Grüneberg.

*Kommunistische Partei Deutschland* (KPD). The German Communist Party. The KPD was reconstituted as the *Sozialistische Einheitspartei Deutschlands* (SED) through the forced merger in April 1946 of the communists' hated rival, the Social Democrats (SPD, or *Sozialdemokratische Partei Deutschland)*, into the KPD.

*Kreis.* Counties subordinate to the *Bezirk* government.

Labor Theory of Value. Marx's *Arbetiswerttheorie,* which held that value is created only by labor. No value is ascribed to capital or to natural resources and raw materials.

*Lößgrenze.* The loess boundary dividing the North German Plain and the Central Uplands. It marks the southern extension of the Vistula Glacier and the border between pine and spruce silviculture.

*Landwirtschaftliche Produktionsgenossenschaften* (LPG). Agricultural Production Collectives: socialist collectives formed to organize farms expropriated in 1945 to further "socialist production relations."

*Menzer Beschluß* (the "Menz Resolution"). The 20 November 1951 decision forbidding clear-cutting and age class management. This resolution was formalized in the management instructions "Replacement of Clear-cut Management with Stocking Maintenance Forest Management."

*Ministerium for Land-, Forst und Nährungsgüterwirtschaft* (MLFN). Ministry for Agriculture, Forestry, and Food Supply.

*Naturgemäße Waldwirtschaft.* "Close-to-nature" forest management. The West German variant of optimal stocking forestry.

NÖS (New Economic System, 1964–67). Implemented through the Second Seven-Year (*Per-*

*spektiv*) Plan (1964–70), the NÖS was introduced at the Sixth Party Congress (15–21 January 1963). The NÖS called for "an optimal system of long-term central planning with an indirect control of enterprises through the use of economic levers (largely in the form of monetary instruments)." Walter Ulbricht and Erich Apel applied cybernetic theory in the NÖS to mimic the control, management, and economic feedbacks and flows of liberal capitalism to allocate resources more efficiently while retaining central control.

*Nutzungssoll.* The production quota set by the central planners. By definition in East Germany it meant the sustained yield harvest.

Optimal Stocking Forestry. *Vorratspflegliche Waldwirtschaft* (VpWw), the postwar variant of *Dauerwald* developed by Herman Krutzsch based on the modified selection method.

ÖSS (Economic System of Socialism, 1967–71). A midcourse correction of the NÖS engineered at the Seventh Party Congress (17–22 April 1967).

*Osthilfe* (Eastern Support, 1930–32). The Brüning government's program of economic aid and resettlement in East Prussia and central Germany.

*Osthilfegesetz.* The "Law on Eastern Support" passed under the Brüning government on 31 March 1931.

*Plenterwald.* Selection forest managed under *Dauerwald* principles.

*Rat für Gegenseitige Wirtschaft* (RGW). The Council for Mutual Economic Assistance (CMEA). The Soviet bloc's answer to the United States' Marshall Plan announced on 25 January 1949 in Moscow.

*Räte der Bezirke.* District councils.

Socialist Spring in the Countryside (*Sozialistische Frühling in der Land*). The forced collectivization of independent small farm and forest land starting in April 1960 to unwind land reform and "create the necessary preconditions for socialism."

SOPADE. The liberal socialist Social Democratic Party (Sozialdemokratische Partei Deutschlands, SPD). The Soviet Military Administration in Germany (SMAD) and the German Communist Party forcibly merged the Soviet zone wing of the SPD into the new *Sozialistische Einheitspartei Deutschlands* (Socialist Unity Party), or SED (see below), in April 1946. Communist repression in the first postwar years focused particularly on the Social Democrats, sending youth and senior leadership into forced labor and concentration camps or driving them into exile. The SPD remained a force in the western zones in opposition to the conservative Christian Democratic Union (CDU, or *Christlich Demokratische Union Deutschlands*) and the Bavarian Christian Social Union (CSU, or *Christlich-Soziale Union*). The SPD was the most consistent and clearest critic of Soviet zone policy, focusing on human rights crimes, on the destruction in the eastern German landscape, on economic problems and Soviet reparations, and on the ravages of Soviet land reform.

*Sowjetische Aktiengesellschaft* (SAG). Expropriated factories run by the Soviet Military Administration for export to the Soviet Union

*Sozialistische Einheitspartei Deutschlands* (SED). The Marxist-Leninist Socialist Unity Party formed from the forced merger in April 1946 of the German Social Democratic Party (SPD) and the German Communist Party (KPD).

*Sowjetische Militäradministration Deutschland* (SMAD). The Soviet Military Administration in Germany, responsible for eastern German government from 1945 up to the formation of East Germany in October 1949.

*Staatliche Forstwirtschaftsbetriebe.* State Forest Districts established in 1952 to manage state forests and eventually all East German forests.

*Staatliche Plankommission.* State Planning Commission, the central economic planning body under the *Ministerrat* (the Council of Ministers) from mid-1961 onwards.

*Staatswald.* State forests of the late 1940s.

*Standortsgemäße Waldwirtschaft.* Site Appropriate Forestry developed by Albert Richter in the 1950s as a more pragmatic evolution of Krutzsch's Optimal Stocking Forestry and *Dauerwald.*

State Committee for Forest Management. (*Staatliches Kommitte für Forstwirtschaft* [*SKF*]). The body responsible for coordinating forest management and reporting to the Council of Ministers.

*Treuhand.* The trustee organization responsible for privatizing East German state property after reunification.

*Volkseigenebetriebe* (VEB). Nationalized People's factories.

*Volks- und Kommunalwirtschaftsunternehmen Wald* (KWU-Wald). People's and communal forests managed under the *Waldgemeinschaften der Vereinigungen der gegenseitigen Bauernhilfe.*

Uneven-Aged Management. Silvicultural treatment related to *Dauerwald* based on natural regeneration. In central Germany it implies mixed hardwoods.

*Valutamark* (VM). East German "foreign currency mark" used to value exports and imports. The rate of exchange was set by the East German government and not published. The exchange rate roughly corresponded to the value of the east mark before the 1961 revaluation. A typical value (in 1982) was 1 VM to .71 DM.

*Vereinigungen der gegenseitigen Bauernhilfe* (VdgB). Unions of Mutual Peasant Aid, the socialist, Party-controlled collectives set up in the 1940s to compete with successful traditional farm cooperatives. Nominally not state controlled, they were directed by the local Party and *Deutsche Bauernpartei* (DBP) organizations. The VdgB's forest complement was the *Waldgemeinschaften der Vereinigungen der gegenseitigen Bauernhilfe*, the Forest Collectives of the Unions of Mutual Peasant Aid.

*Vorratsfestmeter* (Vfm). Standing gross volume of timber measured in cubic meters, usually of trees with diameters in excess of seven cm (*Derbholz*) either with bark (m.R.) or without (o.R.). Later GDR forest inventories included trees with diameters as small as four cm.

*Volkseigene Betriebe* (VEB). People's enterprises.

*Volkseigenengüter* (VEG). "The People's Own Estates" created out of the one million hectares of agricultural land the Party leadership seized in 1945 under the land reform.

*Vertrauliche Verschlußsache* (VVS). GDR's second highest level of security. Environmental and forestry data were classified "VVS" by the Council of Ministers' 16 November 1982 "Law for Securing the Protection of Secrecy in the Environment."

*Volkskammer.* The People's Chamber, the SED's legislative body.

*Volkswald.* People's Forest created by the 15 June 1949 *Verordnung* out of non–land-reform state, community, and copperative forest (*Staatswald, Gemeinde- und Körpershaftswald*) and private forest not distributed to "new peasants" in 1945.

*Volkswirtschaftspläne.* National Economic Plans.

*Vorratspflegliche Waldwirtschaft* (VpWw). Optimal Stocking Forestry, the postwar variant of *Dauerwald* invented by Herman Krutzsch. Based on the modified selection method.

*Waldgemeinschaften der Vereinigungen der gegenseitigen Bauernhilfe.* Forest Collectives of the Unions of Mutual Peasant Aid, socialist forest collectives set up to compete with the successful independent *Waldgenossenschaften.*

*Waldbauergenossenschaften.* Peasant Forest Cooperatives, similar to *Waldgenossenschaften.*

*Waldgenossenschaften.* Traditional forest cooperatives originating in the sixteenth century. *Waldgenossenschaften* were successful in the 1950s but liquidated with forced collectivization in the late 1950s culminating in the Socialist Spring in the Countryside in 1960.

*Waldreinertragslehre.* Forest rent theory of management which emphasizes building up of forest capital. The opposite of the *Bodenreinertragslehre*, it leads to a close-to-nature forest.

# *Bibliography*


Ackermann, Anton, "Gilt es einen besonderen deutschen Weg zum Sozialismus?" *Einheit: Zeitschrift für Theorie und Praxis des Wissenschaftlichen Sozialismus* (February 1946), Sonderheft.

Ager, Derek Victor, *The geology of Europe* (New York: Wiley, 1980).

Anderson, J. G. C., *The structure of western Europe* (New York: Pergamon Press, 1978).

Apel, Erich, "Einige Grundfragen der Leitung unserer sozialistischen Volkswirtschaft," *Einheit: Zeitschrift für Theorie und Praxis des Wissenschaftlichen Sozialismus* 11/12 (1961): 1632.

Apel, Erich, and Günther Mittag, *Ökonomische Gesetze des Sozialismus und neues ökonomisches System der Planung und Leitung der Volkswirtschaft,* 2nd ed. (Berlin: Dietz, 1964).

——, *Wissenschaftlich Führungstätigkeit: Neue Rolle der VVB* (Berlin: Dietz, 1964).

Arbeitsgemeinschaft Naturgemäße Waldwirtschaftswald, "Stellungnahme der Schriftleitung. Erste Norddeutsche Tagung der ANW," *Allgemeine Forstzeitschrift* 5 (1950): 496–99.

Ardagh, John, *Germany and the Germans* (New York: Harper and Row, 1987).

Ascher, William, and Harold Dean Lasswell, *Forecasting: An appraisal for policy-makers and planners* (Baltimore: Johns Hopkins University Press, 1979).

Autorenkollektiv, "Nur für Dienstgebrauch," *Richtlinie für die Bewirtschaftung* (Berlin: SKF [Staatliches Komitee für Forstwirtschaft], 1979).

Backmund, F., "Naturgemäße Waldwirtschaft—ein neues Schlagwort?" *Forstwissenschaftliches Centralblatt* 72 (1953): 144.

Badstubner, Rolf, "'Beratungen' bei J. W. Stalin. Neue Dokumente," in *Utopie kreativ* 7 (1991): 96–16.
Badstubner, Rolf, and Wilfried Loth, eds., *Wilhelm Pieck — Aufzeichnungen zur Deutschlandpolitik, 1945–1953*, trans. Stephen Connors (Berlin: Akademie Verlag, 1994).
Baring, Arnulf, and Volker Zastrow, *Unser neuer Größenwahn. Deutschland zwischen Ost und West*, 3d ed. (Stuttgart: Deutsche Verlags-Anstalt, 1989).
Barth, Roland, R. Kallweit, K. Schikora, and G. Schübel, *Der Wald im Land Brandenburg* (Potsdam: Forsteinrichtungsamtes, 1990).
Bartonek, Rolf, "Umweltkriminalität—in der DDR bislang ein Kavaliersdelikt," *Neues Deutschland* 101 (2 May 1990): 8.
Bath, Matthias, "Es wird weitergeschossen. Zur Normgenese fes 'Schießbefehls,'" *Deutschland Archiv* Analysen und Berichte 18:9 (September 1985): 959.
Bauerkamper, Arnd, "'Loyale Kader'? Neue Eliten und die SED-Gesellschaftspolitik auf dem Lande von 1945 bis zu den fruhen 1960er Jahren," *Archiv für Sozialgeschichte* 39 (1999): 265.
——, "Zwangsmodernisierung und Krisenzyklen: Die Bodenreform und Kollektivierung in Brandenburg 1945–1960/61," *Geschichte und Gesellschaft: Zeitschrift für Historische Sozialwissenschaft* 25:4 (1999): 556–88.
Baylis, Thomas Arthur, "Economic reform as ideology: East Germany's New Economic System," *Comparative Politics* 3:2 (January 1971).
——, *The technical intelligentsia and the East German elite* (Los Angeles: University of California Press, 1974).
Beer, Ferdinand, "Kahlschlag oder Einzelstammentnahmen," *Forst- und Holzwirtschaft* 3:14 (15 July 1949): 11.
——, "Naturgemäßer Waldwirtschaft," *Forst- und Holzwirtschaft* 3:5 (15 March 1951): 65.
——, *Naturverjüngung oder Kahlschlag?* (Berlin, 1947).
——, "Über einige Erfolge und Probleme bei der Aufforstungsarbeiten im Jahre 1949," *Forst- und Holzwirtschaft* 3:21 (1 November 1949): 329.
——, "Unser Aufforstungsplan 1949," *Forst- und Holzwirtschaft* 3:6 (5 March 1949): 1.
Behrens, Fritz, "Zum Problem der Ausnutzung ökonomische Gesetze in der Übergangsperiode," *Zur ökonomische Theorie and Politik in der Übergangsperiode*, 3 Sonderheft, Wirtschaftswissenchaft V (1957): 105–40.
Bell, Daniel, "The dispossessed—1962," *Columbia University Forum* 5:4 (fall 1962): 5–6.
Bemmann, A., "Entwicklung und Nutzung des Waldfonds in der D.D.R.," *Forstarchiv* 61:2 (March/April 1990): 57.
Bennewitz, Inge, and Rainer Potratz, "Forschungen zur DDR-Geschichte," *Zwangsaussiedlungen an der innerdeutschen Grenze: Analysen und Dokumente* (Berlin: Ch. Links, 1994), Bd. 4.
Berend, Ivan T., *The Hungarian economic reforms, 1953–1988* (Cambridge: Cambridge University Press, 1990).
Bergmann, J. H., and W. Flohr, "Zur Wirkung von Fremdstoffen in den Wäldern der DDR unter besonderer Berücksichtigung einer Veränderung der Bodenflora," *Sozialistische Forstwirtschaft* 38:6 (1988): 164–66.

Berlin, Isaiah, *Karl Marx: His life and environment,* 4th ed. (New York: Oxford University Press, 1996).

Beyse, R., "DDR-Bodenreform vor dem Bundesverfassungsgericht," *Holz-Zentralblatt* 117:20 (1991): 305.

Bieberstein, D., *Haupttendenzen und Probleme der Waldfondsentwicklung in der Deutsche Demokratische Republik. 1944–1950,* in "Forestry in East Germany: A special number on the 150th anniversary of the beginning of forestry research at Eberswalde" (in German) *Beiträge für die Forstwirtschaft* 15:3/4 (1981): 144–50.

Bieger, Wilhelm, "Umfang und Ursachen der thüringer Sturmschäden," *Forst- und Holzwirtschaft* 1:2 (15 April 1947).

Binsack, Rudolf, "Zeitläufte: enteignet, deportiert, geflohen: Ein Augenzeuge schildert das Unrecht der Bodenreform in der Sowjetzone," *Die Zeit* (19 March 1998): 3.

Bitschkowski, Horst, and Franz Krahn, "Wirksamer Beitrag zu der höherer Leistungskraft. Gerhard Grüneberg bei Genossen in Schwinkendorf," *Neues Deutschland* 258 (11 January 1980): 3.

Bittighöfer, B., H. Edeling, H. Kulow, "Theoretische und politisch-ideologische Fragen der Beziehung von Mensch und Umwelt," *Deutsche Zeitschrift für Philosophie* 1 (1972): 68.

Blanckmeister, Johannes, "Kurswechsel im Waldbau," *Sozialistische Forstwirtschaft* 12:1 (January 1962): 7.

——, "Wege und Irrwege des Waldbaus in der letzten 150 Jahre," in *Kolloquium anläßlich des 75 Geburtstag von Nationalpreisträger Prof (em.) Dr. ing. Habil. Johannes Blanckmeister* (Tharandt: Technische Universität Dresden, 1973), 12.

Blume, Friedrich, ed., "Mendelssohn, Felix," in *Die Musik in Geschichte und Gegenwart* (Kassel: Bärenreiter, 1961).

Borgeest, Bernhard, "Keine Ruhe nach dem Sturm: Die Übel heißen Fichte, Borkenkäfer und 'der Jagdirrsinn,'" *Die Zeit* 16 (13 April 1990): 81.

Bormann, F. Herbert, and Gene E. Likens, *Pattern and process in a forested ecosystem: Disturbance, development and the steady state based on the Hubbard Brook ecosystem study* (New York: Springer-Verlag, 1981).

Boulton, E. H. B., "The forests of Germany: What they can supply on reparations account for the U. K.," *Timber Trade Journal* 173 (1945).

Boyle, Nicholas, "The poetry of desire," in *Goethe: The poet and the age,* vol. 1 (Oxford: Clarendon Press, 2000).

——, "Revolution and renunciation," in *Goethe: The poet and the age,* vol. 2 (Oxford: Clarendon Press, 2000).

Boys, R. J., D. P. Forster, and P. Jozan, "Mortality from causes amenable and non-amenable to medical care: The experience of eastern Europe," *British Medical Journal* 303:6807 (1991): 879–83.

Bramwell, Anna, *Ecology in the 20th century: A history* (New Haven: Yale University Press, 1989), 34.

Brandt, Willy, "Man darf die Grüne nicht in Schönheit lassen," *Die Welt* (5 February 1987).

Bröll, Werner, "Das sozialistische Wirtschaftssystem," in "*DDR,*" ed. Heinz Rausch and Theo Stammen (München: Verlag C. H. Beck, 1978).

Brüning, Eberhard and Hannes Mayer, *Waldbauliche Terminologie. Fachwörter der forstlichen Produktion* IUFRO-Gruppe Ökosyteme (Vienna: Universität für Bodenkultur, 1980).

Bryson, Phillip J., and Manfred Melzer, *The end of the East German economy: From Honecker to reunification* (New York: Macmillan, 1991).

Brzezinski, Zbigniew, "A common house, a common home," *New York Times* (15 November 1989), A–29.

Bundesminister für innerdeutsche Beziehungen, ed., *Informationen* 24:4 (1988): 4.

Bundesministerium für Ernährung, Landwirtschaft und Forsten, "Bundeswaldinventur 1987: Übersichtstabellen," Entwurf auf den Ergebnissen vom 23.9.90 basierend, *Schriftenreihe des Bundesministerium für Ernährung, Landwirtschaft und Forsten* (1987).

Bundesministerium für gesamtdeutsche Fragen (BGF), *SBZ von A-Z* (Bonn: Deutscher Bundesverlag, 1953).

——, *SBZ von A-Z,* 3rd ed. (Bonn: Deutscher Bundesverlag, 1956).

——, *SBZ von A-Z,* 4th ed. (Bonn: Deutscher Bundesverlag, 1958).

——, *SBZ von A-Z,* 11th ed. (Bonn: Deutscher Bundesverlag, 1969), 760.

Burke, Edmund, "Letter to a Noble Lord" (1796), repr. in *Works,* vol. 5 (London: G. Bell, 1884–1893).

——. *Reflections on the revolution in France* (1790), The Harvard Classics, ed. Charles W. Eliot (New York: P. F. Collier, 1909–14).

Burschel, Peter, "Karl Gayer und der Mischwald," *Allgemeine Forst- und Jagdzeitung* 23 (1987).

Carsten, Francis Ludwig, *A history of the Prussian Junkers* (Aldershot, Hants, Eng.: Gower, 1989).

——, *The origins of Prussia* (Westport, Conn: Greenwood Press, 1981).

——, "Origins of the Junkers," in *Essays in German history* (London: Hambleton Press, 1983).

Carter, Jimmy, "Speech to the nation" (15 July 1979), http://www.pbs.org/wgbh/amex/carter/filmmore/ps_crisis.html (accessed 28 February 2005).

——, "Town meeting remarks, Bardstown, Kentucky" (31 July 1979) *Public papers of the presidents of the United States: Jimmy Carter,* book 2 (Washington, D.C.: Superintendent of Documents, 1979), 1340.

Cervantes, Miguel de Saavedra, *The adventures of Don Quixote de la Mancha,* adapted from the Peter Motteux translation by Leighton Barret (New York: Alfred A. Knopf, 1939).

Chen, Lincoln C., Friederike Wittgenstein, and Elizabeth McKeon, "The upsurge of mortality in Russia: Causes and policy implications" (in Notes and Commentary), *Population and Development Review* 22:3 (September 1996): 517.

Childs, David, "Marxism-Leninism in the German Democratic Republic: The Socialist Unity Party (Party)," review of Martin McCauley's *Marxism-Leninism in the German Democratic Republic: the Socialist Unity Party* (London: Macmillan, 1979) *Soviet Studies* (April 1981): 317.

Childs, David, ed., *Honecker's Germany* (London: Allen & Unwin, 1985).

Childs, David, Thomas A. Baylis, and Marilyn Rueschemeyer, eds., *East Germany in comparative perspective* (London: Routledge, 1989).
Clark, Delbert, "Critic of U.S. leads for German post: Baumgartner of Bavaria, who assailed military regime, may be 2-zone farm chief," *New York Times* (13 July 1947).
Cline, A. C., "A brief view of forest conditions in Europe," *Journal of Forestry* 43 (1945).
Cloos, H., "Ein Blockbild von Deutschland," *Geologische Rundschau* 44:480 (1955).
Collier, Irwin L., Jr., and David H. Papell, "About two marks: The East/West mark exchange rate before the Berlin Wall," *American Economic Review* 78:3 (June 1988): 531–42.
Conference on Security and Co-operation in Europe, *Final Act* Helsinki 1975, section 5, "Environment." Text at Hellenic Resources Network (HR-Net) at http://www.hri.org/docs/Helsinki75.html (accessed 21 September 2003).
Conquest, Robert, "Correspondence," *Soviet Studies* 21:2 (October 1969): 264.
——, "Excess camp deaths and camp numbers: Some comments," *Soviet Studies* 43:5 (1991): 949.
——, *The Great Terror* (New York: Macmillan, 1968).
——, *The harvest of sorrow: Soviet collectivization and the Terror-famine* (New York: Oxford University Press, 1986).
——, "Victims of Stalinism: A comment," *Europe-Asia Studies* 49:7 (November 1997): 1317.
Conquest, Robert, Stephen Cohen, and Stephen G. Wheatcroft, "New demographic evidence on excess collectivization deaths: Further comments on Wheatcroft, Rosefielde, Anderson and Silver," *Slavic Review* 45:2 (summer 1986): 295.
Cornelson, Doris, "Umweltprobleme und Umweltbewußtsein in der DDR," *Gegenwartskunde* 1 (1989): 48.
Cornelson, Doris, et al., *Handbuch DDR-Wirtschaft*, trans. Lux Furtmüller, 4th ed. *Deutsches Institut für Wirtschaftforschung* (DIW) (Farnborough: Saxon House, 1984).
Cornides, Wilhelm, *Wirtschaftsstatistik der deutschen Besatzungszonen, 1945–1948 in Verbindung mit der deutschen Produktionsstatistik der Vorkriegszeit* (Oberursel [Taunus]: Europa-Archiv, 1948).
Cotta, Heinrich, *Anweisung zum Waldbau* (Dresden and Leipzig, 1916).
Courtois, Stephane, *Livre noir du communisme, The black book of communism: Crimes, terror, repression*, trans. Jonathan Murphy, Mark Kramer consulting ed. (Cambridge: Harvard University Press, 1999).
Crafts, N. F. R., "The golden age of economic growth in Western Europe, 1950–1973," *Economic History Review* 48:3 (1995): 429.
Creuzberger, Stefan, "Abschirmungspolitik gegenüber dem westlichen Deutschland im Jahre 1952," in *Die sowjetische Deutschland-Politik in der Ära Adenauer,* ed. Gerhard Wettig (Bonn, 1997), 12–36.
Croan, Melvin, "Soviet uses of the doctrine of the 'parliamentary road' to socialism: East Germany, 1945–1946," *American Slavic and East European Review* 17:3 (October 1958): 302.
Cronon, William, "Modes of prophesy and production: Placing nature in history," *Journal of American History* 76:4 (March 1990): 11220.

——, "A place for stories," *Journal of American History* 78:4 (March 1992): 1347.

——, "The trouble with wilderness; or, getting back to the wrong nature," in *Uncommon ground: Rethinking the human place in nature,* ed. William Cronon (New York: W. W. Norton, 2000), 69.

Damaschke, Adolf, *Die Bodenreform,* 19th ed. (Jena: Gustav Fischer, 1922).

Dannecker, Karl, "Beispielbetriebe der Praxis unter Kritik der Wissenschaft," *Forstwissenschaftliches Centralblatt* 69 (1950): 744–64.

——, "Um das Plenterprinzip in Sudwestdeutschland," *Schweizerische Zeitschrift für Forstwesen* 100:9 (1949): 414–29.

——, "Vom Naturwald zum naturgemäßsen Wirtschaftswald," *Allgemeine Forst- und Jagdzeitung* 121 (1950): 73.

*Darmstädter Echo,* "Report on forestry in eastern Germany," *Darmstädter Echo* (20 January 1949).

Davies, Norman, "Europe's forgotten war crime," *The Sunday Times* (7 April 2002).

DeBaredeleben, Joan, Review M. Feshbach's *Ecocide in the USSR: Health and nature under siege,* M. Turnbull's *Soviet environmental policies and practices: The most critical investment* and Philip R. Pryde's *Environmental management in the Soviet Union, Slavic Review* 52:3 (fall 1993): 593.

Demokratische Aufbau, "Mecklenburg baut für die Neubauern," *Demokratische Aufbau* 1 (April 1946): 28.

Dengler, Alfred, "Zum Jahresschluß," *Zeitschrift für Forst- und Jagdwesen* 72:12 (1940).

Dengler, Alfred, Ernst Röhrig, and A. Gussone, *Waldbau auf ökologischer Grundlage: Baumartenwahl, Bestandesbegründung und Bestandespflege* Bd. 2 (Ulmer [Eugen], 1990).

Dengler, Alfred, Ernst Röhrig, and Norbert Bartsch, *Waldbau auf ökologischer Grundlage: Der Wald als Vegetationsform und seine Bedeutung für den Menschen* Bd.1 (Ulmer [Eugen], 1992).

Dennis, Mike, *German Democratic Republic: Politics, economics and society* (London: Pinter Publishers, 1988).

Department of Propaganda-Agitation of the East German Communist Party, "Einzelbauer Arnold und sein Verhältnis zum Sozialismus. Ein Wort an die Einzelbauern—vor allem an jene, die es bleiben wollen," Department of Propaganda-Agitation of the East German Communist Party, District Office Karl-Marx-Stadt (Bezirk Karl-Marx-Stadt Department for Propaganda-Agitation, 1960) "German Propaganda Archive" trans. Randall Bytwerk, http://www.calvin.edu/academic/cas/gpa/arnold.htm (accessed 11 July 2003).

*Der Deutsche Forstwirt,* "Durchführung kriegswirtschaftlicher Maßnahmen in der Forst- und Holzwirtschaft," *Der Deutsche Forstwirt* 22:69/70 (1939).

*Der Tagesspiegel,* "Sorgen um den Wald in der D.D.R.," *Der Tagesspiegel* 11:408 (6 April 1983): 13.

Deutschen Institut für Zeitgeschichte, *Handbuch der Deutschen Demokratischen Republik* (Leipzig: Staatsverlag der Deutschen Demokratischen Republik, 1963).

——, *Jahrbuch der Deutschen Demokratischen Republik* (Berlin: Verlag der Wirtschaft, 1956, 1957, and 1959).

Deutsches Institut für Wirtschaftsforschung (DIW), "SBZ 1953 inventory" *(DIW)-Wochenbericht* (1954): 96.

Dickens, Arthur Geoffrey, "Letter to the editor," *The Times* (9 April 1947), in response to A. Sudachkov's 5 April 1947 "Letter to the editor," *The Times* (5 April 1947).

*Die Zeit,* "Special section, letters to the editor in response to 'Noch zwanzig Jahre deutscher Wald?'" *Die Zeit* 6 (4 February 1983): 11.

Dittmar, O., "Zur Geschichte des Forstlichen Versuchswesens des Institutes für Forstwissenschaften Eberswalde," in *Ehrenkolloquim anläßlich des 150. Geburtstag von Professor Dr. phil. Max Friedrich Kunze am 12. Oktober 1988*, ed. H. Kurth (Tharandt: Technischsen Universität Dresden, 1988), 12.

Duschek, S., "Wirtschaftspolitische Betrachtungen des deutschen Großgrundbesitzes," *Zeitschrift für Weltforstwirtschaft* 2 (1935): 477.

Eberts, H., "Forstwirtschaft in Ost und West," *Holz-Zentralblatt* (1949): 942.

Eckart, Karl, "Veränderungen in der Landwirtschaft der DDR seit Anfang der siebziger Jahre," *Deutschland Archiv* 18:4 (1985): 396–411.

Economic Commission for Europe, *Economic survey of Europe in 1954* (Geneva, 1955), 49, cited in Joint Session of the Economic Reports, "Trends in economic growth, a comparison of the Western Powers and the Soviet bloc" (Washington, D.C., 1955), 292.

Economist Intelligence Unit (Great Britain), *EIU country report, East Germany,* 4 (London: Economist Intelligence Unit, 1988), 14.

——, *EIU country report, East Germany* (London: Economist Intelligence Unit, 1989).

Edinger, Louis Joachim, *West German politics* (New York: Columbia University Press, 1986), 66.

Elkins, Thomas Henry, *Germany, an introductory geography*, rev. ed. (New York: Praeger, 1968).

Emerson, Ralph Waldo, "The conservative," in *Nature, addresses, and lectures* (Boston: Houghton Mifflin, 1903), 295–96.

——, "Goethe, or the writer," in *Representative men: Seven lectures* (Cambridge: Belknap Press of Harvard University Press, 1987), 161.

Eneyedi, Gyorgy, August J. Gijawijt, and Barbara Rhode, eds., *Environmental policies in East and West* (London: Taylor Graham, 1987).

Enzensberger, Hans Magnus, "The state of Europe," *Granta* 30 (winter 1990): 136.

Faber, Dorothea, "Entwicklung und Lage der Wohnungswirtschaft in der sowjetischen Besatzungszone 1945–1953," *Wirtschaftsarchiv* 8:17 (5 September 1953): 5943.

Fahner, B. G. Weiß, F. Ullmann, and G. Ervert, "Parteitagsdelegierte aus der Forstwirtschaft berichten vom 10. Parteitag der Party: Unter der Führung der Party weiter voran auf den bewährten Kurs der Hauptaufgabe, der Einheit von Wirtschafts- und Sozialpolitik," *Sozialistische Forstwirtschaft* 31:6 (1981): 161.

Fechner, Max, "Klarheit in der Ostfrage!" *Neues Deutschland* (4 September 1946): 1.

Felfe, Werner, "Alles mit den Menschen — Alles für die Menschen. Ausgewählte Reden und Aufsätze," Speech welcoming the Cuban delegation headed by Fidel Castro, 16 June 1972 at the Leuna Chemical Works (Berlin: Dietz Verlag, 1987).

——, "Aus dem Bericht des Politbüros an der 5. Tagung des Zentralkomitees der Party," *Neues Deutschland* (17 December 1987): 6.

Feshbach, Murray, *Ecological disaster: Cleaning up the hidden legacy of the Soviet regime* (New York: Twentieth Century Fund Press, 1995).
——, *Economics of health and environment in the USSR* (Washington, D.C.: Office of Net Assessment, U.S. Department of Defense, forthcoming).
——, "Russia's population meltdown," *Wilson Quarterly* 25:1 (winter 2001).
Feshbach, Murray, ed., *Environmental and health atlas of Russia* (Moscow: PAIMS Publishing House, 1995).
Feshbach, Murray, and Alfred Friendly, Jr., *Ecocide in the USSR: Health and nature under siege* (New York: Basic Books, 1992).
Fischer, Alexander, ed., *Ploetz, die Deutsche Demokratische Republik: Daten, Fakten, Analysen* (Freiburg: Ploetz, 1988), 36.
Fischer, F., ed., *Industriemäßige Produktionsmethoden der Rohholzbereitstellung aus Fichtenvornutzung. Wissenschaftliche Tagung 13. bis 25. Oktober 1974* (Freiberg: Zentralen Druckerei der Bergakademie Freiberg, 1974).
Fischer, George, ed., *Science and ideology in Soviet society* (New York, 1967).
Fisher, Andrew, "Bundesbank warns of tax rise to prop up east German economy," *Financial Times* (19 March 1992), 1.
Fisher, Duncan, "The emergence of the environmental movement in eastern Europe and its role in the revolution of 1989," in *Environmental action in Eastern Europe*, ed. Barbara Jancar-Webster (Armonk, N.Y.: M. E. Sharpe, 1993), 96.
FitzGerald, D. A., "Abstract of the FitzGerald Report alleging short-sighted policies in feeding the world," *New York Times* (28 October 1947).
Flemming, Thomas, *Die Berliner Mauer: Geschichte eines politischen Bauwerks*. Dokumentation Berliner Mauer-Archiv, Hagen Koch (Berlin: Bebra Verlag, 1999).
Food and Agriculture Organization (FAO) of the United Nations, *Forestry and forest products: World situation, 1937–1946* (Stockholm: Stockholms Bokindustri Aktiebolag, 1946).
Forschungsanstalt für Holz- und Forstwirtschaft, "Bericht über den Gesundheitszustand der Wälder der ostdeutscher Bundesländer (DDR) der Bundesrepublik Deutschland," typewritten (Eberswalde-Finow: IFE, 1991).
*Forst- und Holzwirtschaft,* "Die Forstwirtschaft in der Deutschen Demokratischen Republik am Vorabend des III Parteitages der SED," *Forst- und Holzwirtschaft* 4:15 (1 August 1950): 225.
Forstprojektierung Potsdam, Working paper, Archives, handwritten inventory, undated, said to be 1949.
Frazer, Sir James George, *The golden bough; A study in magic and religion*, 3rd ed. (New York: Macmillan, 1935).
Friedrich, Carl J., and Henry Kissinger, eds., *The Soviet zone of Germany* (New Haven: Bechtle, 1956).
Friedrich, Gerd, et al., *Die Volkswirtschaft der DDR,* Akademie für Gesellschaftswissenschaften beim ZK der SED (Berlin: Verlag die Wirtschaft, 1979).
Frowen, Stephen F., "The economy of the German Democratic Republic," in David Childs, *Honecker's Germany* (London: Allen & Unwin, 1985).
Früchtenicht, "Leistungssteigerung im Walde," *Der Deutsche Forstwirt* 22:69/70 (1940).

Fulbrook, Mary, *Anatomy of a dictatorship: Inside the GDR, 1949–989* (London: Oxford University Press, 1995).
Füllenbach, Josef, *European environmental policy: East and West,* trans. Frank Carter and John Manton (London: Butterworths, 1981), 85.
Garton Ash, Timothy, "Big brother isn't watching anymore," *Guardian* (13 March 1999), 1.
——, "East Germany: The solution," *New York Review of Books* 37:7 (26 April 1990): 14.
——, "Ten years after," *New York Review of Books* (18 November 1999): 16.
Garver, John B., Jr., "The military geography of East Europe," in *East Europe: The impact of geographic forces in a strategic region* (Washington, D.C.: Directorate of Intelligence, 1991), 61.
Gessner, Dieter, *Agrarverbände in der Weimarer Republik: wirtschaftliche und soziale Voraussetzungen agrarkonservativer Politik vor 1933* (Düsseldorf: Droste, 1976).
Gleitze, Bruno, "Zielsetzung und Mittel der sowjetzonalen Wirtschaftspolitik bis zur gegenwärtigen Krisensituation." Speech to the working group of German Economists' Research Institute in SOPADE, "Die sowjetzonale Wirtschaftspolitik," *Querschnitt durch Politik und Wirtschaft* #937 (September 1953), 41.
Godzik, Stefan, and Jadwiga Sienkiewicz, "Air pollution and forest health in Central Europe: Poland, Czechoslovakia, and the German Democratic Republic," in Wladyslaw Grodzinski, Ellis B. Cowling, and Alicja I. Breymeyer, eds., *Ecological risks: Perspectives from Poland and the United States,* Polish Academy of Sciences and National Academy of Sciences of the United States of America (Washington, D.C.: National Academy Press, 1990).
Goethe, Johann Wolfgang von, *Faust,* pt. 1, trans. Randall Jarrell (New York: Farrar, Straus and Giroux, 2000), 235–39.
——, letter to Charlotte von Stein, 10 December 1777, in *Briefe und Tagebücher,* vol. 1 (Leipzig: Im Insel-Verlag, 1957), #159.
——, letter to Clausthal, 11 December 1777, in *Briefe und Tagebücher,* vol. 1 (Leipzig: Im Insel-Verlag, 1957), #227.
——, *Tagebücher*, ed. Herbert Nettl (Düsseldorf-Köln: Eugen Diedrichs Verlag, 1957).
Goldenbaum, Ernst, "Demokratische Bodenreform hat unseren Bauern eine gesicherte Zukunft eröffnet," *Neues Deutschland* 204 (28 August 1975): 3.
Golley, Frank Benjamin, *A history of the ecosystem concept in ecology: More than the sum of the parts* (New Haven: Yale University Press, 1993).
Götz, Hans Herbert, "Als der Klassenkampf in der DDR begann. Die Bodenreform vor 40 Jahren," *Frankfurter Allgemeine Zeitung* 206 (6 September 1985): 13.
——, "Eine Landwirtschaft mit Mammut-Betrieben. 30 Jahre Agrarpolitik in der D.D.R.," *Frankfurter Allgemeine Zeitung* 269 (17 November 1979): 15.
Grieder, Peter, *The East German leadership, 1946–73: Conflict and crisis* (Manchester: Manchester University Press, 1999).
——, "The overthrow of Ulbricht in East Germany," *Debatte* 6:1 (1998): 11.
Grill, Bartolomäus, "Deutschland — ein Waldesmärchen. Den Hain im Hirn, den Forst im Volke: Eine Geschichte alter und neurer Mythen," *Die Zeit* 53 (25 December 1987): 3.

Grossman, H., "Forest inventories as a basis for planning, appraisal of plan fulfillment, and permanent verification of the condition of the forest" (in German), *Archiv für Forstwesen* 18:2 (1969): 211–33.

——, "Present position and possibilities in the information supplied by continuous large-scale inventory" (in German), *Sozialistische Forstwirtschaft* 13:2 (1963): 43–45.

——, "Toward improving the growing stock and increment inventory of the German Democratic Republic" (in German), *Sozialistische Forstwirtschaft* 14:6 (1964): 174–75.

——, "Zehn Jahre permanente Grossrauminventur in der D.D.R.," *Sozialistische Forstwirtschaft* 22:3 (March 1972): 74–76.

Grote, Manfred, "The Party under Honecker," *East European Quarterly* 21:1 (1987): 67–78.

Grüneberg, Gerhard, *Auf sozialistische Art leiten, arbeiten und leben* (Berlin: Dietz, 1959).

——, "30 Jahre Marxistisch-Leninistische Agrarpolitik — 30 Jahre Bündnis der Arbeiterklasse mit den Bauern," *Neues Deutschland* 188 (9 August 1975): 3.

——, *Zu einigen Fragen der Agrarpolitik der SED* (Leipzig 1975).

Guratzsch, Dankwart, "Der Schweigebann bricht," *Die Welt* (20 February 1986).

Haaf, Günther, "Noch zwanzig Jahre deutscher Wald?" *Die Zeit* 2 (7 January 1983): 1.

Haden-Guest, Stephen, John Wright, and Wileen M. Teclaff, *A world geography of forest resources* (New York: Ronald Press, 1956).

HA Forstwirtschaft des Ministeriums für Land-, Forst- und Nahrungsgüterswirtschaft, "Statistisches Übersichtsmaterial 1990."

Hager, Kurt, "Die entwickelte sozialistische Gesellschaft," *Einheit: Zeitschrift für Theorie und Praxis des Wissenschaftlichen Sozialismus* 11 (1971): 1214.

Hämmerle, "Das Osthilfegesetz und seine Auswirkungen auf die Forstwirtschaft," *Der Deutsche Forstwirt* 14 (1932): 118.

Hansrath, "Was geschiet mit den Revierförster?" *Allgemeine Forstzeitschrift* 3:24 (15 December 1948): 261.

Hardach, Karl Willy, *The political economy of Germany in the twentieth century,* trans. by author (Berkeley: University of California Press, 1980).

Harmssen, Gustav W., *Am Abend der Demontage: Sechs Jahre Reparationspolitik mit Dokumentenanhang*, the Bremer Ausschuss für Wirtschaftsforschung (Bremen: F. Trüjen, 1951).

——, *Reparationen, Sozialprodukt, Lebensstandard: Versuch einer Wirtschaftsbilanz* (Bremen: F. Trüjen, 1948).

Harris, Chauncy D., and Gabriele Wulker, "The refugee problem of Germany," *Economic Geography* 29:1 (January 1953): 25.

Harrison, Hope Millard, *Driving the Soviets up the wall: Soviet-East German relations, 1953–1961* (Princeton: Princeton University Press, 2003).

——, "New evidence on Khrushchev's 1958 Berlin ultimatum," *Cold War International History Project Bulletin* #4 "Soviet nuclear history" (Washington, D.C.: Woodrow Wilson International Center for Scholars, 1994).

——, "Ulbricht and the concrete rose: New archival evidence on the dynamics of Soviet-

East German relations and the Berlin crisis, 1958–1961," *Cold War International History Project Working Paper* #5 (Washington, D.C.: Woodrow Wilson International Center for Scholars, 1993).

Hasel, Karl, "Die Beziehung zwischen Land- und Forstwirtschaft in der Sicht des Historikers," *Zeitschrift für Agrargeschichte und Agrarsoziologie* 16:2 (October 1968).

——, "Forstbeamte im NS-Staat am Beispiel des ehemaligen Landes Baden," *Schriftenreihe, Landesforstverwaltung und Forstwirtschaft Baden-Württemberg,* #62 (1985).

——, *Forstgeschichte. Ein Grundriß für Studium und Praxis.* Pareys Studientexte 48 (Hamburg: Verlag Paul Parey, 1985).

——, "Wilhelm Pfeil und die Revolution von 1848," *Allgemeine Forst- und Jagdzeitung* 148:5 (1977).

Hassel, Wolfgang, "'Junkerland in Bauernhand' war damals die Kampflosung: Dokumente des Staatsarchivs Magdeburg über die Bodenreform," *Neues Deutschland* 255 (27 October 1984): 13.

Hatzfeldt, Hermann, "Der Baum," *Die Zeit* 39 (21 September 1984).

Haushofer, Heinz, *Ideengeschichte der Agrarwirtschaft und Agrarpolitik im deutschen Sprachgebiet,* vol. 2 (Munich, 1985), 107.

Heger, A., "Aufbau und Leistung von naturnahen Wäldern im Osten und ihre forstwirtschaftliche Behandlung," *Forstwissenschaftliches Centralblatt* 1 (1944): 34.

Heidrich, H., "Die Aufgaben des Betriebsleiters bei der Lösung der ökonomischen Beiträge," *Forst und Jagd* (September 1960): 2.

——, "Die nächsten Aufgaben bei der Entwicklung und Festigung der sozialistischen Forstwirtschaft in der Deutschen Demokratischen Republik nach dem VII Parteitag der SED," *Sozialistische Forstwirtschaft* 17:9 (September 1967): 273.

——, "Die sozialistischer Forstwirtschaft der DDR im 25. Jahr ihrer Entwicklung und ihre weiteren Aufgaben bei der Erfüllung der Beschlüße des VII. Parteitages der Party," in *Industriemäßige Produktionsmethoden der Rohholzbereitstellung aus Fichtenvornutzung. Wissenschaftliche Tagung 13. bis 25. Oktober 1974,* ed. F. Fischer (Freiberg: Zentralen Druckerei der Bergakademie Freiberg, 1974), 14–15.

——, "Schlußwort," in *Die Anwendung des Neuen Ökonomischen Systems im Bereich der VVB Forstwirtschaft Potsdam,* ed. R. Rüthnick (Potsdam-Babelsberg: VVB Forstwirtschaft Potsdam, 1966), 172.

Heine, Heinrich, *Harzreise* (English and German), trans. Charles G. Leland (New York: Marsilio, 1995).

Heitmann, Friedrich, "International Sammlung von Forststatistiken," *Zeitschrift für Weltforstwirtschaft* 10:7/10 (1944): 473.

Helmholz, Friedrich-Karl, and Adolf Sturzbecher, "Wie bei uns Abgeordnete für die Gesunderhaltung des Waldes sorgen," *Neues Deutschland* 116 (19 May 1989): 3.

Heuer, Uwe-Jens, *Demokratie und Recht im Neuen Ökonomischen System der Planung und Leitung der Volkswirtschaft* (Berlin, 1965).

Hilbert, Anton, "Denkschrift über die ostdeutschen Bodenreform," Gräflich Douglas'-sches Archiv Schloß Langenstein (1946) (Anm. 18), cited in H.-G. Merz, "Bodenreform in der SBZ. Ein Bericht aus dem Jahre 1946," *Deutschland Archiv* 11:24 (1991): 1166.

Hildebrandt, G., "Die Forsteinrichtung in der DDR 1950 bis 1965. Ein Beitrag zur jüngeren deutschen Forsteinrichtungsgeschichte aus Anlaß des 80. Geburtstages von Albert Richter in Eberswalde," *Allgemeine Forst- und Jagdzeitung* 160:6 (1989): 123.

Hildebrandt, Rainer, "Die Todesmühlen der SS übertroffen," *Querschnitt durch Politik und Wirtschaft* (July 1949).

Hilf, Hubert Hugo, "Forstwirtschaft zwischen Gestern und Morgen," *Forstarchiv* 31:3 (15 March 1960).

Hinze, Albrecht, "Exportierte Schäden stören die sozialistische Eintracht. Über die Umwelt in der DDR gibt es nur dürre Nachrichten. 'Menschen nicht verunsichern,'" *Süddeutsche Zeitung* (31 December 1986): 1.

Hoffmann, Dieter, and Kristie Macrakis, eds., *Naturwissenschaft und Technik in der DDR* (Berlin: Akademie Verlag, 1997).

Hofmann, Gerhard, "Vergleich der potentiell-natürlichen und der aktuellen Baumartenanteile auf der Waldfläche der D.D.R.," *Hercynia NF* 24 (1987).

Hofmann, Gerhard, and D. Heinsdorf, "Depositionsgeschen und Waldbewirtschaftung," *Der Wald* 40:7 (1990): 208.

Hohenthal, Carl Graf, "Die Umwelt-Last der D.D.R.," *Frankfurter Allgemeine Zeitung* 18 (22 January 1990): 12.

Höhmann, Hans-Hermann, and Gertraud Seidenstecher, eds., *Umweltschutz und ökonomisches System in Oste: Drei Beispiele Sowjetunion, DDR, Ungarn* (Stuttgart: Kohlhammer, 1973).

Holloway, David, "The politics of catastrophe," review of Murray Feshbach's *Ecocide in the USSR*, Grigori Medvedev's *The Truth about Chernobyl* and *No breathing room: The aftermath of Chernobyl*, and Piers Paul Read's *Ablaze: The story of the heroes and victims of Chernobyl, New York Review of Books* 40:11 (16 October 1993): 36.

Honecker, Erich, "Bündnis war, ist und bleibt Eckpfeiler unserer Politik," *Neues Deutschland* (6–7 September 1975).

——, *Die Aufgaben der Partei bei der weiteren Verwirklichung der Beschlüße des IX Parteitages der Partei* (Berlin: Dietz Verlag, 1978).

——, "Fragen von Wissenschaft und Politik," *Einheit: Zeitschrift für Theorie und Praxis des Wissenschaftlichen Sozialismus* 1 (1972): 8.

——, "Gespräch über aktuelle Fragen der Friedenssicherung und Umweltpolitik. Eric Honecker empfing Abordnung des Bundesverbandes Bürgerinitiativen Umweltschutz aus der BRD. Erhalt natürlicher Lebensbedingungen erfordert die Verhinderung eines atomaren Krieges. DDR tut alles, um die Gefahr zu bannen, die durch USA-Erstschlagswaffen in Westeuropa, besonders in der BRD, entstanden ist. Leistungen der DDR beim Umweltschutz gewürdigt," *Neues Deutschland* (6 September 1984): 1.

——, "Program of the Party's Tenth Party Congress" (in German) *Neues Deutschland* (12 April 1981): 3.

Honecker, Erich, and Irma Brandt, "Arbeiter und Bauern schreiben neues Kapitel der Geschichte," *Neues Deutschland* 312 (6 September 1975): 1–4.

Hoover, Calvin B., "The future of the German economy," *American Economic Review* 36:2 (May 1946): 642.

Hoover, Herbert, "Text of Hoover mission's findings on the food requirements of Germany," *New York Times* (28 February 1947).

Hörnle, Edwin, *Volksstimme* (Chemnitz) (3 July 1947).
——, "Wie kann die deutschen Landwirtschaft ihre Aufgabe erfüllen?" *Neues Deutschland* (16 May 1946).
Hörz, Herbert, "Die Wirksamkeit der ideologischen Arbeit erhöhen! Diskussion auf Einladung der Einheit," *Einheit: Zeitschrift für Theorie und Praxis des Wissenschaftlichen Sozialismus* 1 (1972): 21.
Hueck, Kurt, "Aktuelle Aufgaben der Forstwirtschaft," speech by the dean of the Forstfakultät Eberswald at the Agricultural Science Congress, Berlin, 4 February 1947, *Forst- und Holzwirtschaft* 1:1 (1 April 1947): 6.
Hyde-Price, Adrian, *European security beyond the cold war* (London: Sage Publications for the Royal Institute for International Affairs, 1991), 14.
Iklé, Fred Charles, *Every war must end,* 2nd ed. (New York: Columbia University Press, 2005).
Immler, Hans, *Agrarpolitik in der DDR* (Köln: Verlag Wissenschaft und Politik, 1971).
Ivanov, Mikhail, "Brzezinski welcomed to Soviet Diplomatic Academy," *TASS* (1 November 1989).
Jacobson, Dan, "The invention of Orwell," review of Orwell's *The Complete Works,* edited by Peter Davidson (London: Secker and Warburg, 1998) *The Times Literary Supplement* 4977 (21 August 1998), 4.
Jancar-Webster, Barbara, ed., *Environmental action in Eastern Europe* (Armonk, N.Y.: M. E. Sharpe, 1993).
Jeffries, Ian, and Manfred Melzer, "The New Economic System of planning and management 1963–1970 and recentralization in the 1970s," in Jeffries and Melzer, eds., *East German economy,* trans. Eleonore Breuning and Ian Jeffries (London: Croom Helm, 1987), 26.
Johnson, Nevil, Review of Mary Fulbrook's *Anatomy of a dictatorship: Inside the GDR, 1949–1989* (Oxford: Oxford University Press, 1995), *English History Review* 112: 448 (September 1997): 1030–31.
Jones, Merrill E., "Origins of the East German environmental movement," *German Studies Review* (1993).
Kamm, Henry, "A riddle for Communists: Why does the East German economy prosper?" *New York Times* (13 March 1989), A–3.
Kandler, O., "Epidemiologische Bewertung der Waldschädenserhebungen 1983 bis 1987 in der Bundesrepublik Deutschland," *Allgemeine Forst- und Jagdzeitung* 159:9/10 (1988): 179–94.
Karcz, Jerzy F., "The new Soviet agricultural programme," *Soviet Studies* 17:2 (October 1965).
Karlsch, Rainer, "Ein Staat im Staate: Der Uranbergbau der Wismut AG in Sachsen und Thüringen," *Aus Politik und Zeitgeschichte* B 49–50 (1993): 14–22.
Karlsch, Rainer, and Harm Schröter, eds., "Strahlende Vergangenheit," *Studien zur Geschichte des Uranbergbaus der Wismut* (St. Katharinen: Scripta Mercaturae Verag, 1996).
Karutz, Hans-R., "Beteiligt sich Bonn an den Kosten, dann zeigt D.D.R.," *Die Welt* 221 (20 September 1984): 4.

Kaser, Michael Charles, and Edward Albert Radice, eds., *The economic history of Eastern Europe, 1919–1975* (Oxford: Oxford University Press, 1985–86).

Kautsky, Karl, *Die Agrarfrage; eine Übersicht über die Tendenzen des modernen Landwirtschaft und die Agrarpolitik der Sozialdemokratie* (Stuttgart: J. H. W. Dietz Nachf GmbH, 1899).

Keller, Bill, "Clamor in the East: The Kremlin; Making policy of the inevitable, Gorbachev accepts wide change," *New York Times* (12 November 1989), 1.

Kempe, S., and Rudolf Rüffler, "Analyse aus gewahlter Rahmenbedingungen beim Übergang zur eigenverantwortlichen Bewirtschaftung nichtstaatlichen Waldeigentums im Land Brandenburg," *Berichte aus Forschung und Entwicklung* 23:49 (Eberswalde: Forschungsanstalt für Forst- und Holz, 1991).

Kennan, George F., *Democracy and the student left* (New York, 1968), 9–10.

Keren, Michael, "The New Economic System in the GDR: An obituary," *Soviet Studies* 24:4 (April 1973): 556.

Keudell, Walter von, *34 Jahre Hohenlübbichower Waldwirtschaft* (Neudamm, 1936).

Kiechele, Ignaz, "Bundeswaldinventur," *Forstarchiv* 61:5 (September–October 1990): 204–5.

Kienitz, Erwin, *Denkschrift über forstwirtschaftlichsorganisatorischen Reformen, insbesondere des Bauernwäldes der Deutschen Demokratischen Republik: Ein Beitrag zur sozialistischen Umgestaltung der Forstwirtschaft* (Tharandt: Institut für Forstliche Wirtschaftslehre, 1958).

Kindleberger, Charles P., *The German economy, 1945–1947: Charles P. Kindleberger's letters from the field* (Westport, Conn.: Meckler, 1989).

Kircher, Joseph C., "The forests of the U.S. Zone of Germany," *Journal of Forestry* 45:4 (1947): 249–52.

Kirkpatrick, Meredith, *Environmental problems and policies in East Europe and the USSR* (Monticello, Ill.: Council of Planning Librarians, 1978).

Kitchingnam, G. D., "The 1945 census of woodlands in the British zone of Germany," *Empire Forestry Review* 26:2 (1947): 224–27.

Klatt, Werner, "Food and farming in Germany: I. Food and nutrition," *International Affairs* 26:1 (January 1950): 45.

——, "Food and farming in Germany: II. Farming and land reform," *International Affairs* 26:2 (April 1950): 195.

Klaus, Georg, "Kybernetik und ideologischen Klassenkampf," *Einheit: Zeitschrift für Theorie und Praxis des Wissenschaftlichen Sozialismus* 9 (1970): 1180.

Klein, E., "Forsteinrichtung nach der Wiedervereinigung," *Der Wald* 41:2 (1991): 60.

Klemke, Christian, and Jan Lorenzen, *Die sowjetische Militärherrschaft 1945 bis 1994* (Christoph Links Verlag—LinksDruck GmbH, 2002). From the three-part German television series 21, 26, and 28 April 2002, "Roter Stern über Deutschland," "Speziallager," http://www.orb.de/roterstern/content/37Speziallager.html (accessed 5 June 2002).

Knoth, Nikola, "Die Naturschutzgesetzgebung der DDR von 1954," *Zeitschrift für Geschichtswissenschaft* 39:2 (1991).

Koch, H. W., *A history of Prussia* (London: Longman, 1978).

Kohl, Horst, ed., *Ökonomische Geographie der Deutschen Demokratischen Republik*, 3rd ed. (Gotha and Leipzig: VEB Hermann Haack, 1976).

Köhler, W., "Wild game production and harvesting methods in some intensively managed European forests," *Proceedings*, Fifth World Forestry Congress, Seattle (1960), 1801.

Kohlsdorf, Erich, "Denkschrift über der Rauchschaden Situation im Bereich des mittlerer und östlicher Erzgebirges" (1964), typewritten manuscript in Tharandt archives.

Kontorovich, Vladimir, "Lessons of the 1965 Soviet economic reform," *Soviet Studies* 40:2 (April 1988): 308.

Kopstein, Jeffrey, *The politics of economic decline in East Germany, 1945–1989* (Chapel Hill: University of North Carolina Press, 1997).

——, "Ulbricht embattled: The quest for socialist modernity in the light of new sources," *Europe-Asia Studies* 46:4 (Soviet and East European History) (1994).

Kosing, Alfred, "Natur und Gesellschaft," *Einheit: Zeitschrift für Theorie und Praxis des Wissenschaftlichen Sozialismus* 39:11 (1984): 1020.

Kosing, Alfred, and Richard Heinrich, "Natur-Mensch-Gesellschaft: Das Verhältnis der sozialistischen Gesellschaft zur Natur," *Neues Deutschland* 127 (1 June 1987): 3.

Kosing, Alfred, et al., *Sozialistische Gesellschaft und Natur. Wissenschaftlichen Rates für Marxistisch-Leninistisch Philosophie der DDR* (Berlin: Dietz, 1989).

Kramer, Jane, "Living with Berlin," *New Yorker* (5 July 1999): 50.

Kramer, Mark, "The Soviet Union and the founding of the German Democratic Republic: 50 years later—a review," *Europe-Asia Studies* 51:6 (September 99): 1093.

Krause, Klaus Peter, "Begriffsbewirrungen über die 'Bodenreform' zwischen 1945 und 1949," *Frankfurter Allgemeine Zeitung* (2 September 1994): 8.

Krutzsch, Hermann, *Bärenthoren 1924* (Neudamm, 1924).

——, "Vorratspflege," *Forst- und Holzwirtschaft* 3:7 (1 April 1949): 99.

——, *Waldaufbau* (Berlin, 1952).

Krutzsch, Hermann, and Johannes Weck, *Bärenthoren 1934: Der naturgemässe Wirtschaftswald* (Neudamm, 1935).

Kuhrig, Heinz, "Demokratische Bodenreform legte den Grundstein für stetig steigende Agrarproduktion," *Neues Deutschland* 198 (21 August 1975): 3.

Kurjo, Andreas, "Kontroverse um das Umweltbundesamt," *Deutschland Archiv* 8:7 (August 1974): 888.

Kurth, Horst, "Bestandesinventur in der Deutschen Demokratischen Republik," *Sozialistische Forstwirtschaft* 38:4 (1988): 141.

——, "Die Entwicklung der Forstwirtschaft in der DDR," *Allgemeine Forstzeitschrift* 35 (1990): 894.

——, "Die Forsteinrichtung in der Deutschen Demokratischen Republik," IUFRO S.4.04 Conference Report, Bukarest (1983), 81–96.

——, "Entwicklungstendenzen der Forsteinrichtung als Ergebnis des internationalen Forsteinrichtungssymposium," in *Kolloquium anläßlich des 75 Geburtstag von Nationalpreisträger Prof (em.) Dr.-ing. habil Johannes Blanckmeister* (Tharandt: Technische Universität Dresden, 1973).

——, "Max Robert Pressler—ein Pionier intensiver Bestandwirtschaft," *Sozialistische Forstwirtschaft* 36:12 (1986).

——, *Ziele, Aufgaben, Methoden und Arbeitsgefüge der Forsteinrichtung in der Deutschen Demokratischen Republik,* Habilitation dissertation (Diss B.), Technisches Universität Dresden, Sektion Forstwirtschaft (Tharandt, 1969).

Kurth, Horst, ed., *Wissenschaftliche-Technische Darlegung zur Intensivierung der Holzproduktion und zur komplexen und volkswirtschaftlichen effektiven Holzverwertung* (Tharandt: Technische Universität Dresden, 1979).

Laird, Roy D., ed., *Soviet agriculture and peasant affairs* (Lawrence: Kansas University Press, 1963).

Laßmann, G., *Die Rolle und Bedeutung der Forstwirtschaft in System der Volkswirtschaft der Deutschen Demokratischen Republik,* Shriftenreihe für Forstökonomie (Berlin: VEB Deutsche Landwirtschaftsverlag, 1960).

Laufer, Jochen, "'Genossen, wie ist das Gesamtbild?' Ackermann, Ulbricht und Sobottka in Moskau im Juni 1945," *Deutschland Archiv* 29:3 (May-June 1996): 355.

Lehmann, Joachim, "Gedanken zur Anwendung des Produktionsprinzips bei der Leitung unserer staatliche Forstwirtschaftsbetriebe," *Sozialistische Forstwirtschaft* 14:10 (October 1964): 294.

Lehrmann, J., *Wege zur standortsgerechter Forstwirtschaft* (Radebeul, 1956).

Leithäuser, Johannes, "'Aber laßt nicht die Massen dran.' Für die DDR-Funktionäre Jagdreviere so groß wie das Saarland," *Frankfurter Allgemeine Zeitung* 269 (17 November 1990): 7–8.

Lemmel, Hans, "Der Dauerwaldgedanke und 'das eiserne Gesetz des Örtlichen,'" *Der Deutsche Forstwirt* 19 (1937).

——, "Der deutsche Wald in der Bodenreform," *Allgemeine Forst- und Jagdzeitung* 125:3 (1954).

——, *Der Organismusidee in Möllers Dauerwaldgedanken.*

Leonhard, Wolfgang, *Das kurze Leben der DDR. Bericht und Kommentare aus vier Jahrzehnten* (Stuttgart, 1990).

——, *Die Revolution entlässt ihre Kinder* (Cologne: Kiepenheuer & Witsch, 1955).

——, "Es muß demokratisch aussehen," *Die Zeit* (7 May 1965).

——, "Iron Curtain. Episode 2," interview (4 October 1998), U.S. National Security Archive, http://www.gwu.edu/~nsarchiv/coldwar/interviews/episode-2/leonhard2.html (accessed 8 March 2003).

Leptin, Gert, "The GDR," in *The new economic systems of Eastern Europe*, ed. Hans-Hermann Höhmann, Michael Kaser, and Karl C. Thalheim (London: C. Hurst, 1975).

Leuschner, Bruno, *Ökonomie und Klassenkampf. Ausgewählte Reden und Aufsätze 1945–1965.* Institut für Marxismus-Leninismus beim ZK de SED (East Berlin: Dietz, 1984).

Levin, Simon A., *Fragile dominion: Complexity and the commons* (Reading, Mass.: Perseus Books, 1999).

Levine, Herbert S., "Economics," in *Science and ideology in Soviet society,* ed. George Fischer (New York, 1967).

Lietz, B., "Steigerung der Produktion und Effektivität bei Erhöhung der lebender Holzvorräte—Schwerpunkt der Forstwirtschaft in den 80er Jahre," *Sozialistische Forstwirtschaft* 33:9 (1983): 266–71, 274–75.

Lisitzin, Evgeny N., "Collaborative arrangements for environmental protection in European socialist countries," in *Environmental policies in East and West,* ed. Gyorgy Eneyedi, August J. Gijawijt, and Barbara Rhode (London: Taylor Graham, 1987), 352.

Loeser, Franz, "Sind die formalisierten Methoden des marxistischen Gesellschaft Wissenschaften Klassenindifferent?" *Staat und Recht* 3 (1969): 467.

Ludz, Peter Christian, *The changing party elite in East Germany* (Cambridge: MIT Press, 1973), cited in John M. Starrels, "Comparative and elite politics," *World Politics* 29:1 (October 1976): 130.

MacIsaac, David, ed., *The United States strategic bombing survey* (New York: Garland 1976).

Macrakis, Kristie, and Dieter Hoffmann, eds., *Science under socialism: East Germany in comparative perspective* (Cambridge: Harvard University Press, 1999).

Maddison, Angus, *The world economy: A millennial perspective* (Paris: Development Centre of the Organisation for Economic Co-operation and Development, 2001).

Maier, Charles, *Dissolution: The crisis of Communism and the end of East Germany* (Princeton: Princeton University Press, 1997).

Maiziere, Lothar de, "The East German forest" (in German) *Neues Deutschland* (20 April 1990).

Majunke, Curt, ed., "Tätigkeitsbericht 1990," Forschungsanstalt für Holz- und Forstwirtschaft Eberswalde. Abteilung Forstschutz/Hauptstelle für Forstpflanzenschutz, typewritten (Eberswalde-Finow: IFE, 1990).

Mann, Helmut, *Prinzipien der Preisbildung für Rohholz in der Deutschen Demokratischen Republik* (Berlin: Deutscher Bauernverlag, 1958).

Mantel, Kurt, "Forstgeschichte," in "Stand und Ergebnisse der forstlichen Forschung seit 1945," *Schriftenreihe des AID* (1952), 144–53.

Mara, Michael, "DDR Prominenz entspannt sich bei der Jagd," *Der Tagesspiegel* 12:447 (4 September 1986): 12.

——, "Partei befurchtet eine 'grüne Unterwanderung,'" *Der Tagesspiegel* 12:107 (23 July 1985): 3.

——, "Wachsendes Umweltbewußtsein in der D.D.R.," *Der Tagesspiegel* 11:808 (27 July 1984): 3.

Marion, Gräfin Dönhoff, "Dogma oder Weizen?" *Die Zeit* 37 (5 September 1975): 6.

Marx, Karl (A Rhinelander), "Remarks on debates on the law on thefts of wood and the Proceedings of the Sixth Rhine Provincial Assembly," *Rheinische Zeitung* 88 (October 1842). First published in the supplement to the *Rheinische Zeitung*, nos. 298, 300, 303, 305, and 307 (25, 27, and 30 October, 1 and 3 November 1842). Transcribed by director@marx.org, November 1996, http://www.marxists.org/archive/marx/works/1842/10/25.htm (accessed 6 July 2003).

Marx, Karl, "Lohnarbeit und Kapital," *Werke,* vol. 6 (Berlin: Dietz Verlag, 1959), 407.

Marx, Karl, and Friedrich Engels, *The Communist Manifesto* (*Manifest der Kommunistischen Partei*), ed. David McLellan (New York: Oxford University Press, 1992).

Mayr, Ernst, *One long argument: Charles Darwin and the genesis of modern evolutionary thought* (Cambridge: Harvard University Press, 1993).

Maiziere, Lothar de, Article on eastern German forests, *Neues Deutschland* (20 April 1990).

Mazower, Mark, *Dark continent: Europe's twentieth century* (New York: Alfred A. Knopf, 1999).

McAdams, A. James, *East Germany and détente: Building authority after the wall* (Cambridge: Cambridge University Press, 1985).

McCauley, Martin, Review of "Zwischen Plan und Markt: Die Wirtschaftsreform 1963–1970 in der DDR" by Jörg Roesler, *German History* 10:2 (1992).

Mellor, Roy E. H., "The German Democratic Republic," in *Planning in eastern Europe*, ed. Andrew H. Dawson (London: Croon Helm, 1987).

Melzer, Manfred, "The pricing system of the GDR: Principles and problems," in Jeffries and Melzer, eds. *East German economy*, ed. Ian Jeffries and Manfred Melzer, trans. Eleonore Breuning and Ian Jeffries (London: Croon Helm, 1987), 144.

Mendelssohn-Bartholdy, Felix, *Die erste Walpurgisnacht: Ballade für Chor und Orchester*. Gedichtet von Goethe; componirt von Felix Mendelssohn-Bartholdy; op. 60. Partitur (Leipzig: Fr. Kistner, and London: Ewer & Co., 1844).

——, "Erste Walpurgisnacht," Motet op. 69, #1, #3.

——, *Letters from Italy and Switzerland*, trans. Lady Wallace, 3rd ed. (New York: F. W. Christern, 1865).

——, *Selected letters of Mendelssohn*, ed. W. F. Alexander (London: Swan Sonnenschein, 1894).

Mendelssohn-Bartholdy, Karl, *Goethe and Mendelssohn*, trans. M. E. von Glehn (London: Macmillan, 1872).

Menge, Marlis, and Theo Sommer, "Miteinander leben, gut miteinander auskommen," Ein ZEIT-Gespräch mit Erich Honecker, *Die Zeit* 6 (31 January 1986): 3.

Merkel, Konrad, "Agriculture," in *The East German economy*, ed. Ian Jeffries and Manfred Melzer, trans. Eleonore Breuning and Ian Jeffries (London: Croon Helm, 1987), 141.

——, *Die Agrarwirtschaft in Mitteldeutschland: 'Sozialialisierung' und Produktionsergebnisse* (Bonn: Bundesmininsterium für gesamtdeutsche Fragen, 1963).

Merkel, Wilma, and Stephanie Wahl, *Das geplünderte Deutschland. Die wirtschaftliche Entwicklung im östlichen Teil Deutschlands von 1949–1989*, 2nd ed. (Bonn: Instituts für Wirtschaft und Gesellschaft, 1991).

Merker, Paul, Editorial, *Neues Deutschland* (19 December 1946).

Merz, H.-G., "Bodenreform in der SBZ. Ein Bericht aus dem Jahre 1946," *Deutschland Archiv* 11:24 (1991): 1159.

Meyer, Alfred E., "The functions of ideology in the Soviet political system," *Soviet Studies* 17 (January 1966): 273–85.

Milton, John, *Paradise Lost: The first book*, lines 169–78, in John Milton, *Complete poems*, The Harvard Classics, ed. Charles W. Eliot, vol. 4 (New York: P. F. Collier, 1909–14).

Milward, Alan S., *The European rescue of the nation state* (London: Routledge, 1992).

Ministerium für Land- Forst- und Nahrungsgüterwirtschaft, *Umstellung der Kahlschlagwirtschaft auf vorratspflegliche Waldwirtschaft, Anweisung vom 20 November 1951* (Berlin: Hauptabteilung Forstwirtschaft, 1951).

Mittag, Günther, *Fragen der Parteiarbeit nach dem Produktionsprinzip in Industrie und Bauwesen* (Berlin, 1963).

——, "Tenth Party Congress Demands," *Einheit: Zeitschrift für Theorie und Praxis des Wissenschaftlichen Sozialismus* 5 (1982).

Möhring, Caroline, "Bleibt Fontanes Märchenplatz am Ufer des Werbellin?" *Frankfurter Allgemeine Zeitung* 37 (13 February 1990): 3.

Möller, Alfred, *Dauerwaldwirtschaft* (Berlin, 1921).

——, *Der Dauerwaldgedanke: Sein Sinn und seine Bedeutung* (Berlin, 1922).

Morgenstern, Oskar, K. Knorr, and K. P. Heiss, *Long-term projections of power* (Cambridge, Mass.: Ballinger, 1973), 190.

Morrow, Edward A., "Reparations lag in East Germany: Russian officials are warned to let nothing interfere with deliveries to Soviet," *New York Times* (17 October 1948).

Mühlfriedel, Wolfgang, "Der Wirtschaftsplan 1948: Der erste Versuch eines einheitlichen Planes der deutschen Wirtschaftskommission zur ökonomischen Entwicklung der sowjetischen Besatzungszone," *Jahrbuch für Wirtschaftsgeschichte* 3 (1985): 9–26.

Muir, John, *The mountains of California* (New York: Century, 1894).

Müller, K., R. Budzin, and H. Trinks, "Die Berechnung der Lehre von Karl Marx für die gesellschaftliche Nutzung der Naturkräfte des Waldes," *Sozialistische Forstwirtschaft* 18:5 (May 1968).

Munger, Thornton T., ed., "Observations on the results of artificial forestry in Germany: Excerpt from a trade letter written to an American paper company by their German correspondent," *Journal of Forestry* 21 (1923): 719.

Münker, Wilhelm, *Gerichtstag im Walde. Die Waldwesen klagen an* (Bielefeld: Deutsche Heimat-Verlag Ernst Gieseking, 1944).

Münzer, Wilhelm, *Dem Mischwald gehört die Zukunft—100 Stimmen für den Umschwung zum naturgemäßen Wirtschaftswald* (Hilchenbach, Westf., 1950).

——, *Über 200 fachmännische Stimmen für den Umschwung von Nadelreinbestand zum naturgemäßen Wirtschaftswald* (Bielefeld, 1958).

Naimark, Norman M., *The Russians in Germany: A history of the Soviet zone of occupation, 1945–1949* (Cambridge: Harvard University Press, 1995).

National Foreign Assessment Center, *Estimating Soviet and East European hard currency debt,* "USSR: Hard currency debt," table A-1 (Washington, D.C.: Central Intelligence Agency, 1980), 15.

Nayer, Rene, "Der Tod des Technokraten," *Die Zeit*, North American edition (14 December 1965): 3, cited in Baylis, "Economic reform as ideology," 223.

Neubauer, Ralf, "Rückkehr der Junker?" *Die Zeit* 36 (9 September 1994): 10.

Neuberger, Egon, "Libermanism, computopia, and visible hand: The question of informational efficiency" (in *Knowledge, information, and innovation in the Soviet economy*), *American Economic Review* 56:1/2 (March 1966): 131.

*Neuer Vorwärts,* "Waldraubbau-Holzexport," *Neuer Vorwärts* (Hannover) (8 January 1949).

*Neues Deutschland,* "Bauern sichern die Ernährung: Antifascisten aufs Dorf," *Neues Deutschland* (1 June 1946).

——, "Der Weg Schlange-Schöningen: Zweierlei Maß für Umsiedler," *Neues Deutschland* (29 December 1946).

——, "Die SED zur Grenzfrage," *Neues Deutschland* (19 September 1946): 3.

——, "Herr Schlange will den Spuren verwischen," *Neues Deutschland* (24 October 1946).
——, "Nicht nur für den Ofen . . . ," *Neues Deutschland* (23 June 1946).
——, "Nur Demokratie kann die Hungerkrise im Westen überwinden," *Neues Deutschland* (26 November 1946).
——, "One hundred decisive days: How will Berlin get its firewood?" *Neues Deutschland* (27 June 1946).
——, "Schlange-Schöningen muß gehen!" *Neues Deutschland* (30 November 1946).
*New York Times,* "Berlin depots to be renamed," *New York Times* (8 November 1950).
——, "71 billion penalty seen by Germans: Report on reparations set $28,000,000,000 as value of lost territories," *New York Times* (13 February 1948).
——, "Soviet zone adds border guards," *New York Times* (18 June 1949).
——, "The war against hunger," *New York Times* (17 August 1949).
Nichols, A. J., *Freedom with responsibility: The social market economy in Germany, 1918–1963* (New York: Oxford University Press, 1994).
Nick, Harry, "Sozialistische Rationalisierung, wissenschaftlich-technisch Revolution und Effektivität," *Einheit: Zeitschrift für Theorie und Praxis des Wissenschaftlichen Sozialismus* 2 (1971): 169.
Ochs, Martin S., "German red purge sweeps out books," *New York Times* (10 February 1951), 1.
——, "Marx and Engels works revised under East Germany's book purge," *New York Times* (15 March 1952).
Office of the Military Government for Germany (U.S.) (OMGUS), *Government and its administration in the Soviet zone of Germany,* Civil Administration Division, OMGUS (November 1947).
——, *Special report of the Military Governor: The German forest resource survey* (1 October 1948), vol. 17 (Office of the Military Government for Germany [U.S.], 1948).
Ökten, Rita, *Die Bedeutung des Umweltschutzes für die Wirtschaft der Deutsche Demokratische Republik* (Berlin: A. Spitz, 1986).
Oldag, Andreas, "Bauern opfer im Osten," *Süddeutsche Zeitung* (30 June 1994): 4.
Orr, John Boyd, "Program to meet the world's food crisis," *New York Times Magazine* (9 November 1947).
Ostermann, Christian F., "New research on the GDR," *Cold War International History Project Bulletin* 4:34 (Washington, D.C.: Woodrow Wilson International Center for Scholars, 1994): 39–42.
——, " 'This is not a Politbüro, but a madhouse,' the post Stalin succession struggle, Soviet Deutschlandpolitik and the SED: New evidence from Russian, German, and Hungarian archives," *Cold War International History Project Bulletin* #10 (Washington, D.C.: Woodrow Wilson International Center for Scholars, 1998).
Our Berlin Correspondent, "Berlin claustrophobia," *The Times* (18 March 1948), 5.
——, "In the Russian zone," *The Times* (25 September 1947), 5.
Our Correspondent, "Brain washing down on the farm: Recent East German advances in collectivization by consent," *The Times* (30 Mar 1961), 10.
——, "East Berlin as 'capital,' " *The Times* (30 November 1955), 6.

——, "East Germany 'unaffected by attack on personality cult,'" *The Times* (13 November 1961).
——, "Exodus from East Germany," *The Times* (27 July 1961), 8.
——, "Maize sowing lags in East Germany," *The Times* (6 May 1960), 12.
——, "More work for same pay in East Germany," *The Times* (8 September 1961), 11.
——, "Political rumblings in East Germany: Leaders' concern at criticism," *The Times* (19 February 1957): 6.
——, "Right to intervene retained by Soviet government," *The Times* (23 September 1955), 8.
——, "Soviet aims in East German pact," *The Times* (21 September 1955), 8.
Our Correspondent (Berlin), "Youth bored with indoctrination," *The Times* (10 March 1961), 9.
Our Correspondent (Bonn), "Easter exodus from East Germany," *The Times* (26 April 1960), 8.
——, "Mass flight of peasants from East Germany awaited," *The Times* (20 April 1960), 10.
Our Diplomatic Correspondent, "Dearth of food in Soviet zone. Demonstrations and arrests. Stocks reduced by requisitioning for Berlin," *The Times* (4 August 1948).
——, "Fixing the German reparation," *The Times* (19 June 1945).
——, "Moscow," *The Times* (17 September 1955), 5.
——, "Western powers and the Moscow talks: Dearth of food," *The Times* (11 August 1948).
Our Own Correspondent, "Absenteeism in the Soviet zone," *The Times* (29 September 1947).
——, "Arrests by Russians in Germany: Christian youth leaders," *The Times* (24 March 1947), 5.
——, "Border zone in East Germany: Stricter system of passes, cuts in telephones," *The Times* (28 May 1952), 6.
——, "Cutting Germany in two," *The Times* (4 June 1952), 6.
——, "Deportations in East Germany," *The Times* (14 June 1952), 6.
——, "Soviet arrests in Berlin: 'Fascist activities' alleged," *The Times* (29 March 1947).
——, "200,000 a year in migration from East Germany: West Berlin seeks to retain a higher proportion of influx," *The Times* (6 March 1961), 10.
——, "The worst winter," *The Times* (5 February 1947), 4.
——, "Youth group arrests in eastern Berlin: 'Subversive propaganda,'" *The Times* (16 June 1949), 3.
Our Special Correspondent, "Arrests in the Russian zone: German parents' complaints, re-education motive," *The Times* (19 August 1946), 3.
——, "Arrests of German children: Protest by parents," *The Times* (25 June 1946), 4.
——, "Brutality rife," *The Times* (11 September 1945), 4.
——, "International Timber Conference at Marienbad (Marianske Leyne)," *The Times* (12 May 1947).
——, "Mr. Bevin backs American plan for Germany," *The Times* (16 May 1947), 4.
Our Special Correspondent in East Germany, "Living with Pankow and the Wall," *The Times* (21 September 1962), 13.

Panorama DDR, *Agriculture in the German Democratic Republic: Some information about the life and work of the cooperative farmers* (Berlin: Panorama DDR, 1979).

Parchmann, Wechselberger, "Report of Reichsamtsleiter und Ministerialdirektor im Reichsforstamt," *Der Deutsche Forstwirt* 22:69/70 (1940): 533.

——, "Schluß mit dem Liberalismus in der Forst- und Holzwirtschaft," *Der Deutsche Forstwirt* 16 (1934): 409–11.

——, "Von liberalistische zu nationalsozialistische Forst- und Holzwirtschaft," *Der Deutsche Forstwirt* 16 (1934): 941–44.

Parry, Albert, "Science and technology versus Communism," *Russian Review* 25:3 (July 1966): 227.

Paucke, Horst, "The German Democratic Republic," in *Environmental policies in East and West*, ed. Gyorgy Eneyedi, August J. Gijawijt, and Barbara Rhode (London: Taylor Graham, 1987).

——, "Soziologie und Sozialpolitik," *Soziologie und Sozialpolitik* 1:87 (Berlin: Akademie der Wissenschaften der DDR Institut für Soziologie und Sozialpolitik, 1987).

Paul, Frithjof, "Beiträge zu den Grundlagen der Forstökonomik," *Schriftenreihe für Forstökonomie* 1 (1960): 158.

——, "Die Wirkungsweise ökonomischer Gesetze unter den gegenwärtigen Bedingungen der Forstwirtschaft der Deutschen Demokratischen Republik," *Beiträge zum Neuen Ökonomischen System der Planung und Leitung der Volkswirtschaft in der sozialistischen Forstwirtschaft der DDR* 62 (1963): 70.

Petschow, Ulrich, Jürgen Meyerhoff, and Claus Thomasberger, *Umweltreport DDR: Bilanz der Zerstörung, Kosten der Sanierung, Strategien für den ökologischen Umbau.* Eine Studie des Instituts für Ökologische Wirtschaftsforschung (Frankfurt am Main: S. Fischer, 1990).

Pfalzgraf, "Forstwirtschaft contra Holzwirtschaft?" *Forst- und Holzwirtschaft* (1 July 1947).

Pieck, Wilhelm, *Bodenreform: "Junkerland in Bauernhand"* (Berlin: Verlag Neuer Weg GmbH, 1945).

Piesnack, J., "Grundsätzliche Probleme des Waldzustandes und der Erreichung der vollen Leistungsfähigkeit der Wälder," *Sozialistische Forstwirtschaft* 25:2 (1975): 38.

——, "Zur Leitung und Planung der Bestandespflege," *Sozialistische Forstwirtschaft* 7 2 (1986): 193–224.

Piskol, Joachim, " 'Junkerland in Bauernhand': Wie deutsche Antifaschisten die demokratische Bodenreform 1945 vorbereiteten," *Neues Deutschland* 198 (204 August 1985): 13.

Plochmann, Richard, "Forestry in the Federal Republic of Germany," *Journal of Forestry* 79:7 (July 1981).

Podewin, Norbert, "Global denken, lokal handeln. Walter Ulbrichts Modell des Sozialismus. Eine Würdigung" (30 June 2003), http://www.jungewelt.de/2003/06–30/002.php (accessed 21 September 2003).

Przybylski, Peter, *Tatort Politbüro: die Akte Honecker,* vols. 1–2 (Berlin: Rowohlt, 1991).

Puttkammer, "Forstliche Rechts- und Verwaltungsprobleme der Gegenwart," *Forst- und Holzwirtschaft* 2:3 (1948).

Quadt, Ernst, "Der Ofen und das Brennmaterial," *Neues Deutschland* (20 December 1946).
Raab, Friedrich, *Die deutsche Forstwirtschaft im Spiegel der Reichsstatistik* (Berlin: Verlag von Paul Parey, 1931).
Radcliffe, Philip, *The master musicians: Mendelssohn* (New York: Oxford University Press, 2000).
Raymond, Jack, "Allies to tighten patrolling on East Germany's border," *New York Times* (28 June 1952).
——, "German demands timber be saved. Bizonal Economic Council hears report western Allies are destroying all forests," *New York Times* (29 April 1948), 33.
Reagan, Ronald, Speech at Notre Dame University (May 1981), "international relations," *Encyclopaedia Britannica Online*, http://members.eb.com/bol/topic?eu=108380&sctn=1 (accessed 19 December 1999).
Recknagle, A. B., "Some aspects of European forestry: Management of pine in Prussia; Management of spruce in Saxony," *Forestry Quarterly* 11:2 (June 1913).
——, "Some aspects of European forestry: Observations on Prussian forestry," *Forestry Quarterly* 11:1 (1913).
Rees, Goronwy, "From Berlin to Munich," *Encounter* 22:4 (April 1964): 3.
Reichenstein, E., "Die forstwirtschaftliche Lage Deutschland vor und nach dem 2. Weltkrieg," *Forstarchiv* 21:1/3 (1950): 30.
——, "Entwicklung von Vorrat und Zuwachs in den vier Besatzungszonen Deutschland seit 1945," *Weltholzwirtschaft* 1:7/8 (1949).
Reuter, Ernst, "Zur Öffnung der Sektorengrenzen und zur Bedeutung des 17. Juni 1953, 9.7.53 (RIAS Berlin) Nachrichtenmeldung über die Aufhebung der Sperren an den Sektorengrenzen (8 June 1953) DDR-Rundfunk," http://www.17juni53.de/chronik/5307_1.html (accessed 22 March 2004).
Reutter, R., "Volkssolidarität auf dem Lande: Zwei nachahmenswerte Beispiele," *Neues Deutschland* (9 July 1946).
Richardson, Robert D., Jr., *Emerson: The mind on fire* (Los Angeles: University of California Press, 1995), 249.
Richter, Albert, "Aufgaben und Methoden gegenwartsnaher Forsteinrichtung," *Archiv für Forstwesen* 1 (1952): 31–46.
——, *Aufgaben und Methoden standortsgerechter Forsteinrichtung* Conference Report, DAL Berlin 7:9 (1958).
——, "Fragen der Holzvorrats- und Zuwachsinventur im Walde," *Archiv für Forstwesen* 4 (1952): 467–80.
——, "Vom Ende der Forstfakultät Eberswalde 1963 — Ein persönlicher Bericht," *Allgemeine Forst- und Jagdzeitung* 11/12 162 (1991): 229.
Richter, Hans, ed., *Nutzung und Veränderung der Natur* (Leipzig: Geographisches Gesellschaft der Deutschen Demokratischen Republik, 1981).
Riedel, Alexander, "Chancen und Probleme der sachsischen Forstwirtschaft," lecture to graduating students of the Tharandt Forestry School (26 March 1991).
——, Letter to the editor, "Es liegt nicht an der Forstbürokraten der neuen Länder," *Frankfurter Allgemeine Zeitung* (26 February 1993), 8.

——, "Probleme der Umgestaltung der Forstwirtschaft in Sachsen," *Forstarchiv* 62:3 (1990): 91–94.

Ritter, Gert, and Joseph G. Hajdu, "The East-West German boundary," *Geographical Review* 79:3 (July 1989): 326.

Roesler, Jörg, *Das neue Ökonomische System (NÖS): Dekorations- oder Paradigmenwechsel?* 2nd ed., "Hefte zur DDR-Geschichte," vol. 3 (Berlin: H. Meier, 1994).

——, "The rise and fall of the planned economy in the German Democratic Republic, 1945–1989," *German History* 9:1 (February 1991): 46.

——, *Zwischen Plan und Markt: die Wirtschaftsreform in der DDR zwischen 1963 und 1970* (Berlin: Haufe, 1990).

Roesler, Jörg, Veronika Siedt, and Michael Elle, *Wirtschaftswachstum in der Industrie der DDR, 1945–1970,* Forschungen zur Wirtschaftsgeschichte, 0138–5100, vol. 23 (Berlin: Akademie-Verlag, 1986).

Ronzheimer, Manfred, "Umweltforschung in der DDR—eine Bilanz. Besondere aktuelle Bedeutung. Wichtige Erkentnisse trotz politischer Restriktionen," *Der Tagesspiegel* 13:679 (22 September 1990): 18.

——, " 'Wir hatten das nie für möglich gehalten.' Allmählich öffnen sich die Akten der DDR-Umweltpolitisch-Ringvorlesungen an der TU," *Der Tagesspiegel* 13:582 (31 May 1990): 18.

Rostow, W. W., "The world economy since 1945: A stylized historical analysis," *Economic History Review* 38:2 (May 1985): 25.

Rubner, Heinrich, *Deutsche Forstgeschichte—1933–1945: Forstwissenschaft, Jagd und Umwelt im NS-Staat* (St. Katharinen: Scriptae Mercaturae Verlag, 1985).

——, *Forstgeschichte im Zeitalter der industriellen Revolution,* "Schriften zur Wirtschafts- und Sozialgeschichte," vol. 8 (Berlin: Duncker & Humblot, 1967).

——, "Naturschutz, Forstwirtschaft, und Umweltschutz in ihren Wechselbeziehungen," in *Wirtschaftsentwicklung und Umweltbeeinflussen,* ed. Hermann Kellenbenz (Stuttgart, 1982), 105–33.

Ruffer, Horst, and Ekkehard Schwartz, *Die Forstwirtschaft der Deutschen Demokratischen Republik* (Berlin: VEB Deutscher Landwirtschaftsverlag, 1984), 18.

Ruffer, Horst, Kurt Schmidt, and Harry Wersinger, *Neues Ökonomische System der Planung und Leitung. Instrumente zur Aufbau der Sozialismus in der Forstwirtschaft* (Leipzig: Landwirtschafts-Ausstellung der DDR, 1966).

Rüffler, Rudolf, "Die Verantwortung der Forstwissenschaften nach dem 10. Parteitag der SED," *Beiträge für die Forstwirtschaft* 15:2 (1981): 53–54.

——, *Hauptrichtungen der Entwicklung der Forstwirtschaft in dem Mitgliedländer des RGW* (Eberswalde-Finow: Inf. für Leitungskader der Leitstelle für Inf., 1976).

——, "Zur Geschichte des Instituts für Forstwissenschaften Eberswalde," *Beiträge für die Forstwirtschaft* 14:3/4 (1980): 87–100.

Rüffler, Rudolf, and Gerhard Hoffmann, "Der 8. Weltforstkongress forderte: Die Wälder für die Menschen," *Sozialistische Forstwirtschaft* 29:4 (1979): 123–26.

Rüffler, Rudolf, and W. Luthardt, "Forschungsorganizationen im Institut für Forstwissenschaften Eberswalde," *Beiträge für die Forstwirtschaft* 14:3/4 (1980): 112.

Runder Tisch, *Information zur Analyse der Umweltbedingungen in der DDR und zu weiteren Maßnahmen* (Berlin, 1990).

Rüthnick, Rudolf, "Die weiteren Aufgaben bei der Verwirklichen des Neuen Ökonomischen System der Planung und Leitung der Volkswirtschaft im Bereich der VVB Forstwirtschaft Potsdam," in *Die Anwendung des Neuen Ökonomischen Systems im Bereich der VVB Forstwirtschaft Potsdam,* ed. R. Rüthnick (Potsdam-Babelsberg: VVB Forstwirtschaft Potsdam, 1966), 18.

——, *Erste Konferenz der VVB Forstwirtschaft Potsdam vom 15.-17. Juni 1964 in Leipzig-Markkleeburg* (Potsdam, 1964).

——, "Unser Wald im guten Händen. Wie wir ein Stück Verantwortung für heutige und künftige Generationen wahrnehmen," *Neues Deutschland* 225 (23 September 1989): 9.

Rüthnick, Rudolf, ed., *Die Anwendung des Neuen Ökonomischen Systems im Bereich der VVB Forstwirtschaft Potsdam* (Potsdam-Babelsberg: VVB Forstwirtschaft Potsdam, 1966).

Rüthnick, Rudolf, et al., "40 Jahre DDR 40 Jahre Entwicklung zu einer sozialistischen Forstwirtschaft," *Sozialistische Forstwirtschaft* 39:9 (1989): 257–88.

Rutten, Martin Gerard, *The geology of western Europe* (New York: Elsevier, 1969).

Ryle, G. B., "Forestry in western Germany, 1948," *Forestry* 22:2 (1948): 158.

——, "Germany: Military Government, C.C.G., North German timber control (NGTC)," *Empire Forestry Review* 26:2 (1947): 212–23.

Säglitz, J., "Die Forstwirtschaft in Ostdeutschland—Stand, Probleme, Ziele," *Forstarchiv* 61:6 (November-December 1990): 226.

Sanderson, Paul W., "Scientific-technical innovation in East Germany," *Political Science Quarterly* 96:4 (winter 1981–82).

Sarotte, Mary Elise, *Dealing with the devil: East Germany, détente, and Ostpolitik, 1969–1973* (Chapel Hill: University of North Carolina Press, 2001).

Schieferdecker, Helmut, *Umweltpolitik in der DDR Ringvorlesung. TU-Vorlesungsreihe,* May 1990 lectures (Berlin: Technisches Universität, 1990).

Schikora, Klaus, and H.-E. Wünsche, *Der Waldfonds in der Deutschen Demokratischen Republik* (Potsdam: VEB Forstprojektierung Potsdam, 1973, 1977 & 1986).

——, "KOWA" lists for 1986–89, in *Der Waldfonds in der Deutschen Demokratischen Republik* (Potsdam: VEB Forstprojektierung Potsdam, 1989).

Schindler, W., "30 Jahre staatliche Forstwirtschaftsbetriebe—30 Jahre sozialistische Entwicklung in der Forstwirtschaft. Staatliche Forstwirtschaftsbetrieb Löbau, Löbau, D.D.R.," *Sozialistische Forstwirtschaft* 32:11 (1982): 321–23.

Schmalz, Peter, "Das Umweltpflänzchen beginnt zu grünen. Sie galten und sie gelten noch immer als Außenseiter der sozialistischen Gesellschaft: die Öko-Gruppen in der DDR," *Die Welt* (24 May 1984).

Schmidtz, Manfred, "Der ungeteilte Dreck: Saubere Luft braucht die Kooperation von Bundesrepublik und DDR," *Die Zeit* (1987): 42.

Schoeps, Julius Hans, ed., *Enteignet durch die Bundesrepublik Deutschland: der Fall Mendelssohn-Bartholdy: eine Dokumentation* (Bodenheim: Philo Verlagsgesellschaft, 1997).

Schröder, G., "Das bedeutendste Gesetzwerk in der Geschichte Deutschlands," *Forst und Jagd,* part 1 of 2, 9:11 (November 1959): 483.

——, "The identity of value and matter reproduction in the growing stock and their representation through the price of wood and the value of forests" (in German), *Archiv für Forstwesen* 17:6 (1968): 571–96.

——, "Neue Maßtäbe für die Planung und Leitung der Forstwirtschaft. Rückblick und Ausblick nach dem 11. Plenum des ZK der SED," *Sozialistische Forstwirtschaft* 16:3 (March 1965): 67.

——, "Ökonomische Probleme des Zweiten Fünfjahresplans in der Forstwirtschaft," *Forst und Jagd,* part 2 of 4, 7:5 (May 1957): 198.

——, "Zielsetzung und Methode der Sozialistischen Rekonstruktion in der Forstwirtschaft," *Forst und Jagd,* part 2 of 2, 9:12 (December 1959): 524.

——, "Zu einigen Problemen der Forstwissenschaft und Praxis in Prozeß der wissenschaftlich-technischen Revolution," *Sozialistische Forstwirtschaft* 15:11 (1965): 323.

Schröder, Richard, "Dreimal den Krieg verloren? Gibt die Bundesregierung dem Druck der Alteigentümer nach, fördert sie den Glauben, in der vereinten Republik seien die Ostdeutschen immer die Dummen," *Die Zeit* 15 (11 April 1997): 5.

Schrötter, Helmuth, "Zum Begriff der Nachhaltigkeit," *Archiv für Forstwesen* 13:12 (1964): 1280–81.

Schult, W., *Bedeutung und Inhalts eines Zweigsprogrammes und wissenschaftlich-technischen Konzeptionen für die perspektivische Planung in der Forstwirtschaft* (Leipzig-Markkleeburg: Landwirtschaftsaustellung der DDR, 1965).

Schulz, Gerhard, "Die wissenschaftlich-produktive und ihre ideologisch-eziehierische Funktion der politischen Ökonomie des Sozialismus—eine Einheit," *Einheit: Zeitschrift für Theorie und Praxis des Wissenschaftlichen Sozialismus* 1 (1970): 38.

Schulz, Wolfgang, "Wir und Natur," *Mitteilungen aus der Wildforschung* (May 1985): 55.

Schütze, Christian, "Zuflucht für die deutsche Seele. Was ist so Besonderes an unserem Wald?" *Süddeutsche Zeitung* 116 (13, 21, and 22 May 1983): 116.

Schwartz, Ekkehard, "Die demokratische Bodenreform, der Beginn grundlegender Veränderungen der Waldeigentums und der Forstwirtschaft im Gebiet der Deutschen Demokratischen Republik," *Sozialistische Forstwirtschaft* 20:10 (1970): 289.

SED Agitation Department, "Wer die Deutsche Demokratishe Republik verläßt, stellt sich auf die Seite der Kriegstreiber," Notizbuch des Agitators (Berlin: SED Agitation Department, 1955), trans. in, "German Propaganda Archive," http://www.calvin.edu/academic/CAS/gpa/notiz3.htm (accessed 5 August 2003).

Seidel, Gerhard, Kurt Meiner, Bruno Rausch, and Alfonso Thoms, *Die Landwirtschaft der Deutschen Demokratischen Republik,* trans. Gunvor Leeson (Leipzig: VEB Edition, 1962).

Seidenstecher, Gertraud, "DDR," in *Umweltschutz und ökonomisches System in Oste: Drei Beispiele Sowjetunion, DDR, Ungarn,* ed. Hans-Hermann Höhmann and Gertraud Seidenstecher (Stuttgart: Kohlhammer, 1973), 85.

Seidenstecher, Gertraud, *Umweltschutz in der DDR* (Köln: Bundesinstitut für ostwissenschaftliche und internationale Studien, 1973).

Selvage, Douglas E., "The end of the Berlin Crisis: New evidence from the Polish and East German archives," *Cold War International History Project Bulletin* #11 (Washington, D.C.: Woodrow Wilson International Center for Scholars, 1999).

——, "Poland, the German Democratic Republic and the German question, 1955–1967" (Ph.D. diss., Yale University, 1998).

Smith, Jean Edward, *Germany beyond the Wall* (Boston, 1969).

Smolinski, L., and P. Wiles, "The Soviet planning pendulum," *Problems of Communism* (November-December 1963): 21–34.

Sodaro, Michael J., "Ulbricht's grand design: Economics, ideology, and the GDR's response to détente, 1967–1971," *World Affairs* 142:3 (winter 1979–80): 147.

Solsten, Eric, ed., *Germany: A country study* (Washington, D.C.: Federal Research Division, Library of Congress, 1996), 135.

Sommer, Theo, "Ein deutscher Kommunist, ein deutscher Realist. Anmerkungen zu einem ZEIT-Interview mit Erich Honecker," *Die Zeit* 6 (31 January 1986): 1.

Sontheimer, Kurt, "Intellectuals and politics in Western Germany," *West European Politics* 1 (February 1978): 30.

Sontheimer, Kurt, and Wilhelm Bleek, *The government and politics of East Germany,* trans. Ursula Price (New York: St. Martins Press, 1976).

SOPADE, Vorstand der Sozialdemokratischen Partei Deutschlands, "Der Terror in der Sowjetzone geht weiter," *Querschnitt durch Politik und Wirtschaft* #894 (February 1950).

——, "Die Agrarsituation in der Ostzone," *Querschnitt durch Politik und Wirtschaft* (June 1948), 3, 17.

——, "Die Forstwirtschaft in der Sowjetzone," *Denkschriften, Sopadeinformationsdienst* (Bonn: Vorstand der Sozialdemokratischen Partei Deutschlands, 1955).

——, "Farmers in Saxony," *Querschnitt durch Politik und Wirtschaft* #5 (August 1947): 5.

——, "KZs in der Ostzone," *Querschnitt durch Politik und Wirtschaft* #5 (August 1947): 66.

——, "Landwirtschaft in der Ostzone," *Querschnitt durch Politik und Wirtschaft* (February 1948), 40.

——, "Ostzonenreparationen," *Querschnitt durch Politik und Wirtschaft* (1950).

——, "Raubbau an den Ostzonen-Wäldern," *Querschnitt durch Politik und Wirtschaft* (February 1949).

——, "Raubbau an den Ostzonen-Wäldern," *Querschnitt durch Politik und Wirtschaft* (April 1949).

——, *SOPADE-Querschnitt durch Politik und Wirtschaft* (Hannover: Vorstand der Sozialdemokratischen Partei, 1947–1948, 1949/1950–1952/1954).

——, "Von dem Menschenrechten hat Herr Ulbricht noch nichts hehört," *Querschnitt durch Politik und Wirtschaft* (May 1949).

Sozialistische Einheitspartei Deutschlands, *Programm der Sozialistischen Einheitspartei Deutschland. Protokoll des IX Parteitages der Party* 2 (Berlin, 1976), 224.

Sozialistische Forstwirtschaft, "Was wir heute pflanzen ernten wir unter Kommunismus!" *Sozialistische Forstwirtschaft* 1:1 (1962).

Speer, Julius, "Die Forswirtschaft im Wirtschaftsgeschehen des Jahres 1948," *Allgemeine Forstzeitschrift* 4:1 (5 January 1949): 1.

Spelsberg, Gerd, *Rauchplage: Hundert Jahre Saurer Regen* (Aachen: Alano, 1984).

Spencer, Edmund, "An Englishman resident in Germany," *Sketches of Germany and the Germans* (London: Gilbert & Rivington, Printers, 1836).

Spielke, H. O., G. Breithaupt, H. Bruggel, and H. Stand, *Ökonomik der sozialistischen Forstwirtschaft* (Berlin: VEB Deutsche Landwirtschaftsverlag, 1964).

Spitzer, Gretel, "Hint from East Germany of new line on détente," *The Times* (21 June 1971), 6.

Spurr, S. H., "Post-war forestry in Western Europe. Part II," *Journal of Forestry* 51:6 (1953): 415–21.

Staatlichen Zentralverwaltung für Statistik (SZWS), ed., *Statistisches Jahrbuch der Deutsche Demokratische Republik* (Berlin: Staatsverlag der Deutschen Demokratischen Republik, 1955, 1981, 1989 & 1990).

Staatliches Komitee für Forstwirtschaft, "'Diskussionsmaterial'—Die Wege zu Intensivierung des forstlichen Reproduktionsprozess und zur Erhöhung seiner Effektivität," typewritten and stamped "VD" (*Vertrauliche Dienstsache*) (Berlin, 1972).

Staatliches Komitee für Forstwirtschaft, "Forstressourcen-Umfrage," typewritten (Berlin, 1980).

Stahnke, Arthur A., "GDR economic strategy in the 1980s: The 1981–1985 Plan," in *Studies in GDR culture and society: Nine selected papers from the 14th New Hampshire Symposium on the German Democratic Republic,* ed. Margy Gerber (New York: University Press of America, 1983).

Stammen, Theo, "Von der SBZ zur DDR," in *DDR: Das politische, wirtschaftliche und soziale System,* ed. Heinz Rausch and Theo Stammen, 4th ed. (Munich: Verlag C. H. Beck, 1978), 17.

——, "Zur Verfassungsentwicklung," in *DDR: Das politische, wirtschaftliche und soziale System,* ed. Heinz Rausch and Theo Stammen, 4th ed. (Munich: Verlag C. H. Beck, 1978), 196.

Staritz, Dieter, *Die Gründung der DDR: Von der Sowjetischen Besatzungsherrschaft zum sozialistischen Staat,* 2nd ed. (Munich: Deutscher Taschenbuch Verlag, 1987).

——, "Die SED, Stalin und der 'Aufbau des Sozialismus in der DDR.' Aus dem Akten des Zentralen Parteiarchivs," *Deutschland Archiv* 24 (1991): 686–700.

——, "Die SED, Stalin und die Gründung der DDR. Aus dem Akten des Zentralen Parteiarchivs des Instiutu für Geschichte der Arbeiterbewegung," in *Aus Politik und Zeitgeschichte* B (May 1991): 3–16.

Starrels, John M., "Comparative and elite politics," *World Politics* 29:1 (October 1976): 130.

Statistisches Bundesamt, "Bevölkerung und Wirtschaft 1872–1972." (1973).

——, *Statistisches Jahrbuch für die Bundesrepublik Deutschland* (Stuttgart: W. Kohlhammer, 1953, 1992).

Statistisches Reichsamt, *Die besteuerung der Landwirtschaft* (Berlin: R. Hobbing, 1930).

——, *Statistik des deutschen Reiches,* vol. 592 (Berlin: Puttkammer & Mühlbrecht, 1937).

——, *Statistik des deutschen Reiches. Die Ergebnisse der forstwirtschaftlichen Erhebung in Jahre 1927* vol. 386 (Berlin: Puttkammer & Mühlbrecht, 1930).

Staudenmaier, Peter, "Fascist ideology: The 'Green Wing' of the Nazi Party and its historical antecedents," in *Ecofascism: Lessons from the German experience,* ed. Janet Biehl and Peter Staudenmaier (Edinburgh: AK Press, 1995).

Steffens, Rolf, *Wald, Landeskultur und Gesellschaft,* 2nd ed. (Jena: VEB Gustav Fischer Verlag, 1978).

Stent, Angela E., *Russia and Germany reborn: Unification, the Soviet collapse, and the new Europe* (Princeton: Princeton University Press, 1999).

Stern, Fritz Richard, *The politics of cultural despair; A study in the rise of the Germanic ideology,* 2nd ed. (Berkeley: University of California Press, 1974).

Stewart, John Massey, "No place to hide," review of Murray Feshbach's *Environmental and health atlas USSR* and *Ecological disaster, Nature* 381:6579 (16 May 1996): 203.

Stolper, Wolfgang F., "The labor force and industrial development in Soviet Germany," *Quarterly Journal of Economics* 71:4 (November 1957): 534.

Stolper, Wolfgang F., and Karl W. Roskamp, *The structure of the East German economy* (Cambridge: Harvard University Press, 1960).

Stone, Norman, "Property values," review of "Schicksalsbuch des Sächsisch-Thüringischen Adels 1945," "Restitutionsverbot" (Christoph Rechberg, ed.), and "Dokumentation zum neuen Entschädigungsgesetz EALG" (Udo Madaus) *Times Literary Supplement* (10 May 1996), 9.

Süddeutsche Zeitung, "Nach Prüfung aller Unterlagen Bohl: SBZ Enteignungen sind unumkehrbar," *Süddeutsche Zeitung* (3 September 1994).

Szasz, Thomas, *The second sin* (1974), quoted in *The Oxford book of aphorisms,* ed. John Gross (New York: Oxford University Press, 1987), 328.

Tacitus, Cornelius, *Germania,* trans. M. Hutton, rev. by E. H. Warmington (Cambridge: Harvard University Press, 1970), 137.

*TASS,* "Common European home, as seen by Brzezinski," *TASS* (9 November 1989).

——, "Zbigniew Brzezinski on Perestroika — *TASS* interview" (3 November 1989).

Thalheim, Karl Christian, *Stagnation or change in Communist economies?* (London: Center for Research into Communist Economies, 1986).

——, "Volkswirtschaft," in *Ploetz, Die Deutsche Demokratische Republik: Daten, Fakten, Analysen,* ed. Alexander Fischer (Freiburg: Ploetz, 1988).

Theologischen Studienabteilung beim DDR-Kirchenbund, "Ökonomie, Leistung, Persönlichkeit," *Deutschland Archiv* 15:1 (January 1982): 68.

Thomas, Rüdiger, *Modell DDR Die kalkulierte Emanzipation,* 2nd ed. (Munich, 1977), 25–26.

Thomasius, Harald, "Gesetzmäßigkeiten in der historischen Entwicklung des Waldbaus," in *Kolloquium anläßlich des 75 Geburtstag von Nationalpreisträger Prof (em.) Dr.-ing. habil. Johannes Blanckmeister* (Tharandt: Technische Universität Dresden, 1973), 3.

——, "Waldbauliche Auffassungen, Probleme und Wege in der DDR," *Allgemeine Forstzeitschrift* 28–29 (14 July 1990): 726.

Thompson, Wayne C., Susan L. Thompson, and Juliet S. Thompson, *Historical dictionary of Germany* (Metuchen, N.J.: Scarecrow Press, 1994).

*The Times,* "German timber for Britain. Troops at work in famous forests. 'Operation Woodpecker,'" *The Times* (15 April 1947).

——, "Life in the Soviet zone," *The Times* (27 December 1945), 5.

——, "More work for same pay in East Germany: Legitimate to kill," *The Times* (8 September 1961), 11.

——, "No Surrender in Berlin. Mr Bevin's Review. Ernest Bevin's 30 June 1948 speech to Commons," *The Times* (1 July 1948), 4.

——, "Opportunity in Moscow," *The Times* (7 April 1947).

Titel, W., "Umweltschäden in der DDR dient dem Wohl des Menschen," *Marxistische Blätter* 1 (1972): 49.

Toniolo, Gianni, "Europe's Golden Age, 1950–1973: Speculations from a long-run perspective," *Economic History Review* 51:2 (1998): 252.

Töpfer, Klaus, "In zehn Jahren zur deutschen Umweltunion," *Die Welt* 62 (26 March 1990): 8.

——, *Pilotprojekte für Umweltschutz mit der DDR vereinbart: Erklärung von Bundesumweltminister Klaus Töpfer,* 3rd ed. (New York: Harvester Wheatsheaf, 1989).

Tocqueville, Alexis de, *L'Ancien Régime et la Révolution,* ed. François Furet and Françoise Mélonio, trans. Alan S. Kahan (Chicago: University of Chicago Press, 1998).

Tocqueville, Alexis de, *Selected letters on politics and society*, ed. Roger Boesche, trans. James Toupin and Roger Boesche (Berkeley: University of California Press, 1985), 366–68, 373, http://chnm.gmu.edu/revolution/d/590/ (accessed 18 October 2003).

Treml, Vladimir G., "The politics of 'Libermanism,'" *Soviet Studies* 19:4 (April 1968): 569.

Tümmler, Edgar, Konrad Merkel, and Georg Blohm, *Die Agrarpolitik im Mitteldeutschland und ihre Auswirkung auf Produktion und Verbrauch landwirtschaftliche Erzeugnisse* (Berlin: Duncker & Humblot, 1969).

Turner, Henry Ashby, Jr., *Germany from partition to reunification* (New Haven: Yale University Press, 1992).

——, *Two Germanies since 1945* (New Haven: Yale University Press, 1987).

Ulam, Adam Bruno, *Unfinished revolution: An essay on the sources of influence of Marxism and communism* (New York: Random House, 1960).

Ulbricht, Walter, *Dem VI Parteitag entgegen. Referat auf der 17. Tagung des ZK der SED* (Berlin: Dietz Verlag, 1962), 43.

——, "Der Kampf um Deutschland," *Demokratische Aufbau* 7 (October 1946): 193.

——, "Die demokratisches Bodenreform—ein rühmreiches Blatt in den deutschen Geschichte," *Einheit: Zeitschrift für Theorie und Praxis des Wissenschaftlichen Sozialismus* 10 (1955): 849.

——, "Lebendige Demokratie," *Demokratische Aufbau* (November 1947): 321.

——, "Open letter" *Junge Welt,* cited in *The Times*, "More work for same pay in East Germany: Legitimate to kill," *The Times* (8 September 1961), 11.

——, "Schlußwort zur Wirtschaftskonferenz der SED 1961," *Die Wirtschaft* (special edition) (18 October 1961): 3, cited in F. Walter, "Möglichkeiten einer vertieften wirtschaftlichen Rechnungsfühung der Forstwirtschaft mit Hilfe veränderter Finanzierungsmethoden," *Sozialistische Forstwirtschaft* 5 (1965): 1.

——, "2. Gespräch des Staatsratsvorsitzenden Ulbricht mit Präsident Nasser am 28. Februar 1965 von 18.00 Uhr bis 19.40 Uhr," *Vierteljahrshefte für Zeitgeschichte* 46:4 (October 1998): 803.

——, *Zur Geschichte der neuesten Zeit,* vol. 1 (Berlin: 1955).

United Nations Organizations (UNO), *Probleme der Umwelt, Deutsche Demokratische Republik. Nationaler Bericht der Deutsche Demokratische Republik für die UNO-Konferenz über menschliche Umwelt Bedingungen in Stockholm 1972* (Berlin, 1971).

Unsere Jagd, *Jagd und Jäger in der DDR. Die Jagd gehört dem Volke,* Unsere Jagd Extra (Berlin: Deutsche Landwirtschaftsverlag, September, 2000).

Urban, Martin, "Dieser Wald ist ein Teil unserer Heimat," *Süddeutsche Zeitung* (1 February 1983).

U.S. Congress, House Committee on Energy and Commerce, Subcommittee on Health and the Environment, "Acid rain in Europe: A report, March 1985, on the fact-finding excursion," 99th Cong., 1st sess. Com print 99–F (Washington, D.C., 1985), 8.

VEB Forstprojektierung Potsdam, *Forsterhebung 1949. Flächen- und Vorratsgliederung nach Besitzverhältnissen,* vol. 5–161–2 FE-Unterlagen (1949).

——, "Handwritten summary of 1937 Forest Inventory annotated for East German conditions" (in German) (1949).

——, *Übersicht der Forstwirtschaft in der Deutschen Demokratischen Republik* (1986).

Vera, Franciscus Wilhelmus Maria, *Grazing ecology and forest history* (Wallingford, U.K.: CABI Publications, 2000).

Voge, Kurt, et al., "'Unser Wald soll gesund, sauber, ertragreich sein.' Forstwirtschaftsbetrieb Oranienburg an das ZK der Party," *Neues Deutschland* 68 (21 March 1986): 4, 6.

von Berg, Michael, "Umweltschutz in Deutschland. Verwirklichung einer deutschen Umweltunion," *Deutschland Archiv* 23:6 (June 1990): 897.

von Oppen, Beate Ruhm, *Documents on Germany under occupation, 1945–1954* (London: Oxford University Press, 1955), 148.

von Wulffen, Barbara, "Alle Wälder unseres Lebens," *Süddeutsche Zeitung* 122 (29, 30, and 31 May 1982).

Wachs, Philipp-Christian, *Die Bodenreform von 1945: Die zweite Enteignung der Familie Mendelssohn-Bartholdy* (Baden-Baden: Low & Vorderwulbecke, 1994).

Wagenknecht, E.,"Der Waldbau zwischen Heute und Morgen," *Archiv für Forstwesen* 10:4/6 (1961): 366.

Wagenknecht, E., A. Scamoni, A. Richter, and J. Lehmann, *Eberswalde 19536 — Wege zu Sstandortsgerechter Forstwirtschaft* (Neumann, 1956).

Wagner, Christof, *Der Blendersaumschlag und sein System* (Tübingen, 1912).

Walter, F., "Möglichkeiten einer vertieften wirtschaftlichen Rechnungsfühung der Forstwirtschaft mit Hilfe veränderter Finanzierungsmethoden," *Sozialistische Forstwirtschaft* 5 (1965): 1.

Weber, Hermann, *Geschichte der DDR* (Munich: Deutscher Taschenbuch Verlag, 1985).

Wiedemann, E., "Naturgemäßer Wirtschaftswald und nachhaltige Höchleistungswirtschaft," *Allgemeine Forst- und Jagdzeitung* 5 (1950): 157–62.

Weidermann, Klaus, "Abriß der Geschichte der Fakultät für Forstwirtschaft in Tharandt," *Forstarchiv* 15:11/12 (1966): 1253.

Werner, Heinz, "Vorratspfleglicher Waldwirtschaft. Die Bedeutung der Menzer Tagung vom 14–15 Juni 1951," *Forst- und Holzwirtschaft* 9:5 (1951): 257.

Werner, Steffen, *Kybernetik statt Marx? Politische Ökonomie und marxistische Philoso-*

*phie in der DDR unter dem Einfluss der elektronischen Datenverarbeitung*, 39 (Stuttgart: Verlag Bonn Aktuell, 1977).
Whitney, Craig, "German reunification and economics," *New York Times* (17 November 1989).
Wiebecke, C., "Wiedemanns 'Eisernes Gesetz des Örtlichen,'" *Forstarchiv* 61:5 (September-October 1990): 183.
——, "Zum Stand der deutschen Forststatistik," *Forstarchiv* 26:1 (15 January 1955).
——, "Zur 150: Wiederkehr von Goethes Todestag (22. März 1832)," *Forstarchiv* 53:2 (1982): 72.
Wiebecke, Ernst, *Der Dauerwald in 16 Fragen und Antworten für den Gebrauch im Walde dargestellt,* 4th ed. (Stettin-Neutorney, 1924).
Wiedemann, Eilhard, *Die praktischen Erfolge des Kieferndauerwaldes. Untersuchungen in Bärenthoren, Frankfurt a. O. und Eberswalde. Studien über die früheren Dauerwaldversuche und Kiefernurwald* (Braunschweig, 1925).
——, "Naturgemäßer Wirtschaftswald und nachhaltige Höchsleistungswirtschaft," *Allgemeine Forstzeitschrift* 5 (1950): 157–62.
Wiener, Norbert, *Cybernetics* (New York: John Wiley & Sons, 1949), 19.
——, *The human use of human beings; Cybernetics and society* (Boston: Houghton Mifflin, 1954).
Willenstein, Gustav, *Die große Borkenkäferkalamität in Süddwestdeutschland, 1944–1951* (Ulm, 1954).
Wilson, Edmund, "Karl Marx decides to change the world," in *To the Finland Station* (New York: Farrar, Straus and Giroux, 1972).
Witzgall, "Naturgemäßer Waldwirtschaftswald und Holzwirtschaft," *Allgemeine Forstzeitschrift* 5 (1950): 565–66.
Wobst, W., "Waldbau — Ein geistloses Handwerk?" *Allgemeine Forstzeitschrift* (1948).
Wohlfarth, Erich, "Natur und Technik im Waldbau," *Allgemeine Forstzeitschrift Wien* 70:15/16 (1959): 173.
——, "Waldbau oder Waldpflege?" *Allgemeine Forst- und Jagdzeitung* 121 (1950): 114.
——, "Was folgt aus dem Vorstellung vom Wald als Ganzheit für die praktische Waldbehandlung. Vortrag," *Allgemeine Forst- und Jagdzeitung* 132:4 (1961): 96.
——, "Zur Waldbaulichen Lage der Gegenwart," *Forstarchiv* 23:4 (1 May 1952).
Wolf, Michael L., "The history of German game management," *Forest History* 14:3 (October 1970): 16.
Wordie, J. R., "The chronology of English enclosure, 1500–1914," *Economic History Review* 36:4 (November 1983).
Wünsche, H.-E., and Klaus Schikora, "Der Waldfonds der DDR — ausgewählte Fakten über Entwicklung und Zustand," *Sozialistische Forstwirtschaft* 40:3 (1990).
Zank, Wolfgang, "Als Stalin Demokratie befahl," *Die Zeit* 25 (23 June 1995): 75.
——, "'Junkerland in Bauernhand!' 3.3 millionen Hektar Land wurden 1945–49 während der Bodenreform in der Sowjetzone beschlagnahmt," *Die Zeit* 42 (12 October 1990): 49.
Zauberman, Alfred, "Liberman's rules of the game for Soviet industry" (in Notes and Comment), *Slavic Review* 22:4 (December 1963): 734.
Ziegler, E., "Die Wirkung der Industrie-Rauchschäden auf den Wald, ihre Berüchtsichti-

gung bei der Raumplanung und die Notwendigkeit ihrer gesetzlichen Regelung," *Wissenschaftliches Zeitschrift der Technischen Universität Dresden* 6 (1956–57): 777–87.

Zillmann, G., "Fragen des Überganges zu industriegemäßigen Produktionsmethoden in der Forstwirtschaft" *Sozialistische Forstwirtschaft* 14:2 (February 1964): 35.

Zubok, Vladislav M., "Khrushchev and the Berlin crisis, 1958–1962," Cold War International History Project, Working Paper no. 6 (Washington, D.C.: Woodrow Wilson International Center for Scholars) (May 1993), 24.

Zubok, Vladislav M., and Constantine V. Pleshakov, *Inside the Kremlin's Cold War: From Stalin to Khrushchev* (Cambridge: Harvard University Press, 1997).

Zvosec, Christine L., "Environmental deterioration in East Europe," *Survey* 84:28/4 (1984): 117–41.

# Index

*Page numbers in italics indicate figures.*

Absolute truths, 186, 187, 258n51
Accounting, 85, 122, 125–26, 149, 174, 236n38
Ackermann, Anton, 57, 122, 135, 136, 209n17, 226n105
Adenauer, Konrad, 51, 102, 103, 104, 105, 208n130, 229n6
Afforestation: and conifers, 17; and plantation forestry, 43, 85; and First Two-Year Plan, 73, 218n142; and reparations, 74; and gross accounting, 85; in private forests, 86–88, 224nn72,76; fees for, 87–88; and New Economic System, 133, 167; and harvest levels, *166*, 167; and reunification, 175
Agricultural economy: and depression, 28, 195n62; and reparations, 32, 38; and land reform of Soviets, 64, 65, 67, 71, 74, 76, 77, 78, 91; and production quotas, 71–72, 217n128; and peasants' lack of confidence in state, 86; and collectivization, 112–13; and Khrushchev, 116–17, 122; restructuring of, 120; and Marxist-Leninist policy, 129, 146; and Ulbricht, 137; and production, 140, 142, 143, 145, 146, 155, 158; and free markets, 169; and reunification, 174
Agriculture: and food supply during Second World War, 30, 31, 172, 196n5; and land reform of Soviets, 31, 54, 57, 58, 209n17; and reparations, 32, 38; agrarian reform, 52, 58, 71, 192–93n36; and Schlange-Schöningen, 59; and traditional cooperatives, 64, 214n73; and collectivization, 91, 111–12; and socialist farming, 110; farm policy, 129, 139, 146–47, 155, 158; and Soviet agrarian reform, 144; rationalization of, 192–93n36. *See also* Farms and farmers
Allied occupation zones, *35*

Allies: and Soviet reparations demands, 38, 39, 40, 201n53; and German timber harvests, 41–42, 44; and land reform of Soviets, 54, 55, 61, 106; and East-West frontier, 101; and Ulbricht, 106
American zone: costs of reparations, 39, 202n56; and clear-cutting, 43, 204n84; and food supply, 44, 45, 48–49, 208n4; and forest policy, 44, 204–5n87, 216n101; and Germany Treaty, 102
Apel, Erich, 93–95, 120, 122–23, 125–28, 133–35, 143–44, 226nn105,111
Ära Grüneberg. *See* Industrial Production Methods
Arendt, Hannah, 186
Artificial forest: and cultural identity, 20, 21, 28, 187; structure of, 22, 37; and forest decline, 23, 131, 135, 153, 154, 173, 185; replanting of, 74; and materialism, 150
Ash, Timothy Garton, 28
Atlantic Alliance, 55, 100, 102, 103
Autarky, 28, 89, 94, 174, 175

Barciok, Hans, 153, 248n68
Baring, Arnulf, 157
Barth, Roland, 200n38, 238n71
Barzun, Jacques, 119
Berlin, xii–xiii, xx, 1, 8–9, 68, 114, 137, 139
Berlin Crisis of *1948–49*, 65
Berlin, Isaiah, 186, 258n51
Berlin Wall: deaths at, xi; and *Republikflucht*, xii; fall of, 2, 7, 23, 26, 28, 65, 169, 182; building of, 76, 83, 91, 98, 102, 103, 115, 184; and Ulbricht, 91, 103, 115, 116, 137, 229n7; and East-West frontier, 101; and Soviet bloc brutality, 115–16; and Soviet Unions' strategic interests, 137; closing boundaries of, 139; and Grüneberg, 142
Berlioz, Hector, 14, 191n23
Bevin, Ernest, 39–40, 201n53
Biedenkopf, Kurt, 177
Binsack, Rudolf, 30, 31, 32, 197n11
Binsack family, 31–32, 197n11
Blanckmeister, Johannes, 78, 79, 80, 82, 83, 123–24
Bramwell, Anna, 245n31
Brandenburg forest, 3, 7, 172
Brandt, Willy, 11, 75, 137, 229n6, 240n102
Brezhnev, Leonid, 128, 136–38, 148, 179, 180
Brezhnev-Faktion, 137, 138, 143, 178
Britain, 17, 20
British zone: costs of reparations, 39–40, 202n56; and forest operations, 44; and food supply, 48–49, 54, 208nn3,4; and timber reparations, 49, 207n119; and forest working conditions, 50; and Schlange-Schöningen, 58–59; and Germany Treaty, 102; and forest management, 216n101
Brocken, 11–16, 23, 26, 190n10, 193n45
Brüning, Heinrich, 58, 211n33
Brzezinski, Zbigniew, 180, 181–82, 257n34
*Bündnispolitik* (Union of Workers and Peasants): and land reform of Soviets, 59, 60, 89; and collectivization, 71, 73, 90; and ideology, 97; and Ulbricht, 137; and Honecker, 139, 151, 158, 179; and population distribution, 147
Burke, Edmund, 53, 118
Byrnes, James, 55

Capitalism: Marxist-Leninist attacks on, 55–56, 78–79, 129, 147; clear-cutting as, 56, 208n10; and private forest economy, 89; and Socialist Reconstruction, 94; and West Germany, 102; socialism compared to, 107, 117, 129, 139–40, 184; Soviet victory over, 119; and Marx, 145; and environmentalism, 147, 158; and pollution, 147, 151, 245n33

Carter, Jimmy, 140, 180
Castro, Fidel, 151
Catholic Church, 7, 172
Ceauşuescu, Nicolae, 136
Central Europe, xxi, 16, 17, 33, 51, 100, 104
Central German Uplands, 24–25, 26, 27
Central planning: and reparations, 41, 51–52; and forest management, 45, 47, 75, 85, 89, 127, 129, 134, 205n92; and agriculture, 54; and economic growth, 94; and Khrushchev, 118, 120; and Honecker, 139; weakness of, 180
Cervantes, Miguel de, 141
Chemical industries, 20, 155
China, 121–22, 136, 137, 178–79
Christian Democrats, 60–61, 212n49
Churchill, Winston, 39
Class warfare: and Soviet occupation, 33; and land reform of Soviets, 52, 53, 55, 61–62, 64, 71, 73; and collectivization, 92, 111; and farmers, 144
Clay, Lucius, 44–45
Clean Air Act of *1970*, 153
Clean Water Act of *1972*, 153
Clear-cutting: and forest management, 22, 79, 80, 125, 192–93n36; during Second World War, 36; in Soviet zone, 37, 43–44, 73–74, 204nn83,84; as capitalism, 56, 208n10; and peasants, 68; and Iron Law of Site Conditions, 83; and State Forest Districts, 84; and close-to-nature forests, 123; and Plan quotas, 124, 155; and Industrial Production Methods, 143; forbidding of, 175
Close-to-nature forests and forestry: restoration of, xiv, 6, 17, 21, 34, 74–75, 192n36; and forest as organism, 22, 146, 192–93n36, 193n41; and forest management, 22, 23, 28, 79, 96, 161, 170; industrial forest versus, 22, 96, 124; and Nazis, 36, 123; and new forest, 56; and *Dauerwald,* 79, 83; and Menz Resolution, 80; and Iron Law of Site Conditions, 82; Party leadership's attack of, 109, 129; and accounting, 126; and reversing forest decline, 154; and wildlife policies, 169; and reunification, 175; and cultural identity, 193n37
Coal supply, 20, 68–69, 89
Cold war: Soviet advantage in central planning, 41; and forest management, 95, 131; and German landscape, 99; and boundaries, 101; and Information Technology Revolution, 121; and capitalist relations, 147
Collectivization: forced, xii, 91–92, 97, 98, 99, 101, 111, 112–13, 114, 122, 137, 146, 184; and Stalin, xiii–xiv, 64; church foresters spared from, 7; of forests, 45; and land reform of Soviets, 52, 59, 60–61, 63, 64, 71, 89, 111, 211n36; peasants' fear of, 68, 69, 70, 71–72, 86, 90, 216n107; and farms, 71, 91, 92, 99, 101, 111–12, 113, 225n95; and Party leadership, 71, 76, 89–90, 111, 112, 113; and forest landowners, 86, 87, 88, 91, 225nn91,95; as voluntary, 86, 90, 112, 114; resistance to, 110, 111, 113, 142, 150; and Ulbricht, 112–13, 116, 137, 138; and food supply, 113, 114, 122; and reunification, 175, 176. *See also* Socialist Spring in the Countryside
Command economy: principles of, 99; and economic conditions, 100, 140; and Ulbricht, 100, 109, 117, 133; performance of, 117; questioning of, 121; and shortages, 144; and public health, 151; and Gorbachev, 178; weakness of, 180, 183
Conservation: in United States, 12; and Party leadership, 100, 107; and foresters, 109, 125, 127, 169; and Marx, 129; and wildlife policies, 169
Consumption, 9, 159, 251n109
Cotta, Heinrich, 17, 21, 82, 170, 192n36

Cultural identity: and German forest, 11–12, 20, 21, 23, 83; and artificial forest, 20, 21, 28, 187; and forest structure, 21, 193n37; cultural conflict, 22; and land reform of Soviets, 62; Party's destruction of, 107–8; and fear of modernity, 157
Cybermarxism, 98, 100, 117, 119, 136, 144, 145, 184
Cybernetics: and Ulbricht, 81, 93, 95, 119, 120, 122, 123, 136, 137, 144, 184; and Party leadership, 83; and State Forest Districts, 84; and Marxism-Leninism, 98, 100, 117, 119, 137–38, 144; and New Economic System, 100, 119, 128, 184; and IBM's Model *360,* 120; and price/interest rate reform, 125; and Apel, 134; rejection of, 135; and Honecker, 139; and control levels, 186; Soviet definition of, 233n3
Czechoslovakia, 131, 135–36, 153–54, 180, 248n68, 256n28

Damaschke, Adolf, 59
Dannecker, Karl, 79, 219n7
*Dauerwald* (permanent forest): industrial forestry versus, 22, 36, 56, 76, 81–83, 172, 187; and Möller, 22, 189n2; and Party leadership, 22, 82, 83, 107, 108, 109, 125; and foresters, 75, 123, 124; and Krutzsch, 78, 218n2; and close-to-nature forestry, 79, 83; and forest ecology, 80; and Menz Resolution, 123–24, 235n21; effects of, 143; and Schröder, 146; and Göring, 172, 253n1; anomalies of, 187, 258n52; and Green Party, 193n41
Davies, Norman, 196n3
Decentralization, 120, 122, 123–25, 133–34, 158
Degen, J. C., 12
Democratic Socialism, 134, 140, 239n84
Dengler, Alfred, 108, 124
Dessau Forestry Conference, 79, 80
Developed System of Socialism, 135
Dickens, A. G., 39
Dubçek, Alexander, 181

East bloc, 149, 151
East Germany: social scientists' observations of, xi–xii, xiii, 186; propaganda of, xix–xx; historically as central Germany, xxi; as frontline Soviet state, 3; natural forest cover type map of, *18;* current forest cover type map of, *19;* political geography of, 23, 28, 34, 195n61; West Germany's political geography compared to, 23, 26, 28; water supply of, 26, 194n52; Soviet occupation of, 32–33; founding of, 73, 102–3, 229n8; West Germany's economic condition compared to, 92–93, 98, 107, 225n101, 226n106, 227n114; U.S. reconnaissance flights over, 100; Soviet support of, 104–5, 106, 116, 140, 169; economic independence from Soviet Union, 122; political and administrative map of, *130;* as discrete polity, 135; and Soviet oil subsidies, 141; and Helsinki Final Act, 149; emergence of German nation, *164–65;* collapse of, 169, 172
East-West frontier, 100, 101–4, 105, 106, 112
Eberswalde Forest Institute, 2–3, 4, 5, 22, 172, 192–93n36, 205n92
Eberswalde Forestry School, 47, 58, 108–9, 144, 177, 206n104
Eberswalde, Germany, 1, 2, 4–5, 7–8, 95, 124, 172
Economic conditions: and reunification, 5, 75, 182–83, 188, 257nn37,38; and forest structure, 16, 20, 21; and forest decline, 21–22; and land reform of Soviets, 33, 34, 38, 57, 77, 97; and reparations, 33, 34, 38, 77, 78, 97; postwar economic environment, 42, 52; of Soviet zone, 43, 44, 46, 204n84; and food supply, 91, 92–93, 97, 113, 172;

and currency reform of *1957*, 92; of West Germany, 93, 100, 102, 122, 142; and command economy, 100, 140; and Ulbricht, 116, 119; and Soviet bloc, 118, 119, 120, 121, 122, 131, 140; and forest management, 123; and ideology, 134–35; and global inflation and recession, 141; and Honecker, 151, 158, 159, 251n109
Economic System of Socialism (ÖSS), 134–35
Edinger, Louis, 183
Emerson, Ralph Waldo, 12, 13, 77, 186
Enclosure, 17, 20, 21, 67, 145, 192n33
Engels, Friedrich, 59, 107, 129, 147
Environmentalism: and Nazi ideology, 36, 76, 78, 81–82, 147; as core national goal, 135; and Marxism-Leninism, 147–48, 158; and Honecker, 148–49, 151–53; church environmentalism, 150; romantic environmentalism, 157
Enzensberger, Hans Magnus, 157
Erteld, W., 108
Ethnic cleansing, 28, 29, 33, 55, 196n3

Farms and farmers: pollution's damage to farms, xiii, xiv; and land reform, 31, 33, 52, 53, 61–64, 212n50, 213nn57,58; and reparations, 32, 38; expropriated farms, 64, 86, 92, 175–76, 214n78, 255n16; and collectivization, 71, 91, 92, 99, 101, 111–12, 113, 225n95; as factories on land, 73, 145; and Industrial Production Methods, 143; and class warfare, 144; and scientism, 146
Fechner, Max, 54–55, 208n6
Felfe, Werner, 151, 152, 153
Feshbach, Murray, 256n31
Fifth World Forest Conference (1960), 96
First Five-Year Plan (1951–55), 80–81, 83, 84, 85, 87, 88, 89
First Seven-Year Plan (1958–65), 92, 93–94, 97, 98, 122
First Two-Year Plan (1949–50), 73, 218n141
First World War, 21, 44, 55
Food and Agriculture Organization (FAO), 36, 69
Forced labor, 39, 41–42, 46, 49–51, 53, 113, 207nn125,126
Forced labor camps, 51, 72, 112
Forest death (*Waldsterben*): evidence of, 5, 149; and forest management, 23; popular anger over, 132, 150–52, 157, 184–85; and pollution, 142, 151, 153; and Honecker, 152, 185; exaggeration of, 172
Forest decline: and state's instability, xii, xiii, xiv; reversal of, xiv, 154, 161, 187; and Party leadership, 3, 10, 34, 98, 129, 131, 149–50, 153, 185, 198n21; and pollution, 3, 21, 23, 135, 141, 153, 154, 155, 157, 161, 167, 173, 184; in Europe, 22–23; and artificial forest, 23, 131, 135, 153, 154, 173, 185; and forest structure, 23, 74, 95, 97, 160, 161; and industrial forestry, 23, 129, 131, 153, 155, 156, 162; and natural calamities, 37, 160–61, 200n38, 252nn125,126; and reparations, 53, 68, 109; and inventories, 127; and Plan quotas, 128–29; and capitalism, 129, 147; and Industrial Production Methods, 132; and lignite, 142; and wildlife policies, 154, 161, 169, 187; extent of, 155; popular resistance to, 177–78
Forest ecology: and Eberswalde, 2, 7–8, 95, 124, 172; ecological revolution, 22–23; and land reform of Soviets, 68, 74, 77; and close-to-nature forests, 74–75; and *Dauerwald*, 80; and Marxist-Leninist ideology, 83–84, 129; and State Forest Districts, 84, 123, 143; rhetoric of, 99; and Party leadership, 109–10, 123, 124, 170, 185; and inventories, 127; and harvest volumes, 132, 133; and pollution, 142, 154;

Forest ecology (*continued*)
rejection of, 144; and forest decline, 154, 161; and reunification, 170, 175, 177; and Cotta, 192n36
Forest economy: and capitalism, 89; restructuring of, 120; and accounting, 126; and exports for hard currency, 161; as bankrupt, 173; returns on, 174
Forest landowners: and enclosure, 17, 20, 21; and land reform of Soviets, 60, 64, 67; and traditional cooperatives, 68, 87, 88, 89, 90, 224n78, 225n91; and State Forest Service, 85–86; and afforestation, 86–88, 224nn72,76; and collectivization, 86, 87, 88, 91, 225nn91,95; private forest growth, 224n81
Forest management: and modern industrial forest, 16, 17, 20–21, 22, 81, 96–97, 145, 192–93n36; rationalization of, 17, 193n41; and clear-cutting, 22, 79, 80, 125, 192–93n36; and close-to-nature forests, 22, 23, 28, 79, 96, 161, 170; and *Dauerwald* versus industrial forestry, 22, 36, 56, 76, 81–83, 172, 187; during Second World War, 36–37; and National Socialism, 36, 45, 47, 80, 84–85, 123, 199n30, 200n32, 205n93, 216n101; of Nazis, 36, 42, 199n30, 200n32; and reparations, 38; and central planning, 45, 47, 75, 85, 89, 127, 129, 134, 205n92; in Soviet zone, 45–47, 67, 74, 205n92; and factories on land, 73, 145; and materialism, 75, 79, 82, 84, 177, 187; and Menz Resolution, 79–80, 220n14; and Ulbricht, 81, 97, 98, 99, 110, 123, 132, 143, 167; and West Germany, 82, 84, 97, 222n48; and State Forest Districts, 84, 86, 87, 94, 223n50; and production, 85, 124–27, 129, 132, 139, 140, 143–44, 145, 146, 154, 155, 158, 160, 167, 251n117; and cold war, 95, 131; and stocking recovery, 95, 97, 133, 162–63, *163*, 167, 185, 251n117, 253n130; and sustainability, 100, 124; and New Economic System, 122, 123, 127; and Grüneberg, 143–44, 243n5; and aerial fertilization, 155, 156, *156*, 185, 249n86, 250n104; and timber stand improvement, 159; and salvage harvests, 160–61, 173, 252n127; and postreunification, 167; and financial analysis, 192n36; and formal management groups, 222n37. *See also* Industrial Production Methods
Forest policy: and Party's stewardship, xiv, 96, 127, 161–63, 167, 184, 185; in American zone, 44, 204–5n87, 216n101; and State Forest Districts, 85; and peasants, 86; foresters' criticism of, 96, 109; and command economy, 117; and Industrial Production Methods, 129, 132, 139; and Marxism-Leninism, 146; and legislation, 154–55, 174, 249n81; and industrial policy, 155; and reunification, 172
Forest products: free market in, 5–6; Party's focus on, 10; higher-value products from, 20–21; and Stalin's border adjustments, 34; and reparations, 41; pulp and sawlog harvest, 85, 160, *161;* factories, 159; harvest plotted against afforestation and TSI, *166, 167*
Forest structure: history of, xiii; and peasant rights, 16; and cultural identity, 21, 193n37; and artificial forests, 22, 37; and forest decline, 23, 74, 95, 97, 160, 161; inventories of, 34, 36, 37–38, 43, 45, 83, 84, 127, 135, 198–99n24, 236n45; and reparations, 34, 37–39, 45, 74, 109, 162–63, 167, 201nn42,49; and land reform of Soviets, 37, 44, 45, 52, 74, 76; and First Two-Year Plan, 73–74; and fuelwood harvests, 81; and age class distribution, 97, 160, *162*, 163, 185; and sustainability, 124; lack of diversity in, 140, 163, 170, 185; and smoke-tolerant species, 154, 155, 250n104;

and State Forest Districts, 173, 223n50; and reunification, 175
Foresters: and fuelwood, 16; and peasant rights, 20; and forest inventories, 34, 36, 38, 43, 44, 45, 74, 127, 135; Second World War activity of, 37; and war potential considerations, 42, 202n69; Soviet subjugation of German foresters, 45–48, 75–76, 77, 144; Soviet terrorizing of, 49; and land reform of Soviets, 56–57, 67, 74; and afforestation, 73, 74, 133; and close-to-nature forests, 75, 78, 79, 80, 82, 109, 124, 129, 187; and *Dauerwald,* 75, 123, 124; and Party leadership, 79, 95–96, 98, 108–9, 123, 124, 131–33, 135, 144–45, 146, 150, 238n71, 243n7; and Socialist Reconstruction, 93, 95; and Industrial Production Methods, 109, 123, 126, 143; and Plan quotas, 124–25, 127, 159, 160; and Democratic Socialism, 134; and Grüneberg era, 142; and forest decline research, 149–50; and pollution, 150, 151, 154; and environmentalism, 152; and forest industrial plants, 159–60; and wildlife policies, 168, 169; and reunification, 172; staffing of, 174, 254nn4,6
Forestry education: Eberswalde Forestry School/Forest Research Institute, 47, 58, 108–9, 144, 172, 177, 206n104; Tharandt forestry school, 47, 172, 173, 174, 177, 206n104; and Marxism-Leninism, 132, 144
France, 41, 178
Franco-Prussia War, 176
Frankfurt, Germany, 27
Frazer, James George, 11, 190n9
Free German Youth (FDJ), 115, 149
Free markets, 5–6, 88, 89, 147, 169
French zone, 102
Friedrich, Carl J., 40, 65
Fuelwood, 16, 49, 67–69, 70, 74, 81, 88, 202n65, 224n79
Fulda corridor, 27, 194–95n58
Gagarin, Yuri, 119
Galbraith, John Kenneth, 41
Gans, Eduard, 14
German civilians, 39, 44–45, 47–50, 63, 65, 105–6, 206n98. *See also* Rural population
German Communist Party, 54, 57, 60, 61, 71, 78, 212n41
German Democratic Republic. *See* East Germany
German Economic Commission, 49
German forests: characteristics of, 4–5; and cultural identity, 11–12, 20, 21, 23, 83; and coppiced woodland, 15–16, 17, 21, 145; as modern industrial forest, 16, 17, 20–21, 22, 99, 192–93n36; natural regeneration of, 21, 22, 193n37; and Stalin's border adjustments, 34; and land reform of Soviets, 53, 61, 62, 66, 212n50
German landscape: as artificial, 13; and land reform of Soviets, 62, 74, 76, 89, 176, 184; and cold war, 99
German POWs, 41, 42, 51, 53, 104, 229n6
Germany Treaty (1952), 102, 103
Gleitze, Bruno, 40, 43
Goethe, Johann Wolfgang von: and Brocken, 11–16, 23, 190n10; and untamed nature, 11, 12, 13, 14–15, 21, 191nn25,26; custom/modernity conflict, 13, 15, 22, 52, 57, 76; influence of forest on, 17, 190n7; and middle landscape of custom and innovation, 184, 186
Golden Age of Economic Growth (1950–73), 92, 118, 121, 139–40
Goldenbaum, Ernst, 63
Gorbachev, Mikhail, xi, 175, 178, 179–82, 254–55n14, 255n25
Göring, Hermann, 22, 123, 168, 172, 253n1
Graves, Henry, 44
Greeley, William, 44
Green Party, 157, 193n41

Grimm, Jacob, 11, 190n9
Grotewohl, Otto, 58, 73, 104
Grüneberg, Gerhard: and Berlin Wall, 64, 115, 142; and land reform of Soviets, 64, 214n80; and forest management, 85, 96, 142, 143–44, 158, 184, 243n5; and collectivization, 92, 142; and Ulbricht, 138, 142–43; and forestry schools, 144–45; and peace, 158. *See also* Industrial Production Methods

Haaf, Günther, 157
Hallstein Doctrine, 104, 229n6, 240n102
Hartig, G. L., 82, 193n36
Harvey, William, 1
Haussmann, Georges-Eugène, 17, 20
Havel, Vaclav, 180
Hazlitt, William, 28
Hegel, G. F. W., 14, 15, 16, 20
Heger, Alfred, 36
Heidrich, H., 132, 133, 146, 238n71
Heine, Heinrich, 11, 13, 23
Helsinki Final Act (1975), 148–49, 150, 177
Herbert, Zbigniew, 171
Heym, Stefan, 188
Hilbert, Anton, 61, 65, 71
Hildebrandt, Rainer, 51, 95
Himmler, Heinrich, 36
Hinz, Robert, 3, 172
Hitler, Adolf, 33, 58, 61, 63, 107, 113, 181, 211n33, 212n49
Hitler Youth, 72
Holocaust, 176
Honecker, Erich: and West Germany, 93; and refugees, 114–15; and orthodox Marxism-Leninism, 129, 137, 183; and Brezhnev, 136; leadership of, 138–39, 142, 183; and détente, 139, 148, 149; and Marx's historicism, 145–46; and environmentalism, 148–49, 151–53; and pollution, 151, 154; and forest death, 152, 185; and Agricultural Price Reform of *1984*, 158, 159; and consumer demand, 159; and forest management, 167, 241n109; and wildlife policies, 167; and forester staffing, 174; and fortieth anniversary celebration, 178–79, 183, 255n22; and significance of Soviet experience, 211n36; and Ulbricht, 241n109
Hoover, Herbert, 32
Hörnle, Edwin, 66, 73, 216n114
Hueck, Kurt, 37, 56, 206n104
Human rights policies, xii, 51, 149, 180
Humboldt University, 108, 109
Hungarian Revolution (1956), 7, 95, 181

IBM, 119–20
Idealism, 15, 20
Industrial forestry: and forest structure, 21; close-to-nature forests versus, 22, 96, 124; *Dauerwald* versus, 22, 36, 56, 76, 81–83, 172, 187; and forest decline, 23, 129, 131, 153, 155, 156, 162; and spruce plantations, 56–57; and central planning, 89; and Apel, 122; and wildlife policies, 169
Industrial Production Methods: and economic ideology, xiii; and Ulbricht, 72, 137; and human reproduction, 73; and forest management, 96, 143, 145; and foresters, 109, 123, 126, 143; resistance to, 113; and Menz Resolution, 124; and Marxism-Leninism, 129, 145–47; and socialist forestry, 131–32; Grüneberg as champion of, 138, 142, 143, 184; and Honecker, 139; precedence of production, 143–44, 167; and environmentalism, 147–51; and materialism, 150, 184; and pollution, 151–57, 184; and forest products factories, 159–60; and salvage harvests, 160–61, 173, 252n127; and wildlife policies, 161, 167–69, *168*; and investment, 167; ending of, 169–70, 173, 175
Industrial productivity: and Second World War bombing damage, 31; and reparations, 38, 41, 201n48; and food

supply, 66; East and West Germany compared, 92–93, 98, 225n101, 226n106, 227n114; and Socialist Reconstruction, 93–94; forestry's low growth rate, 96; and command economy, 117; as priority, 135; and pollution, 154; and environmentalism, 158; and free markets, 169; and Soviet zone, 196n10
Industrial Revolution, xiii, 16, 17, 20, 145, 157
Information Technology Revolution, 117–21, 128, 137–38
International Union of Forest Research Organizations (IUFRO), 6, 7
Investment, 94, 96, 97, 121, 135
Iranian Revolution, 141
Iron Law of Site Conditions, 82–83, 98

Jefferson, Thomas, 186
Joachim, Hans-Friedrich, 2, 3, 4, 5–9, 172
Junkers, 20, 47, 54, 57–62, 64, 209n17

Kádár, János, 181
Kamm, Henry, 183
Katyn massacre, 181–82, 257n36
Kautsky, Karl, 59
Keats, John, 63
Kennan, George, 187–88
Kennedy, John F., 99
Khrushchev, Nikita: and land reform, 58; and Berlin Wall, 91, 115, 116; and West Germany/Soviet Union relations, 104, 105; and collectivization, 112–13; and *Republikflucht*, 115; and agricultural economy, 116–17, 122; and central planning, 118, 120; images of, 118–19, 233n2; reforms of, 118, 128; and Brezhnev, 136; and Ulbricht, 137; Gorbachev compared to, 178
*Kielwassertheorie* (ship's wake theory), 129
Kienitz, Ignaz, 86, 87, 88, 89–90, 91, 224n78
Kindleberger, Charles, 44–45
Kissinger, Henry, xix, 40, 56, 65
Kohl, Helmut, 170, 175, 188
Kohl, Horst, 173
Kohlsdorf, Erich, 131
Kosygin, Alexei, 116–17
Kramer, Jane, xii–xiii
Krenz, Egon, 149, 179
Krutzsch, Hermann, 22, 56, 76, 78, 79, 218n2
Kurth, Horst, 174, 238n71

Land ownership records, 62, 66, 91
Land Reform Commission, 60, 64, 214n78
Land Reform Conference, 72
Land reform in Germany, history of, 58, 210n28
Land reform of Soviets: and peasants, 31, 53, 59, 60, 62–63, 214n67; and economic conditions, 33, 34, 38, 57, 77, 97; purpose of, 33, 57–58, 62, 71, 89, 91, 146; and forest structure, 37, 44, 45, 52, 74, 76; and food supply, 49, 54, 57, 61, 69–70, 71, 114, 122; and collectivization, 52, 59, 60–61, 63, 64, 71, 89, 111, 211n36; and parliamentary road to socialism, 52, 58; and foresters, 56–57, 67, 74; and division of forestland, 60, 61, 66, 67–68, 212n50, 216n103; and agricultural economy, 64, 65, 67, 71, 74, 76, 77, 78, 91; protests against, 65–66; and village social structure, 73; and New Economic System, 123; effects of, 143, 175; and reunification, 175, 254–55n14, 255n16
Lange, Fritz, 71
Laßmann, G., 84, 90, 91, 96, 109, 231n38
"Law for the Protection of Mothers and Children and for Women's Rights," 73
Law of the Economy of Time, 125–26
Lehmann, Joachim, 239n84
Lemmel, Hans, 58, 71

Lenin, Vladimir, xiii, 59, 71, 73, 120, 134, 136, 140, 144, 211n36
Leonhard, Wolfgang, 57, 210n20
Leuschner, Bruno, 98
Liberalism, 22, 56, 134, 152
Liberal Party, 60–61, 212n49
Liberman, Yevesy, 120
Lignite, 141–42, 242n1
Livestock, 16, 17, 67, 70, 216n114
*Lößgrenze* (loess boundary), 27

Maizière, Lothar de, 175
Maly, Vaclav, 256n28
Mantel, Kurt, 68
Mao Tse-tung, 121–22
Marion, Gräfin Dönhoff, 144
Marx, Karl: and Hegel, 14, 15; and peasant forest rights, 20, 145; and Party's control of publishing, 108; and Labor Theory of Value, 125, 126; and conservation and productivity, 129; and revolutionary power, 143; historicism of, 145–46; and population distribution, 147; and wildlife policies, 168
Marxism-Leninism: and human rights policies, xii; and gross production, xiv; ideological orthodoxy, xx, 90, 95, 137, 138, 139, 140, 145, 167, 169, 181, 186; and scientism, xx, 95, 119, 122, 128–29, 139, 145, 146; and forest decline, 3, 10; forest ecology's triumph over, 9; and war on countryside, 52, 145, 146, 147; attacks on capitalism, 55–56, 78–79, 129, 147; and land reform of Soviets, 62, 63, 64, 71, 73, 74; and forest management, 75, 80–81, 83, 94, 96, 97, 98, 145, 231n38; hostility towards private owners, 86; and Kienitz's analysis, 91; and economic growth, 93, 94; and cybernetics, 98, 100, 117, 119, 137–38, 144; waning popularity of, 106, 180, 181; and foresters, 109, 145; and Information Technology Revolution, 120–21; and accounting reform, 126, 127; and scientific-technical elites, 128; and forestry education, 132; and Party leadership's legitimacy, 134, 138; peasant resistance to, 142; and Honecker, 144; and environmentalism, 147–48, 158; and forest death, 151; and scorn for natural sciences, 157; and orientation to complexity, 185–86; legacies of, 188
Materialism: and Marx, 20, 145; foresters' resistance to, 48, 95, 109; and forest management, 75, 79, 82, 84, 177, 187; and pollution, 135, 150, 184; and Party leadership, 139, 150, 177, 184, 185; and public health, 181; and order and control, 186
Mazower, Mark, xi
Mazowiecki, Tadeusz, 256–57n34
McNamara, Robert, 119
Meadows, 5–6, 8–9, 172
Memory/politics boundaries, 107, 110–11
Mendelssohn, Felix, 13–15, 16, 21, 23, 61, 76, 191nn18,22,23
Menz Resolution, 76, 79–81, 83, 123–24, 220n14, 235n21
Middleton, Drew, 103
Mittag, Günther, 122, 125, 135, 142, 152, 154, 167
Modernity: conflict with custom, 11–12, 13, 15, 16, 22–23, 52, 57, 76, 82, 124; and untamed nature, 11, 12, 14–15, 186; and Haussmann, 20; and antimodernism, 21; and Party's reform rhetoric, 100; and Marxism-Leninism, 121; and United States, 152; fear of, 157–58; assurances of, 187
Möller, Alfred, 5, 7, 22, 56, 79, 107, 172, 189n2
Molotov, Vyacheslav Mikhaylovich, 41
Molotov-Ribbentrop treaty, 33, 55, 106, 181
Montesquieu, Charles Louis, 190n8
Morgenstern, Oscar, 184
Moscow-Bonn Treaty, 139, 240n102

Moscow Treaty (1955), 13, 105–6
Muir, John, 12–13
Münker, Wilhelm, 56–57, 209n15
Müntzer, Thomas, 59, 211n35

Naimark, Norman, 45, 64–65, 70, 206n98, 207n126, 209n17, 212n42
Napoleon Bonaparte, 194–95n58
National Socialism: and *Dauerwald,* 22, 82; and forest management, 36, 45, 47, 80, 84–85, 123, 199n30, 200n32, 205n93, 216n101; German rejection of, 56; green environmentalism of, 76, 78, 81–82, 147; and forest legislation, 154; and expropriated land, 176; and complexity, 185–86. *See also* Nazis
NATO, 27, 176
Nazis: as refugees, 29; and armaments factories, 31; wildlife policies of, 36, 199–200n36; socialism contrasted with, 55; and *Osthilfe* land reform, 58; and land reform of Soviets, 61, 212n52; brutality of Soviets compared to, 65; and race, 186. *See also* National Socialism
Nemchinov, Vasilii Sergeevich, 120, 134, 180
New Economic System (NÖS): and cybernetics, 100, 119, 128, 184; and foresters, 109; structures for, 117; and economic reform, 120; and accounting and control, 122, 125–26, 236n38; and cadres, 122–23, 125, 127, 128, 134, 138, 142; and decentralization, 122, 123–25, 133–34, 158; and price/interest rate reform, 122, 125, 126, 235n30; and State Forest Districts, 122–23, 124, 125, 234n17; and Industrial Production Methods, 132; and afforestation, 133, 167; and command economy, 133; Grüneberg's reversal of, 143; and scientism, 146; development of, 234n15
New Mark, xxi, 30, 177
New Peasants Program, 66–67, 215n95
Nixon, Richard, 153
North German Plain, 26, 28, 78, 156, 193n47
Nuclear weapons, 119, 148, 152, 233n4

Ochs, Martin, 107–8
Oder-Neiße territories: value of, 33, 198n19; recovery of, 53, 54, 55, 176; and Junkers, 62; and Adenauer, 104; and Party leadership, 106–7; and memory/politics frontier, 107; and reunification, 176–77
Optimal Stocking Forestry, 78–83, 97, 98, 123–24
Ortega, Daniel, 179–80
*Osthilfe* land reform, 58, 67, 210n30
Ötken, Rita, 158

Palme, Olof, 152
Paris, 20
Party leadership: and human rights policies, xii; economic ideology of, xiii; forest policy of, xiv, 96, 127, 161–63, 167, 184, 185; power assertion of, xiv, 10, 71, 89, 91, 121, 169, 183; propaganda of, xix–xx, 54, 69, 71, 92, 110, 149, 153; forest created by, 2, 4, 10–11; and forest decline, 3, 10, 34, 129, 131, 149–50, 153, 185, 198n21; and Joachim, 6–7; and *Dauerwald,* 22, 82, 83, 107, 108, 109, 125; and land reform, 32, 54–55, 57–64, 66, 72, 89, 184; and forest management, 45, 47, 50, 67, 68–69, 73, 75, 76, 81, 83–84, 87, 88, 107, 129, 132–33, 139, 167, 205n92; professional elite loyal to, 46; and food supply, 65, 113; and expropriated forestland, 67, 216nn102,103; and collectivization, 71, 76, 89–90, 111, 112, 113; and agricultural economy, 74, 113; and foresters, 79, 95–96, 98, 108–9, 123, 124, 131–33, 135, 144–45, 146, 150, 238n71, 243n7; legitimacy of, 93, 106, 128, 134, 140, 149; and authoritarianism,

Party leadership (*continued*)
97; and war on countryside, 100, 145, 172; and founding of East Germany, 102–3, 229n8; and border defense, 104, 107; and Oder-Neiße territories, 106–7; and Eberwalde Forestry School, 108–9; and forest ecology, 109–10, 123, 124, 170, 185; and work norms, 113; and technology, 119; and New Economic System, 122–23, 125, 127, 128, 134, 138, 142; and accounting reform, 126, 127; and lignite fuel, 141–42; and Grüneberg, 142, 143; and environmentalism, 147, 149, 151–52; and peace, 148; and wildlife policies, 168
Patton, George, 195n58
Paucke, Horst, 154
Paul, Frithjof, 150, 223n58, 243n7
Peace: and Ulbricht, 95; and socialism, 97, 148, 151, 186, 247n61; and Honecker, 152; and forest policy, 158, 159; and Gorbachev, 178, 180
Peasants: and coppiced woodland, 15–16, 17; forest rights of, 16, 17, 20, 145; and land reform of Soviets, 31, 53, 59, 60, 62–63, 214n67; new compared to old, 63–64, 66; and traditional cooperatives, 64, 71, 72, 76, 86, 88, 90, 214n73; New Peasants Program, 66–67; and partitioning of forestland, 66, 67, 68, 216n103; and forest management, 67, 87; fear of collectivization, 68, 69, 70, 71–72, 76, 86, 216n107; and fuelwood harvests, 69, 81; and timber harvest labor, 70; and production quotas, 71–72, 217n128; abandoning of land, 72, 76, 217n129; unrest of, 86, 110, 150; taxation of, 87, 88; protests of, 88, 132; and traditional forms and customs, 110; productivity under collectivization, 113; resistance to collectivization, 142, 150. *See also* Rural population
Peasants' Revolt of *1524–25*, 59
Pfeil, Wilhelm, 82, 83
Physical landscape, 11–12, 13, 14, 119
Pieck, Wilhelm, 59, 60, 102–3, 211–12n38, 212n45, 215n92
Pinchot, Gifford, 44
Pine plantations: formal structure of, 5; and industrial demand, 16; and forest management, 17, 21, 74; and loess boundary, 27; and Stalin's border adjustments, 34; and socialist forestry, 111; and forest decline, 156
Plantation forestry: and Cotta, 17, 21; ecological fragility of, 22, 37, 75; and afforestation, 43, 85; and ecological and economic factors, 82; and reunification, 175
Plochmann, Richard, 2, 172
Poland, 33–34, 92, 94, 116, 122, 154, 176–77, 181
Politburo, 115, 136, 137, 138, 153, 168, 179
Political identity, 11–12, 20
Pollution: extent of, xi, xii, xiv, 131; and forest decline, 3, 21, 23, 135, 141, 153, 154, 155, 157, 161, 167, 173, 184; and accounting, 126; and Party's reactive strategies, 131, 155–56; in West Germany, 142; and capitalism, 147, 151, 245n33; causes of, 147, 151, 245n33; and public health, 149, 150–51, 174; and foresters, 150, 151, 154; popular anger over, 150, 178; transboundary, 152; acid rain, 154, 172
Pomerania, 30, 106
Poplar (*Populus spp.*), 94, 155, 185
Potsdam Agreement: and reparations, 38, 44; and forest surveys, 43; and food supply, 49, 54, 208n4; and Oder-Neiße territories, 55, 104; and land reform, 61; and stocking reduction, 202n69; and forest management, 204n81
Potsdam Conference, 100
Price/interest rate reform, 122, 125, 126, 235n30

Protestant Church, 7, 150, 151, 172
Prussia, xxi, 28, 29–30, 34, 62, 177, 195n61, 210n28
Public health, 149, 150–51, 153, 181

Rau, Heinrich, 47, 206n107
Recknagel, A. B., 22, 96
Red deer (*Cervus elaphus*), 131, 167–68, *168*, 169
Rees, Goronwy, 115–16, 127
Refugees: and Second World War, 29–30, 31, 33, 37; and forest exploitation, 67; and emigration rates, 97–98; and West Germany's strength, 102; and collectivization, 111; and Workers' Uprising in June *1953*, 112; youths as, 114–15; and Honecker, 179. See also *Republikflucht*
Remnick, David, 178
Reparations: and agriculture, 32, 38; and economic conditions, 33, 34, 38, 77, 78, 97; value of, 33, 38–40, 69, 198n19, 201n46, 202n56; and forest structure, 34, 37–39, 45, 74, 109, 162–63, 167, 201nn42,49; and industrial productivity, 38, 201n48; policy on, 38, 40, 41, 43, 45, 51; and forced labor, 39, 51; and central planning, 41, 51–52; and timber harvests, 41–44, 45, 49, 69, 70, *70*, 80, 81, 83, 202n67; manifests of, 42, 203n73; and Soviet calendar, 46; and food supply, 49, 54; and piecework system, 50; and forest decline, 53, 68, 109; harshness of, 58, 101; fuelwood harvests' effect on, 69, *70;* ending of, 81; and private forest harvest, 87; and Socialist Reconstruction, 98; and stocking, 162–63
*Republikflucht:* and Berlin Wall, xii; and economic conditions, 28; and labor shortages, 50; and Party leadership, 72–73, 76; and Socialist Spring in the Countryside, 91, 98; and emigration, 92, 97–98; and foresters, 95–96; extent of, 97, 115, 184; and survival of East Germany, 100, 102, 112, 114; and border guard strength, 101–2; and agricultural economy, 113
Reunification: early days of, 4, 5; and economic conditions, 5, 75, 182–83, 188, 257nn37,38; and Soviet's closing of East-West frontier, 101, 103–4; and Stalin, 102, 229nn7,8; and Soviets' veto power, 105; and Party leadership, 135; and forest management, 167, 169; and forest policy, 172; and forester staffing, 174, 254nn4,6; and plantation forestry, 175; and Oder-Neiße territories, 176–77; force of, 183–84
Rhine River, 11–12, 26
Richter, Albert, 46–47, 79, 82–83, 95–96, 98, 108
Riedel, Alexander, 172, 173–74
Risk, 119, 121, 129, 134, 140, 187
Ritter, Carl, 14
Romania, 136
Roth, Terence, 183
Rüffler, Rudolf, 22, 144
Rural landscape: Community Party's stress on, xiv; and peasant rights, 17; and land reform of Soviets, 74, 77, 78; modernization of, 76; systematic assault on, 83; socialism imposed on, 86; socialist transformation of, 90, 91; and *Republikflucht,* 100; and East-West frontier, 101; and memory/politics boundaries, 110–11; and collectivization, 112; and Marxism-Leninism, 147; and pollution, 154
Rural population: alienation of, 63, 91, 214n70; and Soviet army, 64–65; and land reform of Soviets, 66, 176; migration to cities and western Germany, 67; and timber harvest labor, 70, 216n114; reduction of, 161. *See also* Peasants
Rüthnick, Rudolf, 124, 131, 132, 155, 160, 238nn70,71
Ryle, G. B., 34, 56, 199n30

SAGs (Soviet Joint Stock Companies), 38
Saxony forest, 7, 21–22, 172
Scamoni, Alexis, 94, 108–9
Schabowski, Günter, 169
Schalck-Golodkowski, Alexander, 2, 161
Schikora, Klaus, 34
Schlange-Schöningen, Hans, 58–59, 65, 205n88, 211n34
Schröder, Gerhard, 75, 93, 94, 96, 126, 133, 146
Schröder, Richard, 175
Schumacher, Kurt, 51, 207–8n130
Scientific-technical elites, 122, 128, 137, 180, 237n56, 240n105
Scots pine (*Pinus sylvestris*), 27
Second Berlin Crisis (1958–61), 93
Second Central Forest Conference (1956), 83
Second Five-Year Plan (1956–60), 92, 94
Second Forest Conference (1956), 81
Second Party Conference (1952), 84
Second World War: and Fulda Gap, 27, 195n58; and refugees, 29–30, 31, 33, 37; and Soviet victory, 118, 137; fading memories of, 121
Selvage, Douglas, 115
Semyonov, V. S., 59
Serov, Ivan, 51
Shevardnadze, Eduard, 175
Silesia, 28, 30, 106, 177
Silesia Station, 106–7, 230n29
Silvicultural Guidelines of *1962*, 124
Sixth Central Committee Conference (1963), 98
Snetkov, Boris, 255n22
Social Democratic Party, 59, 64, 152, 153
Socialism: and Soviet occupation, 33; parliamentary road to, 52, 58; and land reform, 53, 60, 62; appeal in Germany, 55–56; legitimacy of, 76, 90, 92, 183; and forest management, 84, 85, 93, 98, 129, 132–33, 144; land tenure structure of, 86; and rural population, 91; and ideology, 95, 97, 151–52; building of, 97; consolidated, 98; advantages of, 107, 117, 132–33; capitalism compared to, 107, 117, 129, 139–40, 184; and IBM's Model *360*, 119–20; special German road to, 122, 136; Democratic Socialism, 134, 140, 239n84; and socialist constitutions, 135; and cybernetics, 138; and environmentalism, 147–48, 149, 158; and Honecker, 179, 183; and reunification, 182; and forest death, 185
Socialist collectives: and peasants, 71, 72, 73, 87, 88, 90, 92, 113; and forest landowners, 86, 91; and farms, 99; purpose of, 245n30. *See also* Collectivization
Socialist forestry: definition of, xx, 134; and Joachim, 7; foresters' resistance to, 48; and State Forest Districts, 93; superiority of, 97; and Party leadership, 110, 111; and Industrial Production Methods, 131–32; and scientism, 146; and stocking increases, 162–63, *163;* ending of, 170, 173; and forest structure, 173, 185; and prices, 173–74
Socialist Reconstruction, 93–98, 110, 122, 133, 226n111
Socialist Spring in the Countryside (1960): and collectivization, 45, 92, 111; and Berlin Wall, 76; and *Republikflucht,* 91, 98; and land reform of Soviets, 106; and Party leadership, 108; and agricultural economy, 112–13, 116; and New Economic System, 117; and Ulbricht, 137; effects of, 143
Socialist Unity Party (SED), 59–60, 78
Sommer, Theo, 152
Sontheimer, Kurt, 155, 250n102
Soviet army: and land reform of Soviets, 3, 64–65; and Brocken, 23, 26; and Fulda Gap, 27; and food supply, 64; and youth leaders, 72; forest encampment of, 100, 101, 102; and border protection, 103
Soviet bloc: and economic conditions, 92,

118, 119, 120, 121, 122, 131, 140; and technological achievements, 93; and pulp imports, 94; as "second world," 100; and East-West frontier, 101; and reunification terms, 102; Berlin Wall as symbol of, 115; and ideological power, 117; and cybernetics, 119; threat from China, 121–22; and hidden capabilities, 128; and pollution, 147, 158; and Helsinki Final Act, 148; and environmentalism, 150; and pollution exports, 154; and wildlife policies, 168; collapse of, 177, 178, 180

Soviet camps, 32, 71, 72, 102

Soviet Communist Party, 120, 136, 144

Soviet Military Administration, 37–38, 228–29n3

Soviet Union: collapse of, xi, xii, 180, 183, 184; power of, xi–xii, 33, 40, 54, 71; propaganda, xix, 46, 52, 53, 54, 63, 102; lands controlled by, xxi; East Germany as frontline state of, 3; soldiers of, 4; ethnic cleansing of, 29, 33, 55, 62, 196n3; pulp imports from, 94–95, 97, 133, 160, 227nn115,119, 238–39n80, 252n123; and East-West frontier, 101, 103–4, 105; economic conditions of, 121, 180; foreign policy of, 137, 138, 139, 148, 180; foresters' cooperation with, 144; environmental destruction in, 181, 256n31; West Germany's relations with, 229n6; agrarian reform of, 244n14; forest legislation, 249n81

Soviet zone: and land reform, 31, 60–61, 63, 65, 212n52; forest inventories of, 36, 37, 43, 83, 203n76, 204n81; clear-cutting in, 37, 43–44, 73–74, 204nn83,84; and central planning, 40–41; and food supply, 42, 47, 48–49, 50, 54, 65–66, 215n87; and forest management, 45–47, 67, 74, 205n92; and plan for reunified Germany, 47–48, 62; and forced labor, 49–51, 207nn125,126; and coal supply, 68–69; rebuilding of forests, 205n87. *See also* Reparations

Spruce plantations: and industrial demand, 16; and forest management, 17, 21, 74; and forest decline, 22, 156; and loess boundary, 27; and Stalin's border adjustments, 34; Soviet reparation quotas for, 46; and industrial forestry, 56–57; replacement of, 94; and socialist forestry, 111; harvest increases, 156–57; and salvage harvests, 161, 252n127; and water supply, 194n55

Spurr, Stephen, 36

Stalin, Joseph: and collectivization, xiii–xiv, 64; border adjustments of, xx, 28, 33, 34, 54–55, 57, 62, 106, 148, 176, 195nn63,64; control of Central Europe, xxi, 33, 51; and land reform, 33, 53, 54, 57, 59, 211n36; and Polish territory, 33–34; and reparations, 40, 43, 51; myth of democratic government, 52; and Germany Treaty, 102; and reunification, 102, 229nn7,8; death of, 104, 118

Stasi, 103, 108, 112

State Forest Districts: and forest management, 84, 86, 87, 94, 223n50; and collectivization, 91; and socialist forestry, 93; and Plan quotas, 97, 160; and foresters, 109; and New Economic System, 122–23, 124, 125, 234n17; and production, 133, 160; and forest structure, 173, 223n50; dissolution of, 174–75, 254n8; staffing of, 174, 254nn4,6; and stocking, 221n31; and sustained yield levels, 254n3

State Forest Service, 85–86, 87, 91, 93

State Planning Commission, 84, 86, 91, 98, 122, 123

Steiner, Rudolf, 21, 78, 210n28

Stern, Fritz, 157

Stettin Station, 106–7, 230n29

Stone, Norman, 176

Sustainability, 21, 75, 83, 96, 100, 124, 222n43
Szasz, Thomas, 153

Tacitus, 10, 11, 189n1, 190n9
Tharandt forestry school, 47, 172, 173, 174, 177, 206n104
Thoreau, Henry David, 10
Tiananmen Square, 178–79
Tocqueville, Alexis de, 178, 181, 188
Treuhand, 175, 182
Truman, Harry, 32

Ulbricht, Walter: and economic conditions, 28, 116, 119; and forced labor camps, 51; and land reform of Soviets, 57, 58, 60, 61, 66, 209n17; and New Peasants Program, 67; and agriculture, 72, 110, 113, 116–17; and government and policy, 78, 215n92; and *Dauerwald,* 80, 82, 123; and First Five-Year Plan, 80, 84, 220n17; and forest management, 81, 97, 98, 99, 110, 123, 132, 143, 167; and Berlin Wall, 91, 103, 115, 116, 137, 229n7; dismissal of, 93, 138, 139, 142–43, 241nn109,115; and investment, 94, 96; and Berlin's boundaries, 95; and command economy, 100, 109, 117, 133; Period of Reforms of 1960s, 100, 119, 120, 122, 139, 146; and Soviet support of East Germany, 104–5, 106, 116; and Moscow Treaty, 105–6; and collectivization, 112–13, 116, 137, 138; and food supply, 114, 132; and accounting, 125, 126; and scientific/technical elites, 128, 237n56; leadership of, 135–36; and Brezhnev, 136, 179; distancing East Germany from Soviet Union, 136, 240n99; and Brandt, 137, 240n102; and counter-détente strategy, 137, 138; and Soviet foreign policy, 137, 138, 139, 148; and Oder-Neiße territories, 176; Gorbachev compared to, 178; and Ackermann, 226n105
U.N. Conference on Environmental Protection, 147
Unification Treaty (1990), 175–76
United Nations, 41
United States: role as power in Europe, 55; reconnaissance flights of, 100; and Information Technology Revolution, 119–20; economic conditions in, 140; and environmentalism, 151, 153; and anti-Americanism, 158. *See also* American zone
U.S. Army, 27, 195n58
U.S. Forest Service, 44
*Urwald* (virgin forest), 16, 75, 79

Versailles Treaty, 38
Vietnam War, 139, 140
Vistula glacier, 12, 26, 27, 28, 30
*Volkseigene Betriebe* (VEB), 84
*Volkseigenengüter* (People's Own Estates, VEG), 61, 63
*Volkseigenewald* (People's Own Forest), 67
*Volkswald* (People's Forest), 2, 45, 79–81, 84–88, 95, 221n31, 224n81
Von Hardenberg, Carl-Hans Graf, 61
Von Humboldt, Wilhelm, 12, 14, 82
VoPos (People's Police), 102–6, 112–14
VVB Forest Management bureaus, 122–23, 124, 125, 143, 234n17

Wagenknecht, Egon, 123
Wagner, Richard, 11
*Waldgenossenschaften* (traditional forest cooperatives), 87, 88
Walpurgisnacht myth, 11, 12, 190–91n12
Walter Ulbricht School of Political and Legal Science, 113
Warsaw Pact, 3, 27, 94, 105, 136, 137, 151, 182, 183
Water supply, xiv, 26, 194nn52,55
Weimer Republic, 58, 211n33, 229n6
West: competitiveness of, 118–19, 120,

121, 131, 134; and Vietnam War, 140; and environmentalism, 147, 148, 152, 153; and socialist industry/agriculture synergy, 155; orientation to complexity, 186
West Germany: forest scientists of, xiii, 2; East Germany's blending with, 4; East Germany's political geography compared to, 23, 26, 28; and effects of Second World War, 32–33; and food supply, 49; and forest management, 82, 84, 97, 222n48; free markets in, 88; East Germany's economic condition compared to, 92–93, 98, 107, 225n101, 226n106, 227n114; industrial production of, 92–93; economic conditions of, 93, 100, 102, 122, 142; legitimacy over all Germany, 102, 104, 135, 229n6; aid to East Germany, 150; and disarmament, 151; and environmentalism, 151, 153, 157; environmental aid of, 152, 248n65; and Honecker, 152; and pollution exports, 154; and forest decline, 157, 172; and stocking increases, 162, 252n129; emergence of German nation, *164–65;* Federal Forest Law, 174, 175, 249n81; and forester staffing, 174, 254n4; and Soviet bloc, 240n102
Whitney, Craig, 183
Wiedemann, Eilhard, 82, 83
Wiener, Norbert, 186, 233n3
Wildlife policies: game wildlife stocks, xiv; of Nazis, 36, 172, 199–200n36, 253n1; of Party leadership, 131; and forest decline, 154, 161, 169, 187; and privileged preserves, 167–69, *168*
Wilson, Edmund, 20
Wismut SAG, 49–50, 51, 206–7n115
Wordsworth, William, 171
Workers' Uprising of 16 June *1953,* 50, 92, 102, 112, 135, 209n17, 226n105
Working Group for Close-to-Nature Forestry, 79

Yeltsin, Boris, 181
Yugoslavia, 92, 229n6

*Zäsur,* 139, 241n109
Zhukov, Georgy Konstantinovich, 30, 42